# THE ANNALS OF
# THE KING'S ROYAL RIFLE CORPS

COLONEL JOHN GALIFFE, C.B.

[Frontispiece.

# THE ANNALS OF
# THE KING'S ROYAL RIFLE CORPS

BY
## LIEUTENANT-COLONEL LEWIS BUTLER

VOLUME II
## "THE GREEN JACKET"

WITH ILLUSTRATIONS AND MAPS

LONDON
JOHN MURRAY, ALBEMARLE STREET,
1923

# PREFACE

TO

## THE SECOND VOLUME

NEARLY ten years have elapsed since the publication of Vol. I. of these Annals, and during the intervening period events among the most momentous in the world's history have absorbed the mind of every one of us. Small wonder, then, that the appearance of the present volume has been delayed; but from the writer's point of view it is unfortunate that these very events can hardly fail to have robbed it of half its interest. After a war involving the employment of men by millions, armed with the deadliest weapons of modern science, the recital of campaigns waged by comparatively insignificant numbers with short-range arms must seem to the present-day reader a very tame affair, even though to our grandfathers a century ago the cause for which they were fighting appeared not less all-important than that for which we have recently been contending.

It is also unfortunate that the lack of regimental record complained of in Vol. I. is accentuated rather than lessened in the present one. When, in 1822, a circular emanated from the A.G. enjoining the completion of the Records from the earliest days of the Regiment, the O.C. 2nd Battalion appears to have taken no notice of it whatever; and although a memorandum of the O.C. 1st Battalion is extant, appointing Captain Charles Leslie to act as historian, and giving the names of

officers to whom application was to be made for information on the part played by the regiment in Wellington's
campaigns, the result was almost nil.

Of the officers who commanded the Rifle Battalion in
the Peninsula, Colonel Galiffe lived till 1847, Major
Davy till 1856, Colonel FitzGerald till 1877 ; yet it does
not seem to have occurred to anyone belonging at that time
to the Regiment to ask them to put on paper their story
of the battalion in the campaigns which should have been of
such deep interest to that generation.[1]  The writer has been
in communication with the descendants of Colonel Galiffe,
and has ascertained that he left a diary which was in all
probability deposited with his other papers in the Museum
at Geneva, where he lived ; but unfortunately the diary
has—at all events for the time being—been lost.

In 1879 the late Major-General Gibbes Rigaud, of our
regiment, published a book entitled 'Celer et Audax,'
which was intended to be a history of the 5th Battalion.
He had access to the papers of Sir William Davy, which were
lent by his son the Rev. Raikes Davy, of Tracey Park, near
Bath, and prints some interesting letters and a little other
useful information.  But while giving General Rigaud all
the credit due to him, it must be confessed that his book
is not a very satisfying work ;[2] a remark which will
however be made, and with justice, by the reader of the
present volume.

The writer has tried in vain to get a glimpse of the

---

[1] Apropos of the last named officer, it seems astonishing that on the death of
Sir W. Beresford, in 1854, FitzGerald, then a General (and afterwards Field
Marshal), was not appointed Colonel-in-Chief of the 60th, in place of Lord Gough,
who had no connection with the 60th and cared nothing about it.

[2] 'Celer et Audax,' must almost certainly have been begun during the lifetime of Sir J. F. FitzGerald ; but if General Rigaud had communicated with him
for the purpose of getting information, it can hardly be doubted that the fact would
have been mentioned in the Preface.

above-named papers.   It would appear that Tracey Park
is still in the possession of the Davy family, but unoccupied ; and a letter to their Solicitors merely elicited a
regret that the fate of the papers was unknown, but that
if still in existence, they had been packed away and were
inaccessible.

Much information which could have been easily obtained
half a century ago is now unattainable, and we must bear
the loss due to the laxity and carelessness of our pre-
decessors.

It is, of course, true that the British Museum and the
Record Office may be searched with advantage ;  and the
writer has spent many hours at both.   Such information
as can be got is largely statistical ;  and even so is often
incomplete and even self-contradictory.   For search of this
kind infinite patience and unlimited leisure are required.
Document after document may be scanned without results ;
and at the close of the day the seeker may depart in deep
disappointment, solaced only by the unconscious humour
provided by some of the official papers.   Thus, in a volume
entitled ' Regimental Losses,' the name  of Colonel Ellis
appears among the deceased ;  and it is only on further
examination that we find it was not the Colonel, but his
horse which had died on the voyage to the West Indies,
and for which (needless to say, in vain) he was demanding
compensation.

In the same volume Ensign Burghhagen is noted as
having died at sea in 1810 ;  yet in the next line but one his
name re-appears as deceased at Calais in January, 1815.
Since, however, in the Army List of 1816, this officer is
shown as still alive we are left with an uncomfortable
suspicion as to the accuracy of W.O. records of that day,
which further inquiry does not tend to diminish.

The object of perusing this volume of Regimental

Losses, was to ascertain how many of our regiment died in the Peninsula. On this subject its information was most defective ; but as showing how search for one thing may lead to discovery of another, the names of ten Riflemen of the 2nd Battalion, said to have been killed in 1831 by the fall of a suspension bridge at Broughton, Manchester, were carefully inscribed. The fact of this accident had been quite unknown to the writer, and the records of the 2nd Battalion make no allusion thereto.[1]

Reverting to the question of Peninsular losses, it would appear that the evidence of the Monthly Returns is the most reliable, but the discrepancy between the figures given therein and those published in the Gazettes is rather remarkable. For instance, during the last three months of the war—February to April, 1814—the Gazettes mention twenty-four Riflemen as being killed in action ; the Monthly Returns give sixty-one.

The writer must express his thanks to many friends who have given him help in the present volume, for which he is deeply grateful. They are too numerous to mention by name, but there are two no longer with us in whose favour an exception must be made. Colonel Henry Donald Browne died before the task of regimental history had been begun. But that it was ever in his mind is shown by the number of important historical memoranda in his own handwriting which were handed over to the author. Valuable service was also rendered by Mr. Walter Courtenay Pepys—once a subaltern in our regiment—in the wearisome task of copying out a great number of Monthly Returns.

The advantage of publishing our Annals by separate volumes at intervals of time is shown by the number of

---

[1] To the courtesy of the Librarian, at the Public Libraries, Manchester, we are indebted for a graphic account of this accident, taken from the *Manchester Guardian* of that date, which, however, states that no Riflemen lost their lives and that only four were badly hurt. Whether the *Guardian* or the W.O. is correct, is unknown.

letters received by the author after the publication of
Vol. I., many of them containing items of information
previously unknown to him.   The pith of such information
is set out in the ' Appendix to Vol. I.' at the end of the
present volume.   Should a new edition of Vol. I. ever be
required, the writer would suggest that this Appendix,
as well as various marginal notes made in his own copy,
be incorporated therein.

He hopes that in the same way the publication of
Vol. II. may produce valuable information in regard to the
doings of the regiment in the Peninsular War.[1]

The late Colonel Willoughby Verner, the distinguished
historian of the Rifle Brigade, had the advantage of many
memoirs written by officers of the old 95th who had served
therein during the Peninsular campaigns.   In the case of
our Regiment such memoirs are practically unknown.
Colonel Charles Leslie, in his 'Military Journal,' makes a few
allusions to the 60th in which he served from 1813 to 1832,
but not many.   If diaries or such like documents exist they
are probably in the possession of descendants of the German
officers of our 5th Battalion and hidden away in some
Prussian or Hanoverian country house.

But it is not only in the dearth of material that we are
at a disadvantage.   It is difficult to avoid the belief that
in the Rifle Brigade there has always been a much greater
atmosphere of historical interest than in our own Regiment.
Every recruit in the R.B. is taught the glories of Tarbes ;
and rightly, for the action was a splendid feat of arms.
How many of us have heard even the name of Vic de
Bigorre ?   It is not, indeed, to be compared with Tarbes
which was fought next day ; for the three Battalions of

---

[1] For the benefit of any reader who may happen to be engaged in military
research, the name of Captain H. D. Challis, 69, Hazlewood Road, Putney, S.W.,
may be mentioned as that of an expert in extracting information, whether from the
Royal United Service Institute, the Public Record Office, or the British Museum.

the R.B. took part in the latter, whereas only three Companies of Riflemen were engaged at Vic de Bigorre, yet the skill and gallantry of these three Companies helped to clear the way for the main body of a Division, and should not even now be entirely forgotten.

In the United States the increasing interest taken in our regiment is very gratifying. The old Regimental Colour of the 1st Battalion presented to the Garrison Chapel in Governor's Island, New York, was installed therein with every possible circumstance of honour and respect; and by way of reciprocal gift a Cochorn, which belonged in the middle of the eighteenth century to ' The Royal Americans,' has been presented to us by the U.S.A. War Department. The writer also receives a considerable number of letters from individuals in the States, on matters connected with our regimental history.

# CONTENTS

# CONTENTS

# CONTENTS xiii

# LIST OF ILLUSTRATIONS

## PORTRAITS

## MAPS

# REFERENCES TO THE REGIMENT

## IN WELLINGTON'S DISPATCHES (EDITION OF 1837) AND SUPPLEMENTARY DISPATCHES; AND IN NAPIER'S 'HISTORY OF THE PENINSULAR WAR,' EDITION OF 1886 (F. WARNE & CO.)

| REFERENCE. | VOL. | PAGE. | DATE. | SUBJECT. |
|---|---|---|---|---|
| S.D. | VI. | — | 3. 8. 08 | Rifles to form Advance Guard. |
| D. | IV. | 75 | ditto | March of Light Brigade. |
| S.D. | VI. | 101 | 7. 8. 08 | Brigading of Force. |
| D. | IV. | 99 | 17. 8. 08 | Roliça. |
| S.D. | XIII. | 294 | 18. 8. 08 | Light Brigade; its new composition. |
| S.D. | VI. | 119 | 18. 8. 08 | Thanked by A. Wellesley. |
| S.D. | XIII. | 294 | 18. 8. 08 | Attachment to Brigades. |
| S.D. | XIII. | 294 | 19. 8. 08 | Position of Rifle Companies. |
| S.D. | VI. | 111 | 21. 8. 08 | Vimieiro. |
| S.D. | VI. | 119 | 21. 8. 08 | Thanked by A. Wellesley. |
| Napier | II. | 430 | | |
| S.D. | XIII. | 325 | 9. 10. 08 | Strength. |
| Napier | II. | 6 | | |
| S.D. | VI. | 211 | 2. 4. 09 | Strength. |
| S.D. | XIII. | 317 | 1. 5. 09 | Strength. |
| S.D. | VI. | 250 | 4. 5. 09 | Brigading of. |
| Napier | II. | 101 | | |
| S.D. | VI. | 253 | 6. 5. 09 | Recommendation of Sir A. Wellesley. |
| S.D. | VI. | 253 | 6. 5. 09 | High Character. |
| D. | IV. | 325 | 12. 5. 09 | Oporto. |
| D. | IV. | 429 | 12. 6. 09 | Recommendation for promotion of Major Davy. |
| D. | IV. | 459 | 23. 6. 09 | Col. Donkin's praise of 5th Battalion. |
| S.D. | XIII. | 319 | 6. 09 | Advance Guard. |
| S.D. | XIII. | 327 | 6. 09 | Brigading. |
| D. | IV. | 484 | 1. 7. 09 | Complaint of Lieutenant —— |
| S.D. | IV. | 481 | 25. 7. 09 | Strength of Talavera. |
| Napier | II. | 168 | | |
| S.D. | XIII. | 344 | 27. 7. 09 | Brigading. |
| D. | IV. | 534 | 29. 7. 09 | Talavera. |
| D. | IV. | 537 | 29. 7. 09 | Talavera. |
| S.D. | VI. | 322 | 29. 7. 09 | Talavera. |
| D. | VI. | 547 | 31. 7. 09 | Information by a Rifleman, taken prisoner and escaped. |

# REGIMENTAL SUCCESSION OF COLONELS-IN-CHIEF, COLONELS-COMMANDANT, AND LIEUTENANT-COLONELS, 1797–1830

## COLONELS-IN-CHIEF

| | |
|---|---|
| H.R.H. The Duke of York . . . . . | Dec. 23, 1797. |
| H.R.H. The Duke of Cambridge . . . . | Jan. 22, 1827. |

## COLONELS-COMMANDANT

| BATT. | NAME. | DATE OF APPOINTMENT. |
|---|---|---|
| 5. | William Morshead . . . . . | Dec. 30, 1797. |
| 1. | William Gardiner . . . . . | Nov. 11, 1798. |
| 6. | Robert Brownrigg . . . . . | July 25, 1799. |
| 5. | Thomas S. Stanwix . . . . . | May 9, 1800. |
| 4. | Lord Charles Fitz-Roy . . . . | June 15, 1804. |
| 6. | Hon. John Hope . . . . . | Oct. 3, 1805. |
| 4. | Edward Morrison . . . . . | Jan. 1, 1806. |
| 6. | Napier Christie Burton . . . . | Jan. 3, 1806. |
| 1. | W. Keppel . . . . . . | April 24, 1806. |
| 5. | Sir George Prevost . . . . | Sept. 8, 1806. |
| 3. | Hon. Edmund Phipps . . . . | Aug. 25, 1807. |
| 1. | Arthur Whetham . . . . . | Feb. 7, 1811. |
| 5. | J. Robinson . . . . . | Jan. 2, 1813. |
| 4. | Hon. Charles Hope . . . . | Feb. 15, 1813. |
| 1. | Sir Henry Clinton, K.B. . . . . | May 20, 1813. |
| 7. | Sir George Murray, K.B. . . . | Aug. 9, 1813. |
| 8. | Sir James Kempt . . . . . | Nov. 4, 1813. |
| 1. | Sir W. P. Ackland, K.C.B. . . . | Aug. 9, 1815. |

On the reduction of the Regiment to two Battalions, General Burton became Colonel-Commandant of the 1st, and General Phipps of the 2nd.

## SUCCESSION OF LIEUTENANT-COLONELS

### 1st *Battalion*

| | |
|---|---|
| J. A. HARRIS | Jan. 16, 1788. |
| W. DOWDESWELL | Jan. 1, 1798. |
| G. RAINEY | April 24, 1803. |
| F. G. VISCOUNT LAKE | May 10, 1808. |
| W. MARLTON | May 26, 1814. |
| A. ANDREWS | Dec. 14, 1815. |

### 2nd *Battalion*

| | |
|---|---|
| D. MACKINTOSH | Sept. 1, 1795. |
| E. CODD | Oct. 26, 1804. |

### 1st *(late 2nd) Battalion*

| | |
|---|---|
| T. BUNBURY | Feb. 6, 1824. |

### 3rd *Battalion*

| | |
|---|---|
| HON. J. ELPHINSTONE | July 20, 1794. |
| G. W. RAMSAY | Dec. 30, 1797. |
| G. MACKIE | Dec. 22, 1808. |

### 2nd *(late 3rd) Battalion*

| | |
|---|---|
| J. GALIFFE | June 16, 1825. |
| H. FITZGERALD | Dec. 25, 1825. |
| HON. A. F. ELLIS | Dec. 18, 1828. |

### 4th *Battalion*

| | |
|---|---|
| G. PREVOST | Aug. 9, 1794. |
| R. LETHBRIDGE | Feb. 10, 1801. |
| G. GORDON | Mar. 9, 1802. |
| T. AUSTIN | June 20, 1805. |
| J. LOMAX | Nov. 16, 1809. |
| W. WOODGATE | June 12, 1814. |

### 5th *Battalion*

| | |
|---|---|
| BARON DE ROTHENBURG | Dec. 30, 1797. |
| W. WILLIAMS | Nov. 15, 1809. |
| J. KEANE | June 25, 1812. |
| J. STOPFORD | Mar. 3, 1815. |

## 6th Battalion

| | |
|---|---|
| L. McLean . . . . . . . . | July 25, 1799. |
| L. von Mosheim . . . . . . . | Sept. 14, 1804. |
| A. Wharton . . . . , . . . | Aug. 16, 1810. |
| A J. Dalrymple . . . . . . . | June 1, 1814. |
| J. Stopford . . . . . . . | Mar. 3, 1815.[1] |

## 7th Battalion

| | |
|---|---|
| H. John . . . . . . . . | Aug. 9, 1813. |

## 8th Battalion

| | |
|---|---|
| J. P. Hunt . . . . . . . | Nov. 11, 1813. |

[1] On the disbandment of the 6th Battalion Colonel Stopford was transferred to the 5th.

# SYNOPSIS OF CONTEMPORARY EVENTS

| | |
|---|---|
| 1803, May 12 . | Rupture of the Peace of Amiens. |
| 1804, Dec. 2 . | Napoleon crowned Emperor of the French. |
| | Third Coalition against France formed by Britain, Russia, and Austria. |
| 1805, Oct. 21 . | Nelson's victory off Cape Trafalgar and death. |
| Dec. 2 . | Austrians and Russians decisively defeated by Napoleon at Austerlitz. |
| 1806, Jan. 23 . | Death of William Pitt, the Prime Minister. |
| Sept. 13 . | Victory of Napoleon over the Russians at Jena. |
| 1807, June . . | Treaty of Tilsit between France and Russia. |
| | The French enter Spain under pretext of coercing Portugal in combination with the Spaniards. |
| Dec. . . | Capture of Lisbon by the French. |
| 1808 . . . | Abdication of King Charles VIII of Spain. |
| | Crown bestowed on Joseph Bonaparte, brother to Napoleon. |
| Aug. . . | Landing in Portugal of a British force under Sir Arthur Wellesley. |
| 1809 . . . | War between Austria and France ending with the defeat of Austria at Wagram and Peace of Schönbrunn (October). |
| | Disastrous British Expedition to Walcheren. |
| 1812, June . . | Napoleon's invasion of Russia, ending in his retreat from Moscow and the destruction of his army. |
| | The United States of America declare war on Great Britain. |
| 1813 . . . | The French driven out of Germany. |
| | The Allies invade France. |
| 1814 . . . | Abdication of Napoleon. |
| | Treaty of Ghent. |
| 1815 . . | Return of the Emperor. |
| | His defeat at Waterloo and surrender to the British. |
| 1820 . . . | Death of King George III. |
| 1821 . . . | Death of Napoleon at St. Helena. |
| 1823 . . . | War between Russia and Turkey. |
| 1830, July . . | Abdication of Charles X of France. |

# THE KING'S ROYAL RIFLE CORPS

## CHAPTER I

In the first volume of these Annals an attempt was made to trace the history of our Regiment from the date of its enrolment, and in doing so to give a sketch of the various circumstances affecting the political geography and development of those vast tracts of North America known respectively as New France and New England. It was endeavoured to show how intimately interwoven with the geography and the politics of those countries were the circumstances which gave birth to The Royal American Regiment, indicated the lines on which it was to be trained, and decided its métier in future history. The lines :

> " Theirs not to reason why,
> Theirs but to do and die,"

would no doubt have been as appropriate a motto for the British Army in general not less in the middle of the eighteenth century than in the days of Tennyson in the nineteenth ; but the genius of Henri Bouquet enabled him to cultivate the individual intelligence of every man in the ranks ; and this difference of system, combined with our colonial origin, served to give a certain distinctiveness to the 60th from the outset of its career.

The fortunes of the newly raised regiment were followed during the campaigns of Abercromby, Wolfe, and Amherst

1

up to the final triumph at Montreal. The anxious vicissitudes of the Indian revolt were noticed ; and it was shewn that, during the American War of Independence, the regiment maintained its reputation even though at the end the battalions had been reduced to a mere skeleton. It was followed through the gloomy days of its residence in the West India Islands ; and the story of the four senior battalions and of the sixth was brought down to the conclusion of the great war with France.

The object of the present volume is to show how a new life was breathed into the Regiment through the raising of three additional battalions and the development of the Light Infantry spirit and traditions ; how it became clothed and equipped in a manner suitable to the duties which it was called upon to perform ; how the genius of Bouquet was revived by a new Commanding Officer who followed the principles of his predecessor and adapted his maxims to the requirements of European warfare.

2

We have already seen that the rifle had been well known to the 60th from its earliest years.

During the war with France and the Red Indians, it was habitually carried by the officers ; and as the weapon, although little known at home, was not only familiar to the Americans, but had been introduced to them by the immigrant Swiss settlers of whom the Regiment was largely composed, it seems probable that many rifles were to be found in the ranks, even when officially armed with the smooth-bore firelock.

In the year 1776, experiments with a rifled weapon were made in the presence of Sir Jeffery Amherst, with a result surprising in accuracy and rapidity of fire, at a

LORD VISCOUNT HOWE.

COLONEL COMMANDANT, 1757.

range then extreme of 200 yards. A Major Fergusson fired at the rate first of four then of six shots a minute. Lying on his back on the ground he hit the bull's eye ; and missed the target only three times.

It was therefore natural that when Sir Jeffery became Commander-in-Chief of the British Army, his thoughts should turn to the improved fire-arm, and particularly to its introduction into his own regiment, raised as we have seen for purposes of fighting as Light troops in broken or wooded ground.

It would appear that in 1794 rifles were issued to the 1st or 4th Battalion, and at the same time to a company of each of the other battalions. The Records of the Royal Gun Wharf, Portsmouth, include the following statement:—

' This Indenture made the Twenty fifth Day of January 1798 in the Thirty eighth year of the Reign of our Sovereign Lord George the Third. . . . Between the most noble Marquis Cornwallis, Master General of His Majesty's Ordnance, and the Principal Officer of the same . . . on the one part ; and Mr. Midhurst of the Inspector General's Department on the other part, Witnesseth That the said Mr. Midhurst hath received out of His Majesty's Stores within in the Office of Ordnance at Portsmouth the particulars undermentioned for use of the Riflemen ordered to St. Domingo on their way to Jamaica. In pursuance of the Board's Order dated 16th inst., viz.—

| | | | | |
|---|---|---|---|---|
| Pigs of lead, 9 ; weight 16 lbs. . | . | | | 1538 |
| Carbine flints | . | . | . | 2000 |

' sd. W. Midhurst.
' Assist. Inspr. General.'

The Riflemen referred to in this document were those of the 4th Battalion.

### 3

There were at this period in the service of the Crown a good many foreign corps, the number of which it was

evidently considered advisable to reduce by the incorpora-
tion of some in the British line.[1]

Among those selected for the purpose were La Tour's
Royal Etranger ; the York Rangers—composed of French
emigrés, and commanded by officers of the old royal army
of France ; Hompesch's Chasseurs ; and two battalions,
one of Chasseurs the other of Fusiliers, raised by the
Prince of Löwenstein-Wertheim—who took his title from
the town in Wurtemburg.  Probably the whole of these
corps—certainly those of Hompesch and Löwenstein—
were armed with a rifle.  All were in a high state of
efficiency, and by their own traditions and the experience
of active service thoroughly practised in the tactics, duties,
and métier of Light Infantry and Scouts.

The rifle being a weapon with which Englishmen in
general were unfamiliar, it was natural enough to utilise
the foreign corps for the purpose of its introduction to the
British Army.

But the movement embraced a far wider scope than the
mere adoption of a more scientific weapon.  At this period
the Light Infantry spirit and tradition, combining the
individual action of the scout with the skill of the marksman, had practically died out in the British Army.  The
Duke of York, who had succeeded to the command of the
Army in 1795, and other reformers grasped the fact that
such spirit must be revived.  But as a pattern could not be
found at home, it was necessary to import one from abroad.
That model was found in the French and German regiments
already mentioned ; and it was only in accordance with the

---

[1] It may also possibly have passed through the mind of the Commander-in
Chief that should it be convenient at any time to bring to England a greater number
of foreigners than the 5000 authorised by the Act of 1794, the violation of parliamentary statute could be made with a better grace, and perhaps with less
likelihood of detection, if they were incorporated in a regiment of the British line
than if known by their original appellations.

circumstances under which the Royal American regiment
had been originally raised and trained that it was chosen
as that with which these Chasseur or Yäger battalions were
to be incorporated.

The remnant of the Light Infantry regiment of Carl
Hompesch was drafted into our 1st Battalion ; the Royal
Etranger into the 3rd, and the Waldstein regiment into the
4th.  The result of these transfers was that the 60th became
one of the first Light Infantry regiments of the British
Army, and its training and regulations the model upon
which subsequent Rifle and Light Infantry battalions were
instructed by Sir John Moore and other General Officers
of the new school.

But a still more important step in the same direction
was about to be taken by the addition to our Regiment of
a 5th battalion armed with rifles, equipped as Yägers and
clothed in green.  The story of its origin will now be told.


4


Baron Carl von Hompesch, whose name has been already
mentioned, appears to have been a Bavarian.  Born about
the year 1757, he entered the Austrian service at the age of
nineteen and served therein as a cavalry officer against
the Prussians and also against the Turks.  Like other
soldiers of fortune Hompesch was not particular as to the
country which he served.  After a few years he entered the
service of Prussia with the rank of Major and the appointment of A.D.C. to the King—presumably Frederick the
Great—and after taking part in the opening scenes of the
war with France was rewarded with the Order of Merit and
the rank of Colonel.

Having, however, some partiality for England he
made overtures in 1793 for a commission in the British

Army, which had recently joined the coalition against the
French republic.  At the instance of Mr. Pitt, the Prime
Minister, Lord Amherst nominated the Baron for a captaincy
in the 60th ; but the Austrian Ambassador objected—what
right he had to interfere is unknown ; still the appointment
was cancelled.  Hompesch was, nevertheless, employed
by the British Government to raise two regiments : one
of Hussars, the other of Chasseurs à pied.  The latter was
almost destroyed at Boxtell in Holland, and the Baron
taken prisoner ; but after some months of captivity he
effected his escape by swimming the Rhine.

After the withdrawal of our army from the Continent
in 1795, his two regiments were ordered to the West Indies,
where over 900 of the officers and men perished, chiefly
of yellow fever.  Hompesch, with the rank of Brigadier-
General, accompanied them in person ; [1] but before em-
barking he undertook to raise yet another regiment
consisting partly of Chasseurs à pied, partly of Mounted
Riflemen, the command of which he gave to his younger
brother Ferdinand.  The quality of the men thus raised
seems to have been higher than the quantity ; and
Ferdinand von Hompesch, either in the wish to complete
its numbers at leisure, or more probably because he still
belonged to his brother's Hussars, handed over the executive
command to Lieutenant-Colonel Baron Francis de Rothenburg, previously second in command of the Hussars.  A
body of 300–400 men were brought from Germany to the
Isle of Wight where they were organised in three companies,
and in January, 1798, formed the nucleus of our 5th
Battalion.[2]

---

[1] He survived, and lived till 1812, by which time he held the rank of Lieutenant-
General.

[2] An interesting relic of the German origin of the battalion remains to the
present day in the regimental practice of marching at ease with ' butts up '; the
custom of German riflemen.

## 5

Concurrently with these proceedings another movement in the same direction was being made in the West Indies. Löwenstein's battalions—weak in numbers—were at this time quartered at Barbados ; and in the last days of 1797, Brigadier-General Cameron, at Martinique, received from England a letter directing him to repair to Barbados for the purpose of superintending the transfer of the two units to the 5th Battalion of the 60th. On January 22, 1798, this body of troops under command of Lieutenant-Colonel von Schlammersdorff was inspected by the Commander-in-Chief in the West Indies, who expressed his entire satisfaction at its appearance, and intimated his intention of reporting favourably thereon to the King.

Thus it happened that two separate formations of the 60th, each termed the 5th Battalion, were being organised in different hemispheres ; but whether it was by inadvertence that the nucleus of two separate battalions received the same numeral, or whether it was from the outset the intention of the Commander-in-Chief to amalgamate them at a future date, cannot at this distance of time be positively ascertained.

Major-General William Morshead who, as an officer in the Coldstream Guards, had seen service in the Low Countries and the West Indies, and in 1800 was employed in the expedition to Ferrol, was appointed Colonel Commandant of the new battalion : and to this extent may be claimed as the first Rifleman of the British army.

The *London Gazette* of January 23, 1798, notified the following appointments to our regiment :—

To be Lieutenant-Colonels

Lieut.-Colonel Frederick de Schlammersdorff, from Löwenstein's
    Chasseurs.

"    "    G. W. Ramsay, from the York Rangers.

"    "    Baron Francis de Rottenburg, from Hompesch's
    Regt.

"    "    Robert Craufurd, from Hompesch's Regt.

"    "    Henry Couper, from The Royal Foreigners
    (Royal Etranger).

Each battalion of the 60th was at this time furnished
with two Lieutenant-Colonels; but to provide for five
was a matter of difficulty. Schlammersdorff, however,
disappears from the scene shortly after the inspection at
Barbados. He appears to have handed over his command
to Major Crizielski, and he died in the following September.
In the interim a vacancy for Ramsay occurred in the 3rd
Battalion. Craufurd was appointed to the Staff so shortly
afterwards that he could hardly have joined the 5th, or
indeed any battalion of the Regiment.

It will be noticed that von Schlammersdorff was senior
to De Rothenburg. If therefore the formation at Barbados
and the Isle of Wight were parts of the same battalion,
Barbados must have been its headquarters. But the
'History of Services of the 1st Battalion,' compiled in
1823-4, during the lifetime of De Rothenburg and many
other original members of the 5th Battalion, distinctly
states that the latter was raised in the Isle of Wight;
and that for this purpose 400 men were drafted from
" the additional companies of Hompesch's Mounted Riflemen and Light Infantry." This statement is confirmed
by the MS. History of the 60th in the War Office Library.
In an official document quoted below, p. 16, n. 2, and dated
February 28, 1798—during Schlammersdorff's lifetime
—De Rothenburg is expressly mentioned as C.O. of the

5th Battalion.   Schlammersdorff is therefore ignored; a fact which gives colour to the view that his command was entirely distinct from that of De Rothenburg.[1]

It is also noticeable that during the short remaining period of von Schlammersdorff's life, Henry Couper signed his name to an official document as *2nd* Lieut.-Colonel; and that the uniform given to the newly raised battalion was green (the colour of Ferdinand Hompesch's Chasseurs) and not grey as had been that worn by Löwenstein's.

Appended are extracts from the earliest Muster Rolls of each of the two formations.

### VON SCHLAMMERSDORFF'S BATTALION

Muster Roll from December 25, 1797, to June, 24, 1798, sworn at Granada (*sic*) October 11, 1798.

| Muster. | Colonel. | Lt.-Colonel. | Majors. | Captains. | Capt.-Lieut. | Lieutenants. | Ensigns. | Paymaster. | Adjutant. | Quartermaster. | Surgeon. | Assistant Surgeon. | Sergeants. | Corporals. | Drummers. | Privates. | Remarks. |
|---|---|---|---|---|---|---|---|---|---|---|---|---|---|---|---|---|---|
| Present . | | 1 | 4 | | 6 | 4 | – | 1 | 1 | 1 | | 1 | 29 | 33 | 18 | 300 | |
| Absent . | 1 | 2 | 10 | 1 | 13 | 11 | | 1 | | | | 3 | 39 | 52 | 4 | 402 | On detachment at Trinidad, etc. |
| Non effective . | 1 | | | | | | | | | | | | 7 | 8 | | 50 | |
| Total . | 1 | 1 | 3 | 14 | 1 | 19 | 15 | – | 2 | 1 | 1 | 4 | 75 | 93 | 22 | 752 | |

[1] It is true that we also find De Rothenburg signing his name as commander of a 'detachment.' But in those days the expression merely meant a portion of a battalion; not a portion detached from the headquarters of a battalion; and in point of fact, at the time when he thus signed his name Von Schlammersdorff was dead and De Rothenburg's own command, even if connected with the force at Barbados, would have become the headquarters of the battalion.

In the absence of Colonel von Schlammersdorff, sick ; and of Major von Dörsner and von Mosheim—the former on leave, the latter at Trinidad—the Roll is signed by Major Crisielsky.

The names of the officers given in this Return are as follows :—

COLONEL

Major-General W. Morshead

LIEUT.-COLONEL

F. von Schlammers-dorff

MAJORS

N. von Dörsner
L. von Mosheim
W. Crisielsky

CAPTAINS

C. Louisenthal
J. Vendt
C. Hurde
L. Montmarin
W. von Donop
C. Seebach
C. Emmenhausen
C. Haberkorn
M. Gandl
G. Braun
F. Gerardy
L. Masburgh
A. Erp
J. Galiffe

CAPTAIN-LIEUTENANT

A. Hellrick

LIEUTENANTS

M. de Vendt
J. Erhardt
C. Lister
J. Franchassin
L. Ratzenhausen
C. Bertner
A. Seeger
G. Langsdorff
H. Krian
J. Hofmann
J. Hausegger
L. Colbert
L. Nesselrode
A. Wolf
F.    (?)
H. Gilse
A.    (?)
A. Munstal
A. Hugon

ENSIGNS

C. Hinckeldey
L. Kellner
C. Kehl
A. Muller
H. Jockell
M. Henkel
W. Koch

ADJUTANTS

Lieutenant F. Ammon
Ens. A. Kemmeter

LIEUT.-COLONEL DE ROTHENBURG'S BATTALION

Monthly Pay and Muster Roll: 25th February to 24th March, 1798.

| Companies. | Sergeants. | Corporals. | Horns. | R. & F. | |
|---|---|---|---|---|---|
| Captain de Gerardy<br>Lieut. Hagenpoet<br>„ de Gilse<br>Ensign G. Zuhleke | 5 | 8 | 2 | 100 | The Field Officers were Lieut.-Colonel de Rothenburg and H. Couper, and Major L. de Verna. |
| Captain Baron d'Erp<br>Lieut. von Wolff<br>„ Blassière<br>Ensign Henckel | 5 | 8 | 2 | 100 | Lieutenant Wolff was Adjutant. |
| Captain de Weise | 3 | 2 | | 3 | |
| Captain Schneider<br>Lieut. W. de Verna<br>„ G. de Verna<br>Ensign Koch | 5 | 8 | 2 | 100 | |
| Total | 18 | 26 | 6 | 303 | |

Von Schlammersdorff's Muster Roll is, as noted above, signed by Major Crizielski and sworn before a magistrate in Grenada. That of De Rothenburg is signed by him and sworn at Newport, Isle of Wight. Neither makes any allusion to the other, and the separate Muster Rolls were continued until May 1799, when the Löwenstein battalion was drafted into the Hompesch.[1]

6

Of Colonel von Schlammersdorff's attainments and ability we know nothing beyond the fact of the compliment paid him in the General Order after the inspection at Barbados.

---

[1] The position of the two formations seems to have been very similar to that of the first and second line Territorials which were known as 1/5th, 2/5th, etc., during the Great War.

Francis, Baron de Rothenburg, on the other hand, was one to whom our Regiment can never adequately express its acknowledgments and whose influence as a trainer of soldiers spread far beyond the limits of his own battalion. De Rothenburg, member of an old family which no doubt took its name from the town on the Oder and a branch of which settled in Austria, was born at Dantzig, in Poland, on November 4, 1757. The events of his early manhood are unknown ; but in March 1782 he received a commission in the French Army, his corps being the German regiment De la Marck known later as the 77th of the Line. The reason that he joined the service of France does not appear ; like Hompesch he was a soldier of fortune and probably adopted the army which seemed to offer the best chance of active service. Anyhow, the archives of the French War Office show that he acquired a reputation as an ' officier exact et intelligent,' and received a gratuity for good service.

In 1787 De Rothenburg was attached as Aide-de-Camp to General Baron de Salis Marcelius, whose daughter he appears to have subsequently married. Baron de Salis was allowed by the French Government to re-organise the forces of the King of Naples and the Sicilies. That duty completed, both officers returned to France. In July 1791 De Rothenburg quitted the French service, probably owing to its slow promotion, for his rank was still only that of ' lieutenant en second,' and we next find him in command of a regiment of Polish infantry. He served under Kosciusko in the war with Russia, and in November 1794 was present at the decisive defeat of the Poles by Suwaroff at Praga, under the walls of Warsaw. His chief had been wounded and taken prisoner a few days previously.

" Freedom shrieked when Kosciusko fell." The Polish

cause was lost, and for a century and a quarter Poland forfeited her existence as an independent nation.

In 1795 De Rothenburg accepted a majority in the regiment of Hussars which bore the name of Count Carl von Hompesch, and he served no doubt in the Low Countries. When this regiment was sent to the West Indies in 1796, he remained in Europe to assist Ferdinand Hompesch in raising the new regiment in which he—De Rothenburg—had the rank of Lieutenant-Colonel, and which in 1798 was merged into the 5th Battalion of the 60th. His commission in the 60th with the same rank is dated, like all those of the newly appointed officers, December 30, 1797. De Rothenburg's life had been so active that he could have had little leisure, yet he had written a literary work which was translated into English in 1798, under the title of ' Regulations for the exercise of Riflemen and Light Infantry,' and was adopted as the text-book on the subject, not only by the 60th, but for the light troops of the British Army, and formed the manual of instruction upon which Sir John Moore trained the ' Light Division.' In a letter written in August 1805 to Colonel Mackenzie, the celebrated Commanding Officer of the 52nd, Sir John writes : '' I mean to make de Rottenburg the groundwork. . . . In reading over his book attentively I perceive much good in it. It only requires to be properly applied.''

This work, which includes the entire drill, manœuvre, and duties of Light Troops and Outposts under all conditions in the field, was commended to the Army at large by the Adjutant-General in a memorandum dated August 1, 1798. But so far as is known the earliest existing copy belongs to the edition of 1808, which is to be found in the library of the Royal United Service Institution. Many of the bugle calls of the present day, and a still greater number of those in use five and twenty years ago, would appear to owe their origin to this book. Writing in September 1805 to the A.G. Horse Guards, Sir J. Moore states that the bugle calls used by the 52nd and the 95th were those of De Rottenburg.

Like Henri Bouquet of the previous generation and Robert Hawley of a future one, in all things connected with the command and interior economy of a battalion, Colonel de Rothenburg [1] was a master of his art.

The later phases of Colonel de Rothenburg's career will be briefly described in due course. Sufficient here to say that the efficiency attained by the regiments under his command as a General Officer, bears no less testimony to his abilities as a trainer of men, than does his popularity (graphically described by a soldier of the 68th) among the men who had served under his command, to the personal affection inspired by his almost parental care and interest in their welfare. [2]

Colonel de Rothenburg seems to have been naturalised as a British subject at the earliest opportunity. His paper of naturalisation states : ' Francis de Rothenburg, Lieutenant-Colonel of the V Battalion of the said Regiment (60th). . . . To be deemed a natural born subject of His Majesty.' The deed was signed by Frederick, Duke of York, Commander-in-Chief, and witnessed by Major-General Robert Brownrigg, his secretary.

De Rothenburg had a warm admirer in the father of Queen Victoria, H.R.H. the Duke of Kent, who writing to Captain de Salaberry, of the 60th, from Kensington Palace in May 1806, says :

---

[1] In the French Army his name was spelt ' Rotenburg.' In the British service it was usually anglicised to ' Rottenburg.'

[2] John Green, of the 68th Light Infantry, in his book ' Vicissitudes of a Soldier's Life,' mentions the loud cheering which greeted General de Rothenburg on revisiting his old Brigade at Brabourne Lees.

' I am delighted to find you met with so kind a reception from that excellent officer and worthy man, Lieut.-Colonel Baron de Rothenburg ; but it was not more than I expected, as I know there was every disposition on his part to show attentions to anyone that comes recommended from me. . . .

' I hope that you will . . . perfect yourself in all the scientific details of the manœuvres of a rifle corps for which he is master unquestionably in our service.'

That De Salaberry profited by the injunctions of the Duke will be seen in the sequel.  De Rothenburg had no finer pupil ; he made De Salaberry his Brigade Major and was closely associated with him during the remainder of his active life.

### 7

From the Foreign Corps were posted  other officers, among them Captain John Galiffe (of the York Rangers), an ex-officer of the French Army in pre-revolutionary days, and destined  to gain renown in the Peninsular War under Wellington.   The officers from Löwenstein's included Lewis von Mosheim, who had begun his career in the Wurtemburg Foot Guards and was fated in after years to become a General Officer in the British service, Michael de Wendt, and Gustavus Braun, both of whom earned distinction in the Peninsula, particularly the second, who for his services in the field received knighthood at the end of the war.   The recruits from Hompesch's also included many who in their respective ranks did well in the Peninsular Campaigns, such as Blassière, de Gilse, Zuhleke, etc.

The connection—such as it was—of Robert Craufurd with our Regiment may be dismissed in very few words. Although a miniature of him in Rifle uniform still exists, it seems hardly possible that he could have actually joined a battalion.  His subsequent career gives evidence of the fact that he was acquainted with De Rothenburg's

system but never assimilated it.   Craufurd is not one of
our regimental heroes, but we respect the memory of the
leader of ' the Light Division.'

## 8

The newly raised battalion was dressed in green uniforms
faced, as became one of a royal regiment, with red.[1]   The
colour thus introduced became the standard uniform for all
future Rifle battalions in the British service ;  and it is to
the 5th Battalion of the 60th that each and all are indebted
for the green jacket which they wear with becoming pride.
The name of the rifle—or as it was usually called ' rifle
gun '—with which the battalion was armed is unknown ;
but it had six grooves and was, therefore, certainly not the
Baker.[2]   The battalion was armed also with a sword
and equipped with a black belt, cartridge box and brown
leather rifle bag, worn slanting over the shoulder, in place
of   the   ordinary   knapsack.[3]   Officers   and   men   wore

[1]  Carl Hompesch's was already dressed in green, with a great deal of red about
it.   The uniform was not like that of the 5th Battalion.   Ferdinand Hompesch's
force was also dressed in green, and the uniform of the 5th Battalion may have been
almost identical.

[2]  The Records of H.M. Gunwharf, Portsmouth, contain the following document :
" This Indenture made the Twenty Second Day of February 1798 in the Thirty
eighth year of the reign of our Sovereign Lord George the Third, etc., between the
Most Noble Marquis Cornwallis, etc., on the one part, and Lieut.-Colonel de
Rottenburg of the 5th Battalion of 60th Regiment, commanded by the said
Lieut.-Colonel Rottenburg on the other part, witnesseth That the said Lieut.-
Colonel hath received out of H.M.'s Stores . . . the particulars undermen-
tioned for use of the said Battalion, agreeable to Major General Vyse's letter dated
the 18th inst.   Carbine flints 678, etc.,

sd. DE ROTTENBURG.<br>Lieut.-Colonel."

[3]  The following advertisement is taken from the Dublin Journal, September 11,
1798 :—

### DESERTED

[4] From the 5th Batt. of the 60th Regt., George Pallas, Rifleman, native of
Hungary, aged 19 years ;  5 feet 2 inches high ;  brown eyes, black hair, small
face, speaks no English, wears green uniform with red facing, his regimental sword
with black belt and cartridge box, a brown leather knapsack, and his rifle gun.'

moustaches ; an innovation, which in conjunction with the green jacket proved so startling that on its first appearance in England the country people are said to have run away, terrified by what they believed to be a foreign invader !

The Act of Parliament under which the battalion was raised confined its service to North America. The Act became a dead letter from the outset. The Riflemen served in Ireland, England, Portugal, Spain, France, Gibraltar, precisely as their services were needed ; but in North America the battalion spent but two of the twenty years of its numerical existence.

## CHAPTER II

THE period at which the Rifle Battalion was raised had become one of extreme gloom and depression. At the outbreak of war with France in 1793, the country had recovered from the blow received by the loss of her American colonies, and had attained a high degree of commercial prosperity. She had thus been lulled into a sense of security; and her pundits, not less sapient in those days than in the present, had convinced themselves that war would be no more, when the French Revolution in 1792, sweeping over the neighbouring country like a tornado, overwhelmed the current of men's thoughts, uprooted ancient landmarks, and in the exuberance of its fanaticism hurled defiance at half Europe.

For the ensuing conflict Britain was totally unprepared, and by 1798 her finances were in disorder; her credit almost exhausted. Her armies had experienced little but disaster; the crews of the navy, essential to her existence, had been driven into mutiny by the abominable treatment of the Admiralty; and the kingdom presented a condition of calamity unparalleled perhaps in her previous history.

It was not unnatural that amid all this turmoil and chaos the condition of affairs in Ireland was giving cause for profound apprehension. In 1798 Ireland had real grievances. Notwithstanding the fact that the large majority of her people belonged to the Church of Rome, her Parliament was exclusively Protestant. But while subjecting Roman Catholics to humiliating disabilities, the British Government, persuaded by the mercantile classes in England, also contrived to tyrannise and offend the Protestants of Ulster by legislation calculated to crush their industries.

The French Revolution raised in some hearts hopes of reform, in others of anarchy. Delegates from Ireland were despatched to Paris and were cordially received by the government of the Reign of Terror. The naval and military assistance of France was promised ; and it was due, not to our Admiralty, but to stress of weather, that an expedition despatched in 1795, under General Hoche, perhaps the finest soldier of the French Republic, failed to effect a landing in Ireland. A storm dispersed Hoche's convoy ; and although his second in command, General Grouchy, reached Bantry Bay with a remnant of the fleet, the latter decided—no doubt wisely—that the force at his disposal was not strong enough to be disembarked. Twenty years later it fell to Grouchy to be once more confronted with a still more serious crisis. At the head of the right wing of the Imperial army he heard in the distance, but did not consider himself warranted by his instructions in marching to, the sound of the guns at Waterloo.

## 2

On the outbreak of war with France in 1793, Great Britain had been, to all intents and purposes, destitute of a military force. Three battalions of the Foot Guards were hurried across in colliers to Holland, at that time threatened with a French invasion. Three battalions of the line followed ; but of the three, two were found unfit

for active service. There was no general system of
administration or training in the army.  Regiments were
for practical purposes the property of their colonel; and
promotion among the regimental officers was—and for
three-quarters of a century continued to be—gained, not
by merit, but by purchase.  The Duke of York proved
unequal to command in the field.  Discipline was at a low
ebb, and the soldiers plundered the inhabitants of the
country in which they were serving; sometimes without
let or hindrance.

The force in Holland, although gradually augmented,
was almost destroyed, not less by cold and hunger than
by the enemy; and our allies, Prussia, Austria, and Russia,
were not in all respects reliable.  At length, in 1795, the
remnant of our troops returned to England under favour
of a convention.  Two years later the naval mutinies at
Spithead and the Nore brought Britain to an even lower
ebb.  The young and vigorous Republic of France appeared
irresistible; and the continual and pressing fear of waking
up any morning to hear that " the French are in the bay "
was no dream but a reality, may be considered the pretext,
if not the excuse, for part of the tyranny practised in
Ireland.

In the last days of 1797, Sir Ralph Abercromby, than
whom no wiser man nor better soldier existed in the
British Army, was appointed to the command in chief in
Ireland.  At his own instance Abercromby was accompanied by Brigadier-General John Moore, the future hero
of Corunna, whom we last saw as Major in our 4th Battalion.
Moore's Diary throws much light upon the situation.  It
appears that the number of men under arms in Ireland
was nearly 77,000 ; but a large proportion belonged to the
Irish militia which was disaffected; and the British
regular troops amounted to only 19,000.  Moore remarks :

' I had opportunities of hearing from Sir Ralph the very defective state in which he found every military preparation. No artillery were in a condition to move ; even the guns attached to the regiments were unprovided with horses. No magazines were formed for the militia regiments ; little or no order and discipline ; and the troops in general were dispersed in small detachments for the protection of individuals. The situation of Commander-in-Chief in Ireland is subservient to the Lord Lieutenant, to whom every application, even of the most trifling kind, must be made, and by him directed. In quiet times a Commander-in-Chief has been little attended to ; and the army has been considered little more than an instrument of corruption in the hands of the Lord Lieutenant and his secretary. So much has this been the custom that even now when the country is undoubtedly in a very alarming state both from internal disaffection and the fear of foreign invasion, it requires, I believe, all Sir Ralph's temper and moderation to carry on the necessary business and to obtain that weight which the situation and the times require.

' The mode which has been followed to quiet the disturbances in this country has been to proclaim the districts in which the people appeared to be most violent ; and to let loose the military, who were encouraged in acts of great violence against all who were supposed to be disaffected. Individuals have been taken up upon suspicion, and without any trial sent out of the country. By these means the disturbances have been quelled and apparent calm produced, but the disaffection has been undoubtedly increased. The gentlemen in general, however, still ask out loud for violent measures as the most proper to be adopted, and a complete line seems to be drawn between the upper and lower orders.'

## 3

The General Officer in command of the southern district of Ireland was Sir James Stewart. Under him General Moore commanded an area with his headquarters at Bandon. He quickly discovered that officers had been appointed to the militia solely for the purposes of gaining their votes at parliamentary elections ; that they were in general profligate, and serving not from a sense of duty, but for the emolument.   They were Protestants, and their men Roman Catholics; and the hatred between these two sections of religious opinion was inveterate to the last degree.   Moore's broad and liberal opinions taught him that the Catholics had a real grievance, and that men denied the privileges of their fellow citizens were rightly filled with a sense of oppression.

Sir Ralph Abercromby issued a strong General Order dealing with the incapacity of the officers, and the irregularities of the troops, which he said proved the army " to be in a state of licentiousness which must render it formidable to every one but the enemy."

The civil government which had secretly encouraged these irregularities at once threw over the Commander-in-Chief, who thereupon resigned his appointment and returned to England.   In place of Abercromby, General Lake was made Commander-in-Chief.

### 4

In the spring of 1798 the four companies of our 5th Battalion at the Isle of Wight were ordered to Ireland under command of Major de Verna.   They arrived on the 20th April and were quartered at Cork.   Colonel de Rothenburg did not follow until three months later.   The disembarkation state showed 1 major, 5 captains, 6 lieutenants, 3 ensigns, 1 adjutant, surgeon, 15 sergeants, 6 buglers, 330 rank and file.   Total 17 officers, 361 other ranks.

The company officers were distributed as follows :—

| Captain | F. de Gerardy | A. Baron d'Erp | L. Schneider | L. de Weise |
|---|---|---|---|---|
| Lieute- | L. Baron de | P. Blassière | W. Verna | (a depôt |
| nant | Hagenpoet | | | company) |
| ,, | F. de Gilse | | C. Verna | |
| Ensign | G. Zuhleke | M. Henckel | C. Koch | |

De Weise's company was merely a cadre.  Lieutenant
J. A. v. Wolff was the adjutant ; but was soon afterwards
succeeded  by  Lieut.  de  Gilse.  The  fifth  captain  was
L. Masburgh.

### 5

During the month of May rebellion openly broke out,
but the movement was partial ; and owing to the precautions previously taken was at first confined chiefly to
the south-eastern counties, Wexford and Wicklow.  Thus
it became comparatively easy, by means of converging
columns, to confine the insurgents to a corner of their
country, and within a few days to break the back of the
revolt.

Sir James Stewart, in accordance with orders from
General Lake, Abercromby's successor, detached a body
of about 1,200 men (including Schneider's company of
the 60th), under Major-General Johnstone, with orders to
march upon Waterford.

On the 10th June, Stewart received further orders
from General Lake to send General Moore eastward by
the Clonmel Road with a Light Infantry battalion and the
two remaining Rifle companies.  Owing to the fact of
orders being countermanded some delay occurred ; but
quitting Cork on the 13th and Fermoy on the 15th, General
Moore, accompanied by Captain Paul Anderson of the
60th as his Brigade Major, soon reached Clogheen, where
he found the High Sheriff zealously engaged in extracting
information by the simple process of flogging.[1]  It was
not by efforts at conciliation or redress of wrongs but by

---

[1] During the remainder of his life, Moore had Anderson at his side, and it was
to him that he gave his dying instructions after Corunna.  The association of the
two is shown among the Moore relics at the R.U.S.I.

the enactment of such barbarities that the wretched people were expected to become imbued with the spirit of loyalty.

On the 17th Moore was at Waterford, and next day at New Ross. Marching out of Ross on the 19th the General, whose force consisted of a troop of Hompesch's mounted men, the Light Infantry, Rifles, and a battery R.A., discovered the insurgents posted upon a hill about a mile and a half distant. He advanced against their left, while General Johnstone attacked their centre. The rebels then retreated, and Moore states that in the pursuit the Riflemen killed 60 or 70 of them—a curious testimony to the effect of their weapon, deplorable though it be that the first employment of a Rifle Battalion in the British Army should have been against British subjects.

On the 20th Moore was ordered to Taghmon, 7 miles from Wexford. The march was delayed till past 3 P.M., and only half a mile had been made when at Langridge— otherwise termed Foulkes' Mill—he encountered insurgents to the number of five or six thousand. The two Rifle Companies were sent to the front to skirmish with the rebels, while two guns and some Light Infantry were advanced to a cross road above Goff's Bridge. The Riflemen rapidly cleared their own front, and then wheeled to the left in aid of the Light Infantry, who were attacked in flank and unable to hold their ground. Sharp fighting ensued, but a counter-attack dispersed the Irish, who, it was said, had been better handled than in any other action. Moore's loss included 10 men killed, 2 of them being Riflemen ; 3 officers and 45 men wounded.[1] The rebels lost heavily. On the 21st, having been reinforced by two battalions, Moore on his own initiative, marched through

---

[1] The names of the two killed—the first Riflemen who fell in action—were Ernst Klitzman and Gerard Osting. Both belonged to Gerardy's company.

Taghmon to Wexford, where he arrived just in time to stop a massacre of a hundred loyalists who had fallen into the hands of the insurgents ; but too late to rescue many others who had already been barbarously murdered.

On the same day Generals Lake and Johnstone with 10,000 men attacked and routed the main body of the rebels, some 15,000 strong, at Vinegar Hill, on which occasion Sergeant Degenhardt of the Rifle Company in Johnstone's force was killed ; Captain Schneider and five of other ranks were wounded.

With the action on Vinegar Hill the rebellion was at an end ; the less said about it the better. Loyalist and rebel vied with each other in barbarity. Moore was almost the only general officer to maintain order and discipline among his troops. He returned to Taghmon on the 27th, and was subsequently employed in extinguishing the embers of revolt in the Wicklow Mountains. During the course of this operation five Riflemen were either killed or reported as ' missing, supposed killed.' Colonel de Rothenburg rejoined the battalion on July 20.

6

Hardly had the task been completed when on August 24, news arrived that a French force had landed in Killala Bay on the coast of Co. Mayo. Moore was at once sent for by General Lord Cornwallis (who had succeeded Lord Camden as Viceroy), and ordered to march with his movable column on Athlone, which was designated as the ultimate point of concentration for the whole available forces ; of which the Viceroy had resolved to take personal command.

Quitting Blessington on the following day, General Moore with the 1st and 2nd Flank Battalions made up

II.                                                                     D

from the 60th, 89th, and 100th, and a detachment of the
Hompesch Cavalry (or, more properly, Mounted Infantry),
reached Athlone, partly by canal boats, partly by march,
on the 27th.   That night came the alarming news that a
force mainly composed of Irish Militia—under General
Lake, the Commander-in-Chief—had been totally defeated
by the French at Castlebar.

In view of the parlous situation Lord Cornwallis was
determined to run no risks.   He moved northward slowly,
and on the 29th met General Lake and his routed force.
Lake reported that the Longford and Kilkenny Militia
had deserted in large numbers to the enemy.   Reports as
to the numerical strength of the latter greatly varied ;
but after a few days it was ascertained that the French
General Humbert, the Commander-in-Chief, had with him
a force of about 1100 French and 5000 Irishmen.

The British army was divided into three brigades, one
of which was commanded by General Hunter of the 60th.
Of these, two formed a division under General Lake ;
while that  of Moore—termed the ' advance  corps '—
was held at Cornwallis' own disposal.

Colonel R. Craufurd with a small mounted force was
despatched in the direction of Castlebar to get news of
the enemy.[1]   On September 3, he rejoined the main
body at Hollymount, and reported that Humbert had not
advanced  beyond  Castlebar.   His  report  was  almost
immediately contradicted by another to the effect that
the enemy had marched eastward early that morning to
Swineford.   Craufurd was directed to return and follow
him.   General Lake was also directed to follow the enemy,
but not to risk an action ;  while Cornwallis himself bore
to the right with a view to intercepting Humbert should

---

[1] Subsequently the well-known commander of the famous Light Division under
Wellington.   At this period Craufurd was an officer in the 60th.

he attempt to cross the Shannon and advance upon Dublin through Longford and Co. Leitrim. At Bally-haunis on the 6th, Moore was despatched in support of Lake to Tobercurry, 19 Irish (*i.e.* nearly 24 English) miles. It soon became evident that Humbert was making for the Shannon at Ballintra; and Moore, recalled by Lord Cornwallis, countermarched and reached Carrick on Shannon on the 8th. Before night came news of the surrender of Humbert and his force at Ballinamuck.

Rumours of another expedition from Brest were, however, still rife, and Moore was directed to concentrate his brigade about Athlone as a central point. The rumours gradually died away; and later on in the year our 5th Battalion returned to Cork. Thus ended this squalid and inglorious yet indispensable campaign.

## CHAPTER III

DURING the month of February 1799 the Riflemen in Ireland, under command of Colonel de Rothenburg, embarked for Martinique.  On April 7, they landed at Fort Royal, where the detachment from Löwenstein's and other corps was drafted into their ranks.  As already mentioned, the Muster Roll for May for the first time included both formations.  It showed 2 lieutenant-colonels (De Rothenburg and Couper), 4 majors (De Verna, Mosheim, Dörsner, and Crizielski), 16 captains, 22 lieutenants, 10 ensigns, 78 sergeants, 81 corporals, 36 buglers, and 891 men.

At Fort Royal a force under General Trigg was being assembled for an attack upon Surinam, better known at the present day as Dutch Guiana, the inhabitants of which, as allies of France, were at war with Great Britain.  Of the 1300 rank and file composing the expeditionary force, about 860 belonged to our 5th Battalion.  On August 16, the squadron convoying the troops anchored in the Surinam river and demanded the surrender of the colony.  The British Ministry had in point of fact already made a secret convention for its surrender, and the Governor at once made arrangements to capitulate.  A detachment of Riflemen having been landed two miles up the river, took possession of the fortifications at Braam's Point, which had already been evacuated by the Dutch.  On the 20th the capitulation actually took place; the 60th took possession of Fort Amsterdam, the Dutch being permitted to march out with the honours of war.

A portion of the garrison consisted of Germans, Austrians, and Hungarians; these were drafted into our battalion, which General Trigg shortly afterwards reported to be ' a very fine body of men, and in every way a most efficient corps.' The Riflemen had now an opportunity of appreciating the readiness with which the Dutch had quitted the colony. They found themselves in a jungle abounding with tigers and in conditions of general misery. An epidemic of home-sickness, culminating in several suicides, broke out; and in order to stop it de Rothenburg had recourse to energetic measures.

At the Peace of Amiens in 1802, Guiana was given back to the Dutch, and the battalion gladly evacuated the country. On November 1 of that year we find it at the island of Trinidad. Writing on January 6, 1803, to the Secretary of State, Lieut.-General Greenfield remarks that 500 of the Riflemen were entitled to their discharge; and expresses his intention of sending the remainder to Antigua *en route* for Halifax, Nova Scotia. While at Trinidad a clothing warrant was issued at home authorising green jackets, pantaloons and black gaiters for the battalion. In October it went on to Halifax, where it remained for two years.

In 1803 the battalion received no doubt a new issue of rifles in accordance with the following Memorandum :—

No. 43.

Counterpart.

' Deliver out of His Majesty's Stores at the *Tower* to Mr. Clarke Carrier, the undermentioned Arms to be by him conveyed to Portsmouth, and there delivered to William Spencer, Esqr., Ordnance Storekeeper, who is to apply to Admiral Gambier to forward them

strength of about 760. At the end of August they proceeded to Ramsgate, where we are told by the local press
that the choruses sung by the Riflemen delighted the
inhabitants. The object of this march was, however, not
music, but active service. They expected to join the
British army under Lord Cathcart, which had invaded
Denmark, on the refusal of that country to surrender to
us her fleet which our Government had reason to believe
was intended for use against England. But the weakness
of the Danish army rendered further reinforcements
unnecessary; and the destination of the battalion was
changed for Cork, where it arrived on September 16. In
December it gained a fresh accession of recruits, bringing
up its strength to 960 rank and file.

of making themselves acquainted with the several duties of the outposts, that they may be enabled to lead and instruct the men intrusted to their care.  He also expects they will seriously consider that any neglect on their part before the enemy may cause the most fatal consequences (from the particular service allotted to the battalion) not only to themselves, but to the whole Army.

' He doubts not their own good sense will point out to them the necessity of impressing upon their minds that vigilance and activity are the first duties required from an Officer on an outpost, and hopes they will not lose the opportunity of profiting by the advice which may be given to them by such of their brother Officers who have had the advantage of acquiring experience upon active service.

' The men are to understand that by the maintenance of order and discipline we can alone look forward to successful opposition to the designs of an enemy; they must on every occasion conform with alacrity to the orders of their Officers ; and as great fatigue is often connected with the duties of Light Troops, they must cheerfully submit, and bear like men the hardships attending a soldier's life.  He feels convinced of their bravery, and is satisfied that they will never yield in that respect to any troops in His Majesty's service.

' The true " Rifleman " will never fire without being sure of his man ; he should if possible make use of forced balls, and only load with cartridges in case of necessity, as when a brisk fire is to be kept up.  And he will recollect that a few direct shots that tell will occasion greater confusion than thousands fired at random and without effect, which will only make the enemy despise our fire, and inspire him with confidence in proportion as he finds us deficient in skill and enterprise.[1]

' It is particularly recommended to the men, and will be strictly enforced, to behave with humanity to the people in an enemy's country, and not to plunder or destroy their houses, or attempt their lives without the most urgent necessity or an order to that effect : interest and humanity both require the maintenance of a strict discipline as the only way to conciliate the minds of the people and to make them our friends.  A contrary conduct besides all other disadvantages attending, will certainly reflect strongly on the credit of the Corps.

[1] It would appear that at this period ' aimed fire ' was considered contrary to the recognised code of military etiquette.  Neither the 95th nor our Riflemen consequently expected quarter from the enemy, but, in point of fact, it does not seem to have ever been refused.

| | | |
|---|---|---|
| Lieutenants | . . | Henry Muller. |
| | | William Linstow. |
| | | F. Baron Eberstein. |
| Ensigns | . . | John Sprecher de Bernegg. |
| | | Isaac Raboteau d'Arcy. |
| | | William Wynne. |
| | | John Joyce. |
| | | Julius von Boeck. |
| | | John Louis Barbaz. |
| | | Charles de Bree. |
| Paymaster | . | George Gilbert. |
| Adjutant | . | Ensign F. de Gilse. |
| Quarter-Master | . | J. A. Kemmeter. |
| Surgeon | . . | M. C. Parke. |
| Assistant Surgeons | . | J. A. du Moulin. |
| | | Charles Wehsarg. |

The following officers did not embark : Lieut.-Colonel de Rothenburg, Captain de Salaberry (Brigade Major to De Rothenburg), Captain J. Macmahon, Captain J. Prevost (A.D.C. to Sir G. Prevost, Commander-in-Chief in Canada),[1] Lieutenants Schriene (attached to the 6th Battalion), Hoffmann (recruiting in Germany), du Chesnay and Ensign W. Morgenthal (D.A.Q.M.G. to Lieut.-General Sir John Moore).

In the absence of Sir Arthur Wellesley, who did not appear until the last moment, command of the force was assumed by Major-General Rowland Hill, who brigaded the battalions ; forming the 60th and a half-battalion of the 95th into a Light Brigade.

## 4

It was not until July 12 that the convoy at last set sail. On the 30th it arrived in Mondego Bay, the spot chosen for disembarkation, but—although under the

---

[1] Colonel Commandant of the 5th Battalion.

circumstances the best available—dangerous on account of the heavy surf, which, indeed, upset several of the boats during the process of landing. The disembarkation began on August 1, and was effected at the village of Lavaos, on the left bank of the Mondego River, the first troops to land being our Rifle Battalion and four companies of the 2nd Battalion 95th—better known at the present day as the Rifle Brigade. The two corps were commanded by Brigadier-General Henry Fane ; and thus began the intimacy which has existed ever since between them. The Riflemen assumed their duties as ' the eyes of the army,' by marching six miles southward the same day and taking up a position to cover the landing of the main body and give timely warning of the enemy's approach.

Sir Arthur, having to the best of his ability sifted the information at his disposal—which for the most part proved very inaccurate—decided to advance upon Lisbon by a route parallel to the sea coast, using the fleet, which at the same time sailed southward, as a movable base of operations. The danger of this project lay in the fact that harbours on the west coast of Portugal were few and far between ; that the coast itself was dangerous, and the general might find himself entirely cut off from the ships and left to the resources of the country, which, although friendly, was poorly furnished with the foodstuffs and supplies necessary to maintain his troops and animals. There was also the contingent peril that Marshal Bessières, posted with a French Army Corps in the province of Leon, might descend on the British rear while Junot contained it in front.[1]

---

[1] Duke of Istria, and one of the eighteen generals included in Napoleon's original creation of Marshals of France, 1804. A cavalry officer of no special ability. Usually commanded the Guard Corps. Killed at Poserna in 1813.

Had Sir Arthur been aware of the actual situation in the Peninsula he would have realised how terribly the British Government had under-rated the task.   In Spain and Portugal the French had about 120,000 men.   They were in possession of the two capitals, Madrid and Lisbon, all the fortresses of Portugal and the greater number of those in Spain.   A strong reserve was massed at Bayonne ;  and the result very shortly showed that the forces already in Spain and Portugal could be trebled with startling rapidity.

On the other hand, the whole British force immediately available for service was only about 40,000 men, which, even if concentrated, would be in the disadvantageous position of acting on exterior lines.

But just at this moment news arrived of a remarkable success gained by the Spanish troops at Baylen in Andalusia, where a whole Army Corps under General Dupont had surrendered to the Spanish General Castaños. The result of this capitulation was that whereas on July 20, Prince Joseph Bonaparte, brother to the Emperor, had entered Madrid as King of Spain, ten days later he recrossed the Somosierra Mountains in headlong flight.

The 30th July happened to be also one of disappointment and annoyance to Sir Arthur Wellesley, for on this day—that of the arrival of the fleet in Mondego Bay— news from England reached him to the effect that the Secretary of State had decided to increase the army in Portugal by two Brigades under Generals Anstruther and Acland, and by a force of about 12,000 men and 42 guns under Sir John Moore, which had just returned to England from an abortive errand on which the British Government had despatched him to Sweden. The army would in consequence be augmented to a strength of about 35,000 sabres and bayonets with 66 guns ; but the command was given to Lieut.-General Sir Hew Dalrymple, and Wellesley would become the junior General of Division, with no less than six General Officers senior to himself. He must indeed have already realised that comparatively junior as he was he could hardly expect to receive the command of so large a force. Had it been given to Sir John Moore he would without doubt have cheerfully acquiesced in the decision ; but Moore himself was only the third in seniority, for although by his attainments and reputation entitled to the supreme command, Sir John had incurred the resentment of the British Ministers by his candid and not altogether complimentary criticisms on their military policy ; and Wellesley could not but feel annoyance that the command of the army should have been entrusted to Dalrymple, an officer practically unknown to the military world, who had seen but little active service, and appeared to have been selected merely from the fact of his being in command of the fortress of Gibraltar.

Undeterred, however, by such considerations, Sir Arthur Wellesley decided to execute his project. He had been informed by the Bishop of Oporto that he would receive the assistance of 10,000 Portuguese, and that the whole of the French force in Portugal did not exceed 15,000 men. A day or two later General Spencer told him that Junot's force amounted to 20,000 ; but Wellesley refused to believe it, although in reality the correct number was still understated. He had, however, ascertained that no part of the enemy's force was at hand to oppose his landing, and on August 1, as already mentioned, began the disembarkation. It was not completed until the 5th, when, in the nick of time, General Spencer arrived in the bay with a division from Gibraltar. The troops were then brigaded as follows :—

| | | | |
|---|---|---|---|
| 1st Brigade . | . Major-General Hill . . | . | 5th Regt. |
| | | | 9th ,, |
| | | | 38th ,, |
| 2nd ,, . | . Major-General Ferguson . | . | 36th ,, |
| | | | 40th ,, |
| | | | 71st ,, |
| 3rd ,, . | . Brig.-General Nightingall . | . | 29th ,, |
| | | | 82nd ,, |
| 4th ,, . | . Brig.-General Bowes . | . | 6th ,, |
| | | | 32nd ,, |
| 5th ,, . | . Brig.-General Catlin Craufurd [1] . | | 45th ,, |
| | | | 50th ,, |
| | | | 91st ,, |
| 6th or Light Brigade Brig.-General H. Fane | . | . | 60th Rifles |
| | | | 4 companies |
| | | | 95th Rifles |

Total strength, about 14,500 of all ranks, with 18 guns.

Wellesley gave orders that the Light Brigade was to be drawn up on one or other of the flanks according to circumstances, and that on the march it was always to form the advance guard. A half-battery R.A. was attached to each Brigade, and a howitzer to the 1st, 2nd, 5th, and Light Brigades.

Arrangements were made to carry bread for seventeen, and meat for five days. The scale of rations was fixed at 1 lb. of bread or biscuit and 1 lb. of meat ; and when the latter was salt, a $\frac{1}{4}$ pint of spirits or a pint of wine was distributed. Six women per cent. were allowed to accompany the column, with their children ; each woman receiving half a ration, and each child a quarter. Orders were issued that on the troops being ordered to march, they were always to receive one day's meat, which was to be immediately cooked and carried by the men next day.

The *rôle* of the Light Brigade was assigned as follows :—

---

[1] Not to be confused with Robert Craufurd, the subsequent commander of the Light Division.

' General Order.                          Lavaos.   3rd August, 1808.

' When the army shall move from its left, the 95th and 5th
battalion 60th will lead the column in the ordinary course.   When
the army shall lead from its right, the 95th and 5th battalion 60th
must form the advance guard, and lead the column from the right
of the two corps. . . .'

Spencer's force was fully landed on the 8th, and at 3 A.M.
next day the advance guard, consisting of 50 Dragoons and
the Light Brigade, one day's march ahead of the main
body, started southward on the Lisbon road.   The weather
was hot and the route lay over scorching sand, and ran
sometimes through pine or olive woods, sometimes along
narrow lanes bordered by prickly pear which by excluding
the air increased the sufferings of the troops.   Sir Arthur's
object was to keep his troops concentrated, and, if possible,
to avoid striking the decisive blow until within the neighbourhood of Lisbon, where its moral effect would be
doubled.   On the 10th he heard of King Joseph's flight
from Madrid, and at once came to the conclusion that
nothing was to be feared from Marshal Bessières, who
under the altered circumstances became fully occupied
with his primary duty, which was that of guarding the
main line of the French communications.

On the 10th the Light Brigade reached Leiria, twenty-
four miles from Lavaos, without opposition.   Wellesley's
next point was Alcobaça, on the Alcoa River, twenty miles
distant ; but in order to shorten the length of his column
he sent the 2nd and 4th Brigades round by Batalha, keeping
the direct road with the remainder.   At Alcobaça, which
the enemy had evacuated only an hour earlier leaving
behind him quantities of meat and other stores, the
army was once more in communication with the fleet
at the port of Nazareth, and received three days' bread
therefrom.

5

Meanwhile General Junot was preparing to make head against the invaders.[1] His army corps comprised about 28,000 men, but of this number a comparatively small portion was available. He had had many difficulties to contend with since the breaking out of the Spanish insurrection : he was cut off from France, and found himself in a position of danger not only in the midst of a hostile population, but menaced by the Spanish Division which had accompanied him to Lisbon. Junot cleverly disarmed five-sixths of this Division ; the remainder escaped. But the news of Dupont's surrender at Baylen and the arrival of two British battalions at the mouth of the Tagus raised to fever heat the excitement of the citizens of Lisbon, already eager for the destruction of the national enemy. As a climax came the news that a British army had landed at the mouth of the Mondego.

Junot considered it indispensable to occupy the Lisbon forts with some 7000 men ; other garrisons absorbed an additional 5000. To oppose Wellesley between 15,000 and 16,000 men only were available. These were organised in three Divisions under Generals Loison, de Laborde, and Kellermann, but posted at a distance from one another. Loison was directed upon Abrantes from Estremos ; de Laborde was ordered to Leiria to act as a retarding force and give time for the concentration of the army. On the 9th the latter was at Candieiros, and advanced to Batalha ; but being anticipated by Wellesley's advance,

---

[1] Duke of Abrantes. Rose from the ranks and attracted the notice of General Bonaparte at Toulon, in 1793. Served with him in Egypt in 1798–9 and in later campaigns. Junot owed his rapid promotion to his early friendship with Napoleon rather than to his own military capacity, which was not high. Although usually termed Marshal Junot by English writers, he never attained a rank higher than that of General of Division. Junot eventually became insane, and committed suicide in 1813.

found it necessary to retire through Alcobaça to the heights of Roliça, whence he despatched a detachment to gain touch with Loison.

## 6

Meanwhile Sir Arthur Wellesley was not without hindrances. A Portuguese Division of 6000 men had joined and accompanied him as far as Leiria, after which its commander insisted on operating on a separate line. Remonstrance was in vain ; the utmost concession made by the Portuguese General was to allow 260 of his cavalry and 1400 of his infantry, under command of Colonel Trant, a British officer in the service of Portugal, to remain with Sir Arthur.

On the 15th Sir Arthur's main body was at Caldas, with the Light Brigade thrown forward at a distance of about three miles in the direction of Obidos. In the evening Sir Arthur, having determined to occupy that village, sent forward for the purpose two companies of the 60th and one of the 95th under Major Travers of the latter regiment. The French piquets were easily driven out of the windmill at Obidos, and abandoned the village ; but Travers, mistaking the *rôle* assigned him, imprudently and (as remarked by Sir A. Wellesley) contrary to orders, followed in pursuit for a distance of nearly three miles. Being counter-attacked on both flanks by the enemy's supports, he effected his retreat with some difficulty, and several of our Riflemen were cut off. It was in this skirmish that the first shots were fired in the Peninsular War. Our loss was one rifleman killed, five wounded, and seven missing.[1] Lieutenant Bunbury of the 95th was killed,

[1] In this skirmish and in the subsequent actions at Roliça and Vimieiro, a much larger number of our Riflemen was reported in the first instance as ' missing.' The great majority of these quickly returned, however, to the battalion.

and Captain Pakenham slightly wounded ; four of their riflemen were reported as missing.

The position taken up by General de Laborde was upon a height sealing the southern end of a valley through which ran the road from Caldas.  The main body of the French was posted on the hills in front of the village of Roliça ; the position covered four or five mountain passes in the rear.  De Laborde's force consisted of about 500 cavalry, 4000 infantry, and 5 guns.

Wellesley knew that on the 16th Loison was at Rio Mayor, 12 miles from Caldas, whence it would be in his power either to descend on the British left and rear, or to join de Laborde by the quickest route.  Sir Arthur, having received information that Loison was taking the latter course, decided on an immediate attack.  He formed his force in three columns.  That on the right, consisting of 1200 Portuguese infantry and the 50 troopers under Trant, was intended to turn the enemy's left and penetrate into the mountains in his rear.  On the other flank a column under General Ferguson, consisting of the 2nd and 4th Brigades, reinforced by three companies of the 60th, a battery, and about forty troopers, was directed to move eastward, and having reached the crest line of the hills bordering the valley, to wheel to the right and move forward thereon parallel with the main road ; then, by means of a *détour*, to drive in the outposts guarding the enemy's right, and afterwards to attack the flank and rear of his main position.  Ferguson was also directed to keep a look-out for Loison.

The centre column, comprising the rest of the cavalry, 2 batteries, with 400 Portuguese light infantry, and the brigades of Hill, Nightingall, Craufurd, and Fane, was ordered to make a frontal attack.

In these three columns at 7 A.M. on the 17th the troops

moved out of Obidos. The Riflemen of the Light Brigade, reduced by the detachment with Ferguson to seven companies of the 60th and four of the 95th, were at once detached into the hills bordering the valley to the east with orders to maintain the communication between the centre and left columns, and to protect the march of the former. The enemy's outposts were soon driven in, and the Light Brigade established itself in the hills threatening the enemy's right; while the brigade of Hill supported by that of Nightingall, with Craufurd's in reserve, menaced the enemy's left centre. On our extreme right Trant and his Portuguese gained ground; on the left, Ferguson was descending the hills and approaching the main body. The French seemed likely to become enveloped; but making a clever although hasty retreat while there was yet time, de Laborde extricated himself and took up a position a mile in rear on what were called the Columbeira Heights.

Several passes led up to the enemy's new position : Trant was directed towards that on our extreme right. On Trant's left the 5th Fusiliers and another battalion of Hill's Brigade were allotted a track the issue from which would have brought them out on the French left; but before they could come into action, the 29th Regiment belonging to Nightingall's Brigade, supported by the 9th from that of Hill, attacked prematurely, and was repulsed with loss.

Meanwhile, on the enemy's right the Light Brigade had begun to make itself felt; but Ferguson, who had been intended once more to outflank the position, wheeled too soon, and in place of coming down on the enemy's rear, found himself frontally in action against de Laborde's right, and was forced to retire in order to regain the proper direction. Thus, on our right the 9th and 29th, and on

7

Sir Arthur had meant to follow up the enemy next day in the direction of Torres Vedras ; but on hearing of the arrival off the coast of Lieut.-General Sir Harry Burrard with two Brigades under Generals Acland and Anstruther, he marched to Lourinha, and on the 19th took post at Vimieiro with the object of covering their landing in Maceira Bay.[1]

On the previous day Sir Arthur had decided that in accordance with the system prevalent on the Continent it was necessary to attach a company of Riflemen to each Brigade.   He paid our Regiment the compliment of selecting it for the purpose, and five of our companies were in consequence attached to the first five Brigades of the Army. Among other duties they invariably performed that of Advance Guards.   Two other companies were attached to the Brigades of Acland and Anstruther as soon as they landed.   To fill their places the 50th Regiment was transferred from the 5th to the Light Brigade.

The Brigade of Anstruther, consisting of the 97th and the 2nd Battalions of the 9th, 43rd, and 52nd Regiments, landed on the evening of the 19th.   That of Acland, made up of the 2nd and 20th Regiments, with two companies of the 1st Battalion 95th and 6 guns, landed on the evening of the 20th, but did not that night join Wellesley's force, which was encamped some two miles from the landing place.   The two companies of the 1st Battalion 95th were transferred on the 22nd to the Light Brigade.   These reinforcements raised Wellesley's numbers to nearly 20,000 men.

On the 20th that General visited Sir Harry Burrard—

---

[1] It may be remembered that Sir H. Burrard had formerly been in the 60th and had served therein during the American War of Independence.

his senior—on board ship and explained his project of advancing early next morning upon Mafra with a view to turning the enemy's position south of Torres Vedras and reaching a point within twenty miles of Lisbon. Sir Harry considered the plan too hazardous, and ordered Sir Arthur to halt pending the expected arrival of Sir John Moore's Division. Wellesley returned to his camp, but at about 11 P.M. was awakened by a messenger bringing news of the approach of the enemy. The in-lying piquets were put under arms, and Ferguson's Brigade ordered to occupy the heights on the further bank of the Maceira stream. Nothing further took place, but in accordance with British custom the force was under arms an hour before daybreak on the 21st, and at 6 A.M. Acland's Brigade marched into camp. The parade was dismissed, and it is evident that Wellesley gave no credit to his information, for writing at 6 A.M. he stated that the enemy was at Torres Vedras, nine miles distant. But an hour later a sudden cry 'Stand to arms!' proved its accuracy; and the enemy's cavalry was observed at a distance of about a mile, moving in an easterly direction with the evident intention of threatening the British left.

The fact was that General Junot had quitted Lisbon on the 15th, and at Torres Vedras had concentrated his Army Corps, consisting of Margeron's cavalry and the infantry divisions of de Laborde, Loison, and Kellermann.[1] It had, however, been found necessary, as already noticed, to leave behind no less than 7000 men for the purpose of maintaining order in Lisbon, as well as some garrisons, such as that of Elvas; and the force collected to attack Wellesley consisted of not much above 13,000 men with

---

[1] Son of the Marshal, who had been a General in the Royal Army and repulsed the Allied forces at Valmy, in 1792. Kellermann, junior, although commanding an infantry division in Portugal was essentially a cavalry officer, and as such served with distinction at Marengo, Austerlitz, and Waterloo.

23 guns.  Sir Arthur's army—nearly 20,000 of all ranks with 18 guns—was posted with a view to water supply rather than for tactical considerations.

'On the right,' says Napier, 'a mountain ridge trending from the sea inland, ended abruptly on a small plain in which the village of Vimieiro is situated; the greater part of the Army was heaped on the summit.  On the other side of the plain the same line was continued by a ridge of less elevation, narrow yet protected by a ravine almost impassable; and being without water had only one regiment and some piquets posted thereon.  In front of the gap between these heights, and within cannon-shot, was an isolated hill of inferior elevation yet of good strength, masking the village and plain of Vimieiro, and leaving only narrow egress from the latter on the right.'

On this hill six guns and two brigades of infantry, those of Fane[1] and Anstruther, were posted; the former on the left.  Ferguson's Brigade was drawn up astride the Maceira, behind the village of Vimieiro.  The remainder of the army occupied the ridge on the left bank of the river.  Sir Arthur's position faced about south-east.  That the high ground beyond the right bank of the Maceira was almost denuded of troops was evident at a glance to Junot on the opposite heights.  Ignorant of the arrival of Acland and Anstruther, and without attempting proper reconnaissance, the French General determined to fall on the British left and centre.  To this intent he directed Brennier's Brigade of de Laborde's Division to turn, and that of Solignac—belonging to Loison—to envelope, the British left.  But by this action Junot dislocated his own organisation; and Loison and de Laborde were hurled against the British centre, each having in hand but one weak brigade.  Kellermann, who also had only one brigade present, was held in reserve.

Wearing, on account of the heat of the weather, white

---

[1] In Fane's Brigade the 95th took the right, and the 60th the left of the line.

linen frocks and cap covers, the French infantry was unusually conspicuous, and Sir A. Wellesley quickly grasped his adversary's purpose. Leaving Hill's Brigade to guard his right, and Fane and Anstruther in advance of the village of Vimieiro, he supported Ferguson with the brigades of Nightingall, Bowes, Catlin Craufurd, and Acland ; and directed the first-named to move eastward to his left and to form line against the threatened attack of Solignac and Brennier. But before Ferguson's Brigade could reach its ground, de Laborde and Loison had launched their attack upon Vimieiro. Loison was easily repulsed by Anstruther ; but the attack of de Laborde upon the Light Brigade, whose left rested on a churchyard, led to desperate fighting. After a vigorous resistance, and disputing every inch of ground, the Riflemen who had found the outposts were driven in ; but General Fane, in accordance with the discretionary power given him by Wellesley, called up Colonel Robe, C.R.A., with the reserve battery of artillery, which poured in a terrible salvo of grape and shrapnel upon the enemy in front, while Acland's guns, moving to support Ferguson and happening to be passing at the moment, unlimbered and struck him in flank. Yet the French column still pressed gallantly forward to the summit of the ridge. Here it was met by a volley from the 50th Regiment under Colonel Walker (who had served in the 60th from 1791 to 1798), charged in flank with the bayonet by the whole brigade, and driven in confusion to the bottom of the hill.

Observing the crisis, Junot brought up Kellermann, who with two battalions re-inforced Loison and with the remaining two attempted not only to pass Fane's left but to penetrate into the village of Vimieiro with a view to cutting the British line in two. The manœuvre was an able one and nearly successful, for Fane had no troops in

II.                                                                 F

French army being unmolested, were cleverly reunited by the Chief of the Staff, and in a circuitous route regained the road to Torres Vedras.   Their loss exceeded 2000 men, and included 14 guns.   The British casualties amounted to 135 officers and men killed, and 585 wounded and missing.   Of our own Regiment 14 Riflemen were killed, Lieutenants Koch and Ritter, 1 sergeant and 22 men wounded, and 1 Rifleman was missing.   The three officers of our 5th Battalion wounded at Roliça were present at Vimieiro, and the list of officers engaged was consequently identical with that in the former action.   In his General Order of the same day Sir Arthur

'congratulates the troops on the signal victory they have this day obtained over the enemy . . . and has in particular to notice the distinguished behaviour of the Royal Artillery, 20th Light Dragoons, 36th, 6th, 43rd, 50th, 52nd, 5th Bn. 60th, 71st, 82nd, 95th, and 97th Regiments.'

By their skill in marksmanship and light infantry tactics the Riflemen had indeed vindicated their *raison d'être*.   A French officer, taken prisoner, quoted by G. Simmons in ' A Rifle Man,' says :—

'I was sent out to skirmish against some of them in green,—grasshoppers I call them, you call them Riflemen. They were behind every bush and stone and soon made sad havoc among my men ; killing all the officers of my company, and wounding myself without being able to do them any injury.'

Referring to the actions of Roliça and Vimieiro, the writer of the ' History of Services of the 1st Battalion ' —probably Colonel Charles Leslie—remarks : ' They,' *i.e.* the Riflemen of the 5th Battalion, ' displayed that steadiness and cool courage so essentially requisite to a Rifleman, in not hurriedly throwing away their fire until there was a positive chance of their fire taking effect."

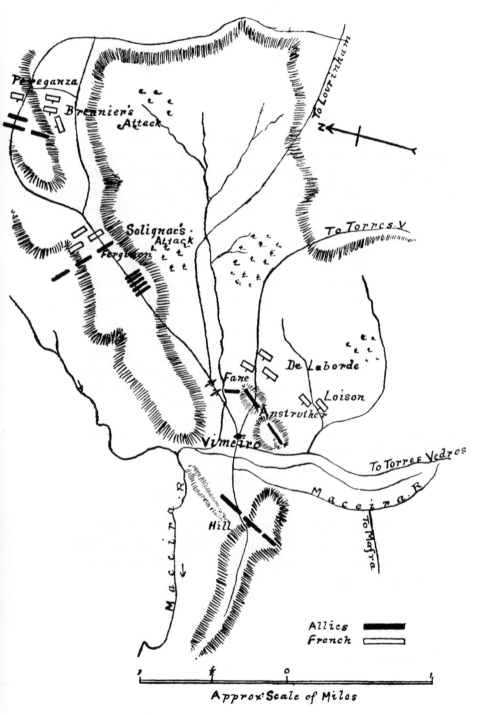

Pereganza

Brennier's Attack

To Lourinha

Solignac's Attack

Ferguson

To Torres V

De Laborde

Fane

Loison

Anstruther

Vimeiro

To Torres Vedres

Maceira R.

Maceira R.

To Mafra

Hill

Allies ▬▬▬▬

French ▭▭▭▭

1    ½    0    1

Approx. Scale of Miles

SKETCH OF THE BATTLE OF VIMIEIRO.

In other words, aimed fire was the essence of the training of each individual Rifleman.

9

Although Sir Harry Burrard had not proved himself a great commander he was the most loyal of men. Being entirely free from the vice of jealousy he was loud in his appreciation of Sir Arthur and the army. Lord Castlereagh, the Secretary of State, was also most complimentary ; and in forwarding copies of their letters a few weeks later to Major Davy, General Fane kindly added, ' I am very happy in being afforded this opportunity of congratulating you any my friends of the 60th upon the flattering approbation our services have met with from our Sovereign.' Baron de Rothenburg naturally wrote in terms of delight :—

' Ashford in Kent, October 12th, 1808.

' MY DEAR DAVY,

'. . . You may easily conceive how highly I have been gratified by reading in the papers of your distinguished conduct in the field. The Duke of York spoke to me in high terms of the Battalion, and I must sincerely congratulate you on the honour and glory you have acquired. I always told you that you might depend upon the bravery of my disciples. . . . My Compliments to Majors Woodgate, Galiffe and all our gentlemen.

' Believe me,
' Yours faithfully,
' FRANCIS DE ROTHENBURG.'

In accordance with G.O. of 21.8.1808, a Rifle Company was attached to each of the newly arrived 7th and 8th Brigades, leaving three only with Battalion Headquarters in the Light Brigade.

## CHAPTER V

SIR HARRY BURRARD's period of command was short, for on the day after the battle of Vimieiro Sir Hew Dalrymple arrived from Gibraltar. Sir Hew so far concurred in Wellesley's opinion that he gave orders for an immediate advance upon Torres Vedras; but before the army could be put in motion an envoy from the French Commander appeared in the person of General Kellermann, who had been deputed by his chief to arrange terms for the evacuation of Portugal. Negotiations ensued which resulted in the arrangement inaccurately known as ' The Convention of Cintra,' [1] under the terms of which the French Army, amounting to 25,747 officers and men, with 30 guns, evacuated Portugal by the end of September, and was conveyed back to France in British vessels. With the furious controversy which arose in England over the terms of this Convention we need not concern ourselves; suffice it to say at this distance of time, most people would probably agree that its objectionable points were greatly outweighed by the solid advantages gained.

At this period there was unfortunately a want of harmony in the higher command of the British Army. Lord Castlereagh, with perfect truth but also perhaps with a certain lack of tact, had recommended Wellesley to Sir Hew as an officer of great ability and worthy of every confidence; and it was in accordance with human

---

[1] ' Inaccurately,' because it was signed at a place far from Cintra.

at Corunna, and to join Sir John at his option either at that port or on the line of march.   Moore chose the latter alternative.

Sir William Napier sums up the situation as follows :—

' No General in Chief was appointed to command the Spanish Armies, nor was Moore referred to any person with whom he could communicate, much less concert a plan of operations.   He was unacquainted with the views of the Spanish Government, and was alike uninformed of the numbers, composition and situation of the armies with which he was to act and those with which he was to contend.   His own genius and £25,000 in the military chest con- stituted his resources for a campaign which was to lead him far from the coast with all its means of supply.   He was to unite his forces by a winter march of 300 miles ;  another 300 were to be passed before he reached the Ebro ;  he was to concert a plan of operations with Generals, jealous, quarrelsome and independent ; their positions extending from the northern sea coast to Saragossa ; their men insubordinate, differing in customs, discipline, language and religion from the English, and despising all foreigners.   And all this was to be accomplished in time to defeat an enemy already in the field, accustomed to great movements, and conducted by the most rapid and decided of men."

Notwithstanding all difficulties, Moore quickly put his forces in motion.   Within five days of the receipt of Lord Castlereagh's letter the leading troops were on the march ; but at this critical moment it was found that the roads north of the Tagus were impracticable for artillery.   General Junot had, it is true, brought his guns over them in the previous year, but the carriages had been broken to pieces. There was, indeed, a southern line of advance through Estremadura, but difficulties of supply rendered this route impracticable for the army as a whole.   It was eventually decided that General Hope, with the Cavalry Brigade, five batteries R.A., and a brigade of infantry should move by Elvas, Talavera, the Escorial and Arevalo to Salamanca, which was named as the point of concentration for the whole of Moore's army. The brigades of Fane and Beresford, with one battery, were directed upon Salamanca, via Coimbra and Almeida ; those of Hill and Lord William Bentinck via Guarda ; and two brigades under Edward Paget, by way of Alcantara and Ciudad Rodrigo.

The 5th (Rifle Battalion) had already been divided into two wings, one of which, comprising the 3rd, 4th, 5th, 9th, and 10th Companies—the last-named commanded, in the absence of Captain Prevost, by Lieutenant Zuhleke— had been doing duty under Major Woodgate with General Hope's Division in the neighbourhood of Elvas. This detachment—whose company officers were Captain Andrews, Ensigns D'Arcy and Debree ; Captain Hames, Ensigns Sprecher and Joyce ; Captain de Wendt and Lieutenant Muller ; Captain Wolff and Lieutenant Sawatzky, and Lieutenant Zuhleke with Ensign de Boeck— was now attached to Lord W. Bentinck's Brigade, and comprised 23 Sergeants, 10 Buglers, 26 Corporals, and 396 Riflemen. The Headquarter half-battalion under Davy, included 3 Captains, 5 Lieutenants, 3 Ensigns, 26 Sergeants, 10 Buglers, 21 Corporals, and 372 Riflemen. It was attached to Beresford's Brigade.

On October 16 and 17 the Headquarter wing was at Rio Mayor, two days later at Leiria. From the 23rd to the 26th it halted at Coimbra ; on November 4 it reached Trancoza. On the 9th the C.O. issued the following Battalion Order :—

' The Major as yet has had much reason to be satisfied with the behaviour of the men on the march, and only regrets that the misconduct of a few individuals should have reflected on the character of the whole. He strongly recommends the continuance of the meritorious conduct which has procured them the good wishes of the inhabitants of the towns of Portugal through which they have passed.'

On the 13th the Battalion Headquarters were at Villa de Pedro Alonzo, when the following Regimental Order was issued :—

'It is with deep concern that the Major discharges a most painful part of his duty—that of announcing to the Battalion that in consequence of the misbehaviour of the five Companies detached, they have been sent back to Lisbon. Under these distressing circumstances the Major calls upon every individual of the Regiment to use his utmost efforts to vindicate and maintain the well-merited reputation that the Battalion has acquired in the field : and doubts not that with the cordial co-operation of the whole they will be able to do away that disgrace which must otherwise for ever reflect upon the character of the Battalion. This order to be read in German.'

The battalion had embarked from England short of eighteen or twenty officers, but in other respects Davy had no one to thank but himself. It was not unnatural that as soon as the men recruited from Junot's army got an opportunity they deserted to their old friends. On the 16th the Headquarter Wing was at Salamanca, and here Davy had the mortification of being ordered to concentrate the whole battalion in Portugal. The move was not, however, actually made until the end of the month : as it turned out, the delay was unfortunate. By this time Moore's army was in the presence of the enemy, and General La Houssaye, in command of a brigade of French dragoons, was able to inform the Emperor of Moore's position from intelligence brought 'by deserters from the regiment which had returned to Portugal in charge of a convoy.'

This fact was of course unknown to the Rifle officers, who, on December 13, forwarded to Sir John Moore from Ciudad Rodrigo a respectful memorial on the subject of being ordered back to Portugal. The following letter was received in reply :—

'Head Quarters, Castro Nuovo: Dec. 19th, 1808.

' Sir,—I have had the honour to lay your letter of the 12th inst. and the memorial of the Officers of the detachment of the 5th battalion 60th Regiment under your Command, before the Commander of the Forces; and I am directed to desire that you will inform them that his Excellency's motives for sending the detachment back arose from the inexpediency of keeping the battalion divided, and not from any idea of the nature of which you apprehend.

' While his Excellency laments that circumstances have rendered it necessary to remove the detachment from the present scene of action, he desires that you will observe to the Officers that their station in Portugal is by no means uninteresting, and that their exertions there may be most usefully directed. I have, etc.

'H. Clinton, A.G.

' To Major Davy, Commanding 5th Battalion 60th Regiment.'

Good will sometimes come out of evil, and deeply as every Rifleman must regret this episode, the result was that the 5th Battalion shares with the 40th—nowadays known as the 1st Battalion South Lancashire Regiment—and the 45th Regiment (now the 1st Battalion of the Sherwood Foresters) the honour of having served in the Peninsular campaigns from the day that Wellington landed in Mondego Bay up to the cessation of hostilities in April 1814.

On December 24 Major Davy dated his regimental orders from Lamego in Portugal; between February 13 and 14, 1809, from Oporto; and on March 7 from Belem, close to Lisbon, where Woodgate with the other wing had arrived on the previous day.

An immediate report of these occurrences had, of course, been made to H.R.H. the Duke of York, who at once sent orders to Sir John Cradock (Burrard's successor as Commander-in-Chief in Portugal) that such men of the 5th Battalion as had been taken from prison ships, and were considered unfit to be trusted, were to be sent at once to England with a view to their being posted to the Rifle Companies of other battalions of the Regiment.

Sir John appreciated the situation.  In a letter to the Secretary of State dated March 6, 1814, he says :—

' The 5th battalion of the 60th which was sent back by Moore in some disgrace is collected (*i.e.* concentrated).  Their services as light troops are so valuable that I shall endeavour as much as possible to re-establish their efficiency, and by getting rid of some of the worst characters to restore them to the service.'

On arrival of the headquarter wing of the battalion at Belem, Davy set himself to carry out these orders, and to restore the efficiency of his battalion.  He writes as follows to the General Commanding his Brigade :—

' Belem, March 7, 1809.

' SIR,—I have the honour to acquaint you that the five Companies of the 5th Battalion 60th Regiment under Command of Major Woodgate have yesterday joined the other five Companies here.  It is my duty to state to you that from immediately after the action of Roliça on the 17th August last the battalion has been divided into detachments which were attached to the several brigades of the Army, and have thus gone through the arduous services of Riflemen during the campaign ; and since that period the battalion has never been united.

' The arms, accoutrements, appointments, clothing, etc., have suffered materially by the nature of the services peculiar to Light troops ; and the interior economy has necessarily been deranged by so many detachments.

' To establish that regularity without which no regiment can serve with credit, and to supply those deficiencies which have taken place, I think it necessary to request that some time may be allowed before the Battalion is sent on active service.  I shall not lose a moment to effect these objects, but I think one month or at least three weeks will be required.  I have, etc.

' W. G. DAVY.

' To Brigadier General Sontag.'

'Lisbon, March 20th, 1809.

' Agreeably to the commands of 7th December 1808, of H.R.H. the Commander-in-Chief, communicated to me through you by letter bearing date 8th January, 1809, 44 men of suspected character have been selected and sent on board ships to be conveyed to England.

' The number of deserters, prisoners on board ships, amounts to 110, amongst which by Major Woodgate's report (93 of them belonging to the Companies detached under his Command on the 21st December, 1808) there are some he has reason to believe of good character who may have been misled by the bad conduct of others ; but in order to hold forth to the Battalion a striking example (at this moment so particularly necessary) of the consequences attending upon desertion, I have refrained from making any application to his Excellency to have them given up to the Regiment.' [1]

## 4

Our steps must now be retraced a little in order to pick up the thread of the general history.

In obedience to instructions from home, Sir Hew Dalrymple resigned the command to Sir H. Burrard on October 3, 1808. Three days afterwards Burrard was virtually superseded by the appointment of Sir John Moore to the command of the army destined to advance into Spain ; for Moore was to take with him the great bulk of the troops, leaving Burrard only some 8000 men for the defence of Portugal. With a laudable absence of jealousy, Sir Harry gave Moore his best assistance in making the necessary preparations ; but during the month of November was also summoned home ; and the force in Portugal was left without a permanent commander until the arrival in the following month of Lieut.-General

---

[1] The Monthly Returns, October 1808 to April 1809, are a little puzzling. The number of deserters during that period is stated at 301. On the other hand, 242 men are shewn as having ' joined from desertion ' and 27 others as ' joined from missing, prisoners of war.' It would therefore appear that men momentarily absent were struck off as deserters, but rejoined almost immediately.

Sir John Craddock, who, like Burrard, was senior in the service to Sir John Moore.

Meanwhile Moore had experienced the fallacy of relying upon Spanish support. The Spanish armies were routed, Madrid was captured by the French; and the Emperor Napoleon, who had recently assumed personal command, despatched his advance and flank guards to threaten the frontier of Portugal and the southern provinces of Spain. Sir John Moore, even after forming his junction with the reinforcements despatched from England under Sir David Baird, had under his command a force of less than 30,000 men, almost unsupported by Spanish troops, and pitted against Napoleon's army of 300,000. It was under these circumstances that with brilliant strategy and splendid audacity he struck at Napoleon's line of communications, and forced the Emperor to recall his advanced Army Corps and to concentrate all the troops at hand for the purpose of overwhelming his bold assailant.[1]

[1] State of the French Army in Spain.   October 10, 1808.

|  | Commander | Strength |
|---|---|---|
| 1st Army Corps . . | Marshal Victor [1] . . . . | 34,000 |
| 2nd ,, ,, . . | ,, Soult [2] . . . . | 33,000 |
| 3rd ,, ,, . . | ,, Moncey [3] . . . . | 37,700 |
| 4th ,, ,, . . | ,, Lefébre [4] . . . . | 26,000 |
| 5th ,, ,, . . | ,, Mortier [5] . . . . | 26,700 |
| 6th ,, ,, . . | ,, Ney [6] . . . . | 38,000 |
| 7th ,, ,, . . | General G. St. Cyr [7] . . . | 42,000 |
| 8th ,, ,, . . | ,, Junot . . . . | 25,700 |
| Reserve . . . | . . . . . | 42,400 |
|  | Total . . | 305,500 |

[1] Duc de Bellune.

[2] Duc de Dalmatie : a good and able soldier under forty years of age.

[3] An officer of the royal army before the Revolution. When, in 1840, the remains of Napoleon were brought to Paris from St. Helena, they were received at Les Invalides by Marshal Moncey, Duc de Conegliano, then in extreme old age.

[4] Duc de Dantzig, a person of no great ability.

[5] Duc de Trévise, a good soldier.

[6] Afterwards Prince of the Maskowa, not a man of great ability, excepting in command of a rearguard during a retreat.

[7] An indolent man, but of great military talent.

Moore evaded the stroke, and, pursued by the Army Corps under Marshals Soult and Ney, retired through Astorga and Lugo to Coruña, where he repulsed the enemy's attack upon his position. Sir John lost his own life in the action ; but he had saved Spain. Among those who were with him at the last was his old and intimate friend, Paul Anderson of our Regiment, whose sash, stained with the General's blood, is still preserved in the Museum at the Royal United Service Institution.

Moore's Army embarked for England. Soult was now free to attack Portugal from the north. Unexpected hindrances however occurred, and it was not until March 28, 1809, that he succeeded in capturing Oporto. The Emperor himself had returned to France early in January.

Meanwhile the position of Sir John Cradock had been one of considerable difficulty. The force remaining in Portugal after Moore's departure consisted of two squadrons of Light Dragoons, ten British and four K.G.L. battalions, with thirty guns, of which, however, six only were horsed. Five of the battalions were sent to reinforce Moore : of the remainder, one battalion garrisoned Almeida, and another Elvas. Three British and four German battalions occupied the neighbourhood of Lisbon, where the mob displayed so much ill-feeling that it seemed likely to be our first assailant. At the end of the year dispatches from Sir John Moore showed the peril of the general situation, which was increased by news of the approach of the French 4th Army Corps along the valley of the Tagus. But as the year expired the effect of Moore's advance began to tell. The 4th Army Corps was halted and recalled to Placentia. Thus the pressure was relieved, and the gradual arrival of further reinforcements raised Cradock's numbers to about 14,000.

Early in March, Major-General William Beresford (at

II.                                                          G a later date Colonel-in-Chief of our Regiment) arrived in Lisbon in response to the request of the Portuguese Government that a British officer might be sent to train and organise its forces.  He was assisted by a considerable number of British officers, who were posted to the Portuguese Army with a step of local rank in the British, and a second step in the Portuguese Service ; Captains in His Majesty's Army thus becoming Lieut.-Colonels in that of Portugal.  A good many Rifle officers applied for service in the Portuguese Army ; among others, Gustavus Braun, von Linstow, Zuhleke, etc.

During this period the British Government had been drifting along apparently in two minds : at one moment the Ministers seemed ready to abandon Portugal, and the return to England of Moore's worn and ragged troops after the Battle of Corunna served to confirm them in this opinion ; at another they furnished Sir John Cradock with reinforcements.  While in the latter mood they despatched Major-General Hill with six battalions, which landed in Lisbon in April, raising Cradock's force to about 19,000 officers and men.  This reinforcement enabled him to make head against Soult.  Cradock accordingly advanced to Leiria, but at this point was relieved by Sir Arthur Wellesley, who had been sent from England to supersede him.

## CHAPTER VI

THE French Emperor had by this time brought into Spain an army of some 330,000 men. Marshal Soult was not the only enemy in the field. He, indeed, was threatening Portugal from the north, but Marshal Victor with the 1st French Army Corps was also menacing it by the valley of the Tagus. The question was, which of the two must be attacked first? Sir Arthur, taking advantage of his interior lines, decided to attack Soult with his main body, leaving Major-General Mackenzie with 13,000 men—9,000 of whom were Portuguese—to hold the right bank of the Tagus between Abrantes and Santarem, and act as a retarding force in the event of Victor's advance.

On April 24 Sir Arthur assumed command, and the General Orders of the 27th notify the appointment of two officers of our Regiment to the Staff : viz. Lieut.-Colonel James Bathurst, as Military Secretary, and Captain Francis Cockburne, D.A.A.G. In the G.O. of May 7 Ensign Morgenthal, whom we last saw as D.A.A.M.G. in Sir John Moore's Corps, ' is continued as attached to the Q.M.G.'s Department.' His special *métier* was that of a draftsman.

Preparations were made for an immediate advance, but before it took place the following General Order was issued :—

'Coimbra. May 4, 1809.

' The Light Infantry Companies belonging to, and the Rifle-men attached to each Brigade of Infantry are to be formed together on the left of the Brigade under the command of a Field Officer or Captain of Light Infantry of the Brigade to be fixed upon by the Officer who commands it. Upon all occasions in which the Brigade may be formed in line or in column, when the Brigade shall be formed for the purpose of opposing an enemy, the Light Infantry Companies and Riflemen will be of course in the front, flank or rear according to the circumstances of the ground and the end of the operations to be performed."

The British Army was organised as follows :—

### CAVALRY DIVISION
*Lieut.-General Payne*

|  |  |  | Sabres. |
|---|---|---|---|
| 1st Brigade | Major-Gen. H. Fane | 3rd Dragoon Guards, 4th Dragoons  .. | 1314 |
| 2nd  ,, | ,, Stapleton Cotton | 14th, 16th, 20th Light Dragoons . | 1432 |

Royal and K.G.L. Artillery, Brig.-General Howarth, 30 guns.

### INFANTRY BRIGADES

|  |  |  | Rank & File. |
|---|---|---|---|
| Brigade of Guards | Brig.-Gen. H. F. Campbell | Coldstream Guards, 3rd Guards, No. 6 Coy. 60th Riflemen | 2292 |
| 1st Brigade | Major-Gen. Hill  . | . 3rd, 48th, 66th ; No. 5 Coy. 60th | . 2022 |
| 2nd  ,, | ,, Mackenzie | . 24th, 27th, 31st, 45th  .   .  . | 2653 |
| 3rd  ,, | ,, Tilson  . | . Headquarters, and Nos. 2, 3, 4, 9, 10 Coys. 60th  .  . | 1771 |
|  |  | 87th, 88th, 1st Portuguese Grenadiers |  |
| 4th  ,, | Brig.-Gen. Sontag  . | . 97th, 2nd Bn. Detachments, 2nd Bn. 16th Portuguese, No. 7 Coy. 60th  .  . | 1674 |
| 5th  ,, | ,, A. Campbell | . 7th, 53rd, 1st Bn. 16th Portuguese, No. 1 Coy. 60th  .  . | 1590 |

Rank & File.

| | | | |
|---|---|---|---|
| 6th Brigade | Brig.-Gen. Stewart . | . 29th, 1st Bn. De-<br>tachments, 1st Bn.<br>16th Portuguese | . 1702 |
| 7th ,, | ,, Cameron. | . 9th, 83rd, 2nd Bn.<br>10th Portuguese, No.<br>8 Coy. 60th . | . 1617 |
| 1st K.G.L. | Major-Gen. J. Murray | . 1st and 7th Bns.<br>K.G.L. . . | . 1662 |
| 2nd ,, | Brig.-Gen. Drieburg . | . 2nd and 5th Bns.<br>K.G.L. . . | . 1402 |

In accordance with Battalion Orders, dated Coimbra, May 4, the Rifle Companies were distributed as follows :—

Major Davy with Bn. H.Q. was in General Tilson's—the 3rd Brigade. The five Companies present were :—

   No. 2. Captain T. Macmahon.
   ,, 3.  ,,  A. Andrews.
   ,, 4.  ,,  P. Blassière.
   ,, 9.  ,,  J. Schöedde.
   ,, 10. Lieut. F. Steitz.
   10 Officers, 306 Other Ranks.

Captain de Salaberry being, as already mentioned, Brigade Major to General de Rothenburg, never served in the Peninsula.

The remaining companies were attached to Brigades :—

| | | | | | |
|---|---|---|---|---|---|
| Brigade of<br> Guards . | No. 6 Coy. | Capt. T. Hames | . 2 Officers | | 61 O.R. |
| 1st Brigade | ,, 5 ,, | ,, M. De Wendt | . 2 | ,, | 61 ,, |
| 4th ,, | ,, 7 ,, | Lieut. G. H. Zuhleke | . 2 | ,, | 61 ,, |
| 5th ,, | ,, 1 ,, | Bt.-Major J. Galiffe | . – | | 64 ,, |
| 7th ,, | ,, 8 ,, | Capt. J. A. Wolff | . 2 | ,, | 60 ,, |

These five companies were under the superintendence of Major Woodgate, who took his instructions directly from the Adjutant-General. The Battalion mustered in all 26 officers and 617 O.R.   Of the latter 56 were, however, sick and 4 ' on command.'

Every care was taken by Major Davy for the maintenance of an efficient system of discipline and interior economy. Daily parades in camp for purposes of inspection were ordered whether the Battalion was on the march or halted. Supplies of clothing were, however, evidently insufficient, so much so that Major Davy found himself obliged to authorise the men to cut away the skirts of their jackets and patch up the upper part. On halting days officers commanding companies were directed to inspect their men at 11 o'clock roll-call, and to pay particular attention to the state of the arms, ammunition, and accoutrements.

' The blue pantaloons which have been torn on the march are to be immediately repaired, and those men who are under the absolute necessity of wearing their white drawers must appear with them well washed and clean (!).

' The stocks which are worn and the rosettes and bugles which have been torn out of the caps are to be replaced.'

On May 6 the issue of the following General Order gave evidence that the confidence of Sir A. Wellesley in his Riflemen was in no way diminished by the cloud which had momentarily shadowed the battalion :—

' The Commander of the Forces recommends the Companies of the 5th Battalion of the 60th Regiment to the particular care and attention of the General Officers commanding the Brigades of Infantry to which they are attached. They will find them to be most useful, active and brave troops in the field, and that they will add essentially to the strength of the Brigades.

' Major Davy will continue to superintend the economy and discipline of the whole Battalion, and for this purpose will remain with that part of the Army which will be most convenient to him with that object.'

In Battalion Orders dated Coimbra, May 5, 1809, the
C.O., after referring to the detail to be observed by the
detached companies, called upon every officer

'to keep up the interior economy and discipline of the Battalion
under circumstances in which, he is well aware, there are great
difficulties to be encountered ; but he trusts their zeal and attention
will surmount them.

'5. Ensign Furst will act as Adjutant to Major Woodgate's
detail.

'6. Lance-Sergeant Schwalbach to be attached to Captain
Wendt's Company.'

The non-commissioned officer mentioned in the last
paragraph deserves something more than a passing mention.
In the muster roll, December 1809, there is a note against
his name to the effect that he had been discharged on
June 14 by order of the Lieut.-General Commanding.
It was intended to utilise his services in a higher sphere.
Schwalbach, born at Tréves in 1774, had already dis-
tinguished himself in the Regiment, and was rewarded
by being given a commission in one of the Caçadores
(*i.e.* Light Infantry) Regiments of the Portuguese Army.
Thenceforward he adopted Portugal as his country.  He
was engaged in the battles of Bussaco, Fuentes de Oñoro,
Vitoria, and the Pyrenees, where he was severely wounded.
After the termination of the war Schwalbach was appointed
Military Governor of the Province of Alemtejo and afterwards of Lower Beira.   In 1839, having attained the rank
of Major-General, he was employed as Special Commis-
sioner in Brazil ;  and eight years later we find him a
General Officer in the Royal Army and ennobled with the
title of Viscount de Setabal.

A portrait of Schwalbach may be found in the *Illustrated London News* of January 2, 1847, accompanied by
the following remarks :—

' Johan Schwalbach is one of the Commanders of the Royal troops at present operating with Marshal Saldanha against Das Antas and the insurgents.  As is evident from his name he is a German, but he has been for many years in the Portuguese service : he was a favourite officer of Dom Pedro,[1] who raised him from the rank of Lieutenant to that of General, and gave him his title.  His latest achievement is the occupation of that place, an important town, from which he takes his title.  He is much dreaded by the Democrats, and they evacuated it—Popular Forces, Junta and all— three hours before he arrived. . . . Schwalbach is a brave soldier and an able General ; he is a military Commander and nothing else, never having mixed himself up with politics at all.  It may be observed in passing that both Marshal Saldanha and his opponent, General Das Antas, were under Lord Wellington in the Peninsula.'

General Schwalbach died in May 1847.

## 2

Sir Arthur quickly decided on his plan of campaign. Feeling that Soult must first be dealt with, and taking advantage of interior lines, he determined to attack that Marshal with the bulk of his force.  Leaving, therefore, General Mackenzie with his own Brigade, Fane's Cavalry Brigade, and 9000 Portuguese to watch Marshal Victor, who had concentrated 30,000 men on the Guadiana with headquarters at Merida, Wellesley concentrated the re- mainder of his army during the first days of May at Coimbra.

During all this time Marshal Soult remained stationary at Oporto, some eighty-five miles north of Coimbra.  His outposts were, it is true, pushed forward to the bank of the Vouga, about fifty miles south of Oporto, but the wave of invasion which had brought him from Coruña appeared to have spent itself, and his power to resume the offensive seemed exhausted.  Hemmed in as the Marshal had been between the sea and the Tamega, his most recent

---

[1] Emperor of Brazil.

MAJOR-GENERAL JOHANN SCHWALBACH.
(By kind permission of *The Illustrated London News*.)

operations had been directed to getting possession of that river, and by so doing to obtain a new line of retreat in case of need by way of the bridge of Amaranté. His divisional General, Loison, had been detached for the purpose, and captured the bridge. Soult's remaining Divisions were commanded by Generals Mermet and Merle.

Although the official unit or organisation did not go beyond the Brigade, Wellesley quickly formed what were to all intents and purposes four Divisions, by allotting the Guards, with the 4th and 5th Brigades, to General Sherbrooke, and the 6th, with the 1st and 2nd Brigades K.G.L., to Lieut.-General Edward Paget. The 1st and 7th Brigades were given to General Hill; and the 3rd Brigade was placed under the orders of Beresford, who had also with him about 6000 Portuguese and 30 guns.

On May 7 Sir Arthur began his advance on Oporto. Cotton's Cavalry Brigade, forming with Paget's Division the advance guard, led the way; Sherbrooke followed; Hill marched by the sea road via Aveiro. Beresford moved simultaneously upon Lamego, about eighty miles distant, which it was hoped would be reached on the 10th or 11th, although the loss of the bridge over the Tamega precluded any great hope of cutting off the French retreat. The Portuguese being but half trained, he was ordered to run no risks.

On the morning of the 10th General Hill had not only reached Ovar, but turned the flank of General Franceschi, who was in command of Soult's advanced cavalry posts. But for an accident Franceschi would hardly have made good his retreat. As it was, he retired upon Mermet's Division at Grijo. The following evening both Generals retreated across the Douro into Oporto, and the same day Beresford drove Loison's Division from the Upper Douro to the Tamega.

Feeling himself unequal to further operations on the
offensive, Marshal Soult had already been contemplating
a retreat on Braganza.   On hearing of the British advance
he decided to wait no longer.   Amaranté was the key to
his position, and Loison was directed to hold the bridge
to the last.   General Mermet's Division was directed on
Vallongo ; the bridge of boats over the Douro was broken,
and all barges and other vessels were conveyed to the
right bank.   The report received from Franceschi led the
Marshal to believe that Hill's Division had embarked on
board ship, and might attempt a landing at the mouth of
the river.   He therefore took up a position on high ground
overlooking it.   A frontal attack from the further side
of a river, deep, rapid, and 300 yards broad, appeared out
of the question.   Nevertheless, even in the face of 10,000
men, Sir Arthur Wellesley was resolved to force the passage.
He had little hope of intercepting the enemy's retreat,
but feared that in retiring the French might overwhelm
Beresford.   Sir Arthur therefore despatched half of the
K.G.L. Brigades with two squadrons and two guns, under
General Murray, to Arnetas, three or four miles higher
up the river, and added Hill's Division to Paget's remaining
troops.   Murray, finding boats, proceeded to cross the
river.   At 10 A.M. Wellesley heard of the fact, and about
the same time Colonel Waters, A.A.G., most cleverly
secured unobserved four barges moored on the further
side and brought them across to the left bank.   A party
of men belonging to the 1st Brigade thereupon crossed,
and occupied a large empty building named ' The Seminary '
on the opposite side.   Its presence was soon discovered
by the enemy, and the British troops with difficulty
maintained their ground.   Reinforcements were sent over
with as little delay as possible, but the contest was doubtful
until Murray was observed advancing from Arnetas by

the right bank. Then the French, seeing their line of retreat menaced, retired in confusion along the road to Amaranté. It was thought that Murray might have done something to obstruct their retreat, but he remained inert. Sir Arthur, perhaps unfortunately, made no further attempt to pursue them. The passage of the Douro had cost him but 20 killed and 95 wounded ; 60 guns were captured in Oporto.[1]

Marshal Soult quickly restored order in his army, but on the 13th received the crushing news that General Loison had abandoned Amaranté. His position consequently became one of the utmost peril. But Soult's talents always shone most brightly in adversity. Quitting the main road he took to a mountain track, and after many hair-breadth adventures succeeded on the 19th in escaping from his pursuers, and crossing the frontier into Spain.

### 3

Hardly had Sir Arthur completed the expulsion of the French from Portugal when news arrived that on May 14 Marshal Victor had advanced against Alcantara, had

---

[1] Napier states that the party which first crossed the Douro consisted of one officer and twenty-five men. He does not state to what regiment they belonged. Later writers say this party belonged to the Buffs, but quote no authority for the statement. Captain Barghan, A.D.C. to Sir Arthur, in a private contemporary letter gives the strength of the party as ' forty men,' but also mentions no regiment. The little detachment may quite possibly have been taken from the Buffs, but it would appear at first sight to be more probably the rifle company attached to the Brigade, one of whose primary functions was to furnish advance guards. Colour is lent to this surmise by General Hill's order to his Brigade next day, in which, after ' acknowledging the highly distinguished conduct of the Brigade under his command in the action of yesterday,' he continues, ' The Major-General feels the highest satisfaction in returning his best thanks to Lieut.-Colonel Drummond, and *Captain de Wendt and the Company of the 60th attached to his Brigade, and begs that the same will be conveyed to the Officers and men of this gallant Corps.*' General Hill would hardly have spoken of his Riflemen in such high terms had they not performed some special feat. The only fact however positively known is that a Rifleman shot a French officer on a wall ! ! !

forced the bridge over the Tagus, and occupied the town. Wellesley at once put his troops in motion southward to reinforce Mackenzie, and by June 8 was at Abrantes; but the alarm proved groundless. Victor showed but little enterprise, and halted between the Tagus and the Guadiana. Wellesley also rested; for though anxious to advance against Victor, he was short of ready money, and had also difficulty in arranging a satisfactory plan of campaign with Don Gregorio Cuesta, who commanded the Spanish troops on the Upper Tagus. Time, too, was required for organising supply. Almost the only means of transport was by pack mule; the consequence being that on service the army was followed by a string of mules many miles long.

Technical instruction was required for Staff and regimental officers; and the administration of the army had to be arranged on a sound and practical basis.

As to the men, and particularly the light troops, the French General Foy, who for years was opposed to the British Army in the Peninsula, writing shortly after the close of the war, makes an elaborate comparison in his 'Memoirs' between the English and French troops; and speaking of our own particular arm, says that Napoleon added to each regiment a picked company, called Voltigeurs, composed of men of small stature but active and intelligent. These Voltigeurs constituted the Light Infantry of the French Army, and habitually acted as skirmishers which harassed the enemy, escaped by their speed from his troops in force, and from his artillery by its extended order. It was, he says, thought

'that the English soldier had not enough intelligence and smartness to combine with the regular duties of the Line the individual action of the skirmisher. When the need of a special light infantry began to be felt the best marksmen of different corps were at first selected;

but it was afterwards found advisable to make the 8 Battalions of
the 60th, and 3 Battalions of the 95th and some of the foreign Corps
exclusively into skirmishers.   These troops were armed with the
rifle.   During the late war Companies of these Riflemen were always
attached to the various Brigades.   The echoing sound of their
horns answered the double purpose of directing their own movements
and of signalling such movements of the enemy as would otherwise
have escaped the notice of the General in command.'

Marshal Soult's remarks on the Riflemen of the 60th
will be shown in due course.   French writers never speak
of them excepting in terms of respect.

Meanwhile the British Commander was planning a
sudden descent upon the Tagus, with the object of seizing
the bridges at Almaraz and severing Marshal Victor's
communications with Madrid.   But Victor evaded the
blow by a movement in retreat, and the sole immediate
result of his threatened attack was the destruction of the
Alcantara bridge, which was blown up by our Engineers
in a moment—as it turned out—of groundless alarm.

The time spent at Abrantes was, however, by no means
wasted.   The improvement of staff work and organisation
occupied every spare moment of the Commander-in-Chief.
The office of Chief of the Staff was at that time unknown
in the British Army, and the work of such an officer was
performed by Wellesley himself.   The Adjutant-General, Sir
Charles Stewart, was chiefly useful as being an intermediary
with his brother Lord Castlereagh, Secretary of State.
In other respects he left much to be desired ; being deaf,
short-sighted, ignorant of his work, and deficient in loyalty
to Sir Arthur.   The Q.M.G., Colonel George Murray, had
all the makings of an excellent Staff officer ; and with
tuition and experience gradually became invaluable to
his chief, who also, later on, found an able assistant in
his Military Secretary, Colonel FitzRoy Somerset, known

in the evening of life as Lord Raglan, Commander-in-Chief of the British Army in the Crimea.

The Commissariat and Transport services were re-organised, and by degrees the army became a mobile machine.

## 4

The situation of the French Army in Spain was as follows :—

The 7th Army Corps was engaged in Catalonia and the 3rd in Aragon. These two Corps, something over 50,000 strong, were outside the sphere of Wellesley's operations.

By the express orders of the Emperor, the 2nd, 5th, and 6th Corps, 54,000 strong and 107 guns, had been formed into an army under Marshal Soult.

The 4th Corps, now commanded by General Sebastiani, in La Mancha, was in communication with the 1st Army Corps under Marshal Victor, now operating on the Tagus ; but its primary objective was a Spanish army commanded by General Venegas, which threatened a march northward on Madrid.

By the end of June Sir Arthur was ready to advance by the line of the Tagus against Madrid. Strict orders were issued for the march. The troops were always to start at daylight in order that their halting place might be reached as early in the day as possible. They were to halt for five minutes every hour and a half. Entrenching tools were to be carried in the proportion of eight spades, eight shovels, four picks and four axes for each regiment of cavalry ; and five spades, five shovels, and five axes for each infantry battalion—a ridiculously small supply according to modern notions !

With the exception of the Rifle Regiments, the infantry was armed with the clumsy musket known as the ' Brown

Bess,' whose accurate range hardly exceeded one hundred yards. The heaviest field gun was the 9-pounder; the 6- or even 3-pounder being more common. The effective range of artillery fire did not exceed 600 or 800 yards. The guns were mostly drawn by teams of oxen, and it was only in course of time that all were regularly horsed, with eight in pairs, driven by men of the driver corps. Each battery consisted of five guns and a 5·5 howitzer, firing a 24-pound shell.

While at Abrantes Sir Arthur Wellesley organised the British Army in permanent Divisions as follows :—

### CAVALRY DIVISION
*Lieut.-General Payne*

|  |  |  | Rank & File. |  |
|---|---|---|---|---|
| 1st Bde. | Brig.-Gen. Fane | 3rd Dragoon Guards | 525 | |
| | | 4th Dragoons | 545 | 1070 |
| 2nd ,, | Maj.-Gen. Cotton | 14th Light Dragoons | 464 | |
| | | 16th ,, ,, | 525 | 989 |
| 3rd ,, | Col. Anson | 23rd Light Dragoons | 459 | |
| | | 1st Light Dragoons K.G.L. | 451 | 910 |
| | | | | 2969 |

One 3-pounder battery R.A.

### 1ST INFANTRY DIVISION
*Lieut.-General Sir John Sherbrooke, K.B.*

| Brigade of Guards | Br.-Gen. H. Campbell | 1st Bn. Coldstream | 970 | |
|---|---|---|---|---|
| | | 1st Bn. 3rd Guards | 1019 | |
| | | No. 6 Coy. 60th Riflemen (Capt. Hames) | 56 | 2045 |

7th Inf.                                                    Rank & File.
  Bde.    . Br.-Gen. Cameron   . 1st Bn. 40th Regt. .  745
                              2nd Bn. 83rd  ,,   .  535
                              1st Bn. 61st  ,,   .  778
                              No. 8 Coy. Rifles
                              (Capt. Wolff)   .   51
                                                   ——
                                                       2109
1st Bde.      ,,    Langwerth . 1st Bn. K.G.L.   .  604
  K.G.L. .                  5th ,,    ,,    .  610
                                                       ——
                                                       1214
2nd Bde.      ,,    Lowe     . 2nd Bn. K.G.L.   .  678
  K.G.L. .                  7th ,,    ,,    .  557
                              1st, 2nd Lt. Bns.   .  106
                                                     ——
                                                       1341

                                                       6709

One 6-pounder battery R.A.
Two 6-pounder batteries K.G.L.

### 2ND DIVISION
*Maj.-Gen. Rowland Hill*

1st Bde.  . Maj.-Gen. Tilson    . 1st Bn. The Buffs  .  746
                              2nd Bn. 48th .   .  567
                              2nd Bn. 66th .   .  526
                              No. 5 Coy. Rifles
                              (Capt. De Wendt)  52
                                                   ——
                                                       1891
6th Bde.  . Br.-Gen. R. Stewart . 29th     .     .     .  598
                              1st Bn. 48th .   .  807
                              [1]1st Bn. Detachments 609
                                                    ——
                                                        2014

                                                       3905

### 3RD DIVISION
*Maj.-Gen. Mackenzie*

2nd Bde.  . Maj.-Gen. Mackenzie  2nd Bn. 24th .   .  787
                                  2nd Bn. 31st .   .  733
                                  1st Bn. 45th .   .  756
                                                   ——
                                                       2276

[1] The Battalions of Detachments were composed of men left behind in Portugal sick when Sir J. Moore advanced into Spain, or of stragglers from his army.

Rank & File.

3rd Bde. . Col. Donkin    .    . 2nd Bn. 87th .    . 599
                                 1st Bn. 88th .    . 599
                                 Nos. 2, 3, 4, 9, 10
                                 Coys. Rifles (Major
                                 Davy)        .    . 273
                                                      ——  1471

                                                          3747

One 6-pounder Battery R.A.


### 4TH DIVISION
*Brig.-Gen. A. Campbell (acting)*

4th Bde.    Br.-Gen. Sontag    . 97th    .    .    . 502
                                 2nd Bn. Detach-
                                 ments .    .    . 625
                                 No. 7 Coy. Rifles
                                 (Capt. Prevost)    . 56
                                                      ——  1183
5th Bde. .      ,,     A. Campbell 2nd Bn. 7th Fusiliers 431
                                 2nd Bn. 53rd .    . 537
                                 No. 1 Coy. Rifles
                                 (Bt.-Maj. Galiffe) . 64
                                                      ——  1032

                                                          2215

One 6-pounder Battery R.A.


### SUMMARY

Cavalry   .        .        .        .    2,969 Sabres
Infantry  .        .        .        .    16,576 R. & F.
Artillery .        .        .        .    1,011 Men, 30 guns
R.E.      .        .        .        .        22 Sappers
Royal Staff Corps  .        .              63
                                          ————
                                          20,641 men, 30 guns.

The total of all ranks may be estimated at about 23,000.

The Rifle officers actually present appear to have been distributed as follows : Major Davy, commanding ; Lieut.

II.                                                    H

De Gilse, Adjutant; Lieut. J. A. Kemmeter, Quarter-master.

|  | 1st Division |  | 2nd Division |
|---|---|---|---|
| No. 6 Coy. | Capt. Hames. | No. 5 | Capt. de Wendt |
|  | Ensign Joyce |  | Lieut. Muller |
|  | ,,    Altenstein |  |  |
| ,, 8 ,, | Capt. Wolff |  |  |
|  | Lieut. Friess |  |  |
|  | ,,    Sawatzky |  |  |

|  | 3rd Division |  | 4th Division |
|---|---|---|---|
| No. 2 Coy. | Lieut. Koch | No. 1 | Brevet-Major Galiffe |
|  | ,,    Eberstein |  | Lieut. Ritter |
| ,, 3 ,, | Capt. Andrews | ,, 7 | Lieut. Zuhleke |
| ,, 4 ,, | ,,    Blassière |  | ,,    Du Châtelet |
|  | Lieut. Holmes |  | ,,    Mitchell |
|  | ,,    Franchini |  |  |
| ,, 9 ,, | Capt. Schöedde |  |  |
|  | Lieut. Furst |  |  |
|  | Ensn. Wynne |  |  |
| ,, 10 ,, | Lieut. Steitz |  |  |
|  | Ensn. Barbaz |  |  |

The Spanish forces under General Cuesta, destined to co-operate with the British amounted to about 38,000 men and 70 guns, and the force under General Venegas comprised 25,000 men.

Opposed to the Allies were the 1st and 4th French Army Corps, amounting to 46,000 men and 78 guns; the Reserve Division, under General Dessolles, 7000 men; and King Joseph's Guards, about 5000 strong. These bodies of troops, although not yet concentrated, were combined in an army under command of King Joseph in person, who had with him as C.G.S. Marshal Jourdan. Their base of operations was Madrid. Wellesley was unaware of the fact that, in addition, Marshal Soult had direct orders from the Emperor Napoleon to march southward from Salamanca with the 2nd Corps, and assail the British left and rear. Other miscalculations underrating the strength of the French were also made, and the Allied commanders believed that it would be possible by combined action to drive the enemy from Madrid and force him to evacuate the southern provinces of Spain. But the French were operating on short interior lines—the distance between the King at Talavera and Sebastiani at Toledo being not much above fifty miles—while the Allies were on exterior lines with divergent bases. At Abrantes Wellesley was 180 miles distant in a straight line from Almaraz; and even after effecting a junction with Cuesta, would still be considerably more than 100 miles from Venegas at Ciudad Real.

## 5

At daylight on June 27 Sir Arthur marched from Abrantes: two squadrons of cavalry and the Headquarter Wing of our 5th Battalion forming the vanguard of the army in its advance. On the 30th he reached Castello Branco. On July 10 he had concentrated his force at Plasencia, where, from want of supplies, he was detained until the 17th. On realising the British advance Marshal Victor had fallen back, and on June 28 had taken up a position behind the Alberche, a few miles east of Talavera de la Reyna. Wellesley effected his junction with Cuesta.

Marshal Victor was still in position as the Allies approached. He had not been joined by the King or the 4th Corps, and his isolated position invited attack; but warned in time, the Marshal retired behind the Guadarama River covering Toledo, and effected the concentration of the whole army. Cuesta, following him up, was roughly handled by the French, and barely rescued from disaster on the 26th by the arrival of the 1st and 3rd British

Divisions, which covered his retreat. The repulse of the Spaniards was not without its advantage, for their General allowed Sir Arthur to take command of both armies ; and the latter took up a position for battle with his right resting on the Tagus at Talavera and his left on high ground three miles northward. The Spanish Army was posted on the right on such strong ground as to be practically unassailable. On its left came the 4th and 1st British Divisions in the order named. The 2nd Division carried on the line, but failed to occupy a ' round hill ' on which the left flank of the army was intended to rest. The 3rd Division was to be posted in second line. The line of battle was protected by a stream which on the left deepened into a formidable obstacle : the ground, excepting that occupied by the 2nd Division, was level and open. The French Army may be taken at 50,000 men and 80 guns. The Allies had concentrated 57,000 men and 100 guns ; but in the ensuing battle the 23,000 men and 30 guns of Wellesley's army bore the whole brunt of the French attack.

6

The dispatches of Wellington give us pretty clearly his opinion of the Rifle Battalion. Writing from Abrantes on June 17 to the Military Secretary at the Horse Guards, he says :—

' I enclose a memorial which has been given into my hands by Major Davy which I beg you will lay before the Commander in Chief.

' I believe that H.R.H. the late C.-in-C. had intended to promote all the Majors commanding battalions in the late service in Portugal [1] ; and certainly if the services of any battalion could

---

[1] *I.e.* from Mondego Bay to Vimieiro.

give to their C.O. a claim to promotion, the conduct and services of the 5th Bn. 60th Foot entitled their C.O. to this advantage. I have had every reason to be satisfied with their conduct again upon this occasion, and I shall be very much obliged to you if you will recommend Major Davy to the favourable consideration of the C.-in-C.'

And, a day or two later, to Colonel Donkin, Commander of the 3rd Brigade :—

'Abrantes, 23rd June, 1809.

' MY DEAR COLONEL,
        ' I have received your letter of the 21st.[1]  I am delighted with your account of the 5th Bn. 60th Regiment.  Indeed everything that I have seen and known of that excellent corps has borne the same stamp.'

It would be difficult to find any regiment to which the great Duke alludes in terms of higher praise than those which he invariably uses towards our Rifle Battalion ; but on this occasion his tribute is particularly gratifying as showing his unshaken confidence despite the temporary cloud already noticed as the result of enlisting French prisoners of war.

This confidence was now destined to undergo a high test.  While Wellesley was taking up his position for battle at Talavera he left the 3rd Division with Anson's Cavalry Brigade on outpost duty about a mile west of the Alberche.  In its rear the plain was studded with cork and olive trees.  In its front the trees growing more thickly formed a wood which needed careful patrolling, a duty which appears to have been entirely neglected by Anson and his cavalry.  It so happened also that the Division in retiring west of the Alberche after covering Cuesta's retreat set fire to some French huts on the further bank,

_____
[1] This letter is, unfortunately, not to be found at Apsley House among the large collection of papers, documents, etc., bearing on Wellington's campaigns.

and the wind being from the east, the smoke of the burning huts obscured the view of the British sentries.   The 2nd Brigade was holding the right and Donkin's the left of the outpost line.   At 3 P.M. on July 27 the post was surprised by a sudden appearance in force of the Divisions of Lapisse and Ruffin belonging to Victor's Corps and marching in three columns upon parallel roads leading to Talavera.   Two young battalions in the 2nd Brigade fell back, one with heavy loss ; but the 45th, a veteran regiment of perhaps 700 bayonets, held its ground.   In Donkin's Brigade matters were even worse.   Two battalions, also of young soldiers, and neither in particularly good order, fell into confusion, fired upon one another and broke in disorder.   The five Headquarter Companies of our 5th Battalion under Davy—only about 250 rifles— alone remained to stem the tide of attack.

Sir Arthur Wellesley was close at hand.   Accounts vary in detail, but it would seem that on observing the confusion he moved into a large low house named the Palacio de Salinas, and mounting to the roof grasped the situation.   The weight of the attack had compelled the 45th and Riflemen to fall back, albeit in perfect order. Hardly had Wellesley reached the roof when he found himself in danger of being cut off.   But the Riflemen proved worthy of their charge, and holding their ground gave the General time to escape.   Sir Arthur, being young and active, dropped from a window in the house, mounted his horse, and took personal command of the two battalions which, despite the pressure of the advancing masses of the enemy whose dark uniforms appeared like a huge black cloud ready each moment to burst and overwhelm them, continued to retire slowly and in good order for nearly six miles across the plain, supported by the cavalry and by some of the troops which had given way in the first instance. By the resolute bearing of the veteran battalions the danger was averted, and time sorely needed was gained for the occupation of the battle position at Talavera.

Colonel J. Campbell, who was present, states in his book, ' The British Army,' that ' the well-executed retirement which so far as permitted by the nature of the ground was made by Wings and Companies was ascertained to have excited the admiration of the French.'

Another contemporary writer says : ' A disaster was averted only by the steady conduct of the 60th Rifles and the 45th. There is no doubt that their conduct was magnificent.'

And in his subsequent G.O. and Dispatch Wellesley notices ' the gallantry, steadiness and discipline ' of the two battalions, which he states ' were conspicuous, and begs Major Davy and Col. Guard of the 45th to accept his particular thanks.'

During this episode the casualties of the 3rd Division amounted to about 400, but those of the 45th and 60th— the mainstay of the retirement—were very small. It was, perhaps, fortunate that this crucial test of steadiness and discipline fell on the two regiments which had been longest in the country. Near sunset the battle position at Talavera was at length reached, and the 2nd Brigade formed up as intended in second line behind the 1st Division ; but Colonel Donkin, whose line of retreat led him to the left flank of the position, noticing that the round hill on the left of the line was still unoccupied by the 2nd Division, seized it with his own Brigade ; whereupon the French following seized another round hill opposite it, and Marshal Victor, observing the weakness of Donkin's Brigade, attacked that officer's hill with the divisions of Ruffin and Villatte. Donkin repulsed the frontal attack, but the enemy's right turned his left, and gained the summit of the hill. Nevertheless, despite the fatigue of his men, who had been under arms for fifteen hours, Donkin, aided perhaps by the fact that it was now nearly dark, maintained his position until reinforced by three battalions of the 2nd Division; but the fight vehemently continued: the two lines were hardly twenty yards apart, and it was not till after a vigorous struggle that the French were driven off the hill.

The 1st Division on the right of the 3rd was also heavily attacked and the two Brigades K.G.L. were roughly handled. Darkness stopped the fighting. It was no doubt during this attack that Captain Wolff was wounded.

Although July 27, 1809, is a notable day in the history of our Regiment, its casualties were surprisingly small. Only three Riflemen were returned as killed, and four as wounded; but a bugler and eighteen men ' missing ' were no doubt for the most part dead or injured; and Captain Wolff, as already mentioned, was severely wounded.

On the 28th, soon after daybreak, Ruffin, supported by the division of General Villatte, once more attacked the front and flank of the British left: once more the fight was desperate; but the French, having lost 1500 men in forty minutes, broke and retired.

A lull now occurred; but at 1.30 P.M. King Joseph attacked all along the line: the Corps of Victor on the right, Sebastiani's on the left, Desolles in reserve. Wellesley surveyed the field of battle from Donkin's hill, which was evidently the key to the British position. The Spaniards were too strongly posted to be attacked, and the brunt of the battle fell upon the British line. Campbell's Division on our right, Hill's on our left, repulsed every attack; but in the centre the line was momentarily broken by the

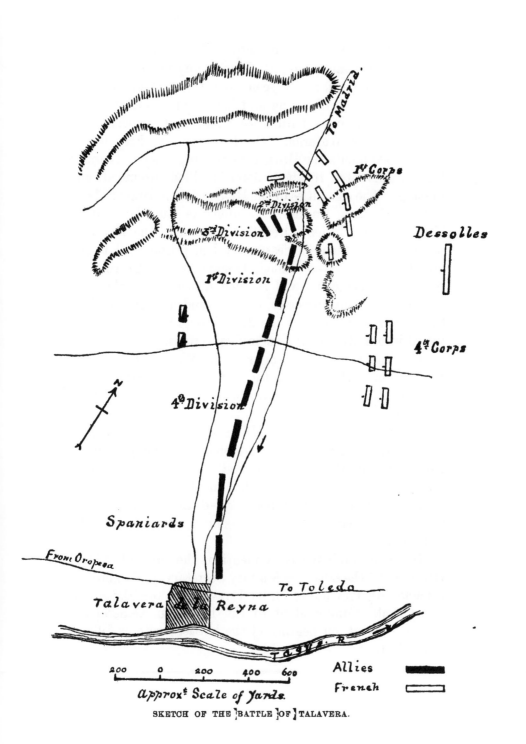

To Madrid.

1st Corps

2nd Division

3rd Division

1st Division

Dessolles

4th Corps

N

4th Division

Spaniards

From Oropesa

To Toledo

Talavera de la Reyna

Talva R.

200    0    200    400    600

Approxt Scale of Yards.

Allies

French

SKETCH OF THE BATTLE OF TALAVERA.

weight of Sebastiani's Army Corps. Timely reinforcements restored order, and at nightfall the enemy retired. Harder fighting had been rarely seen, and the British infantry, largely composed of young and half-trained soldiers, gave proofs of steadiness hardly less than that displayed six years later at Waterloo. On this day the French lost nearly 7500 men ; the British losses, proportionately much higher, amounted to about 5400 in killed, wounded, and missing. The casualties of the Riflemen were Brevet-Major Galiffe, Captain Andrews, Lieutenants Zuhleke, Ritter, Mitchell, and Ensign Altenstein wounded—the last four severely ; Lieutenant Friess was taken prisoner ; 1 bugler and 6 riflemen were killed, 1 sergeant and 24 riflemen wounded, 2 sergeants and 10 men missing. Of the last-named one, made prisoner, escaped and as mentioned by Sir Arthur in a letter of the 31st, brought to the C.-in-C. useful information of the enemy's positions.

It is mentioned in ' Stories of the Wars of the French Revolution ' ' that during the battle of the 28th, a hare was put up on the plain below our left, and was shot by a Rifleman.'

After the action the 3rd Division resumed its outpost duty. Next morning it was reinforced by a brigade under Robert Craufurd, already noticed as having been for a few years an officer in our Regiment. Mackenzie, the Divisional General, had been killed at Talavera ; and Craufurd, as senior Brigadier, assumed command of the 3rd Division.

In his dispatch Wellesley mentions the name of Colonel Bathurst, of the 60th, his Military Secretary, and expresses his acknowledgments to Major Davy of the 5th Battalion. The Colonel Commandant of the Battalion, General Sir George Prevost, on hearing of the battle, wrote as follows to the latter :—

'Halifax, N.S.    October 16th, 1809.

' DEAR SIR,—I received a few days ago your letter from Abrantes
of the 25th June, and at the same time the newspapers announced
the great and glorious victory of Talavera.   From the report of
Sir Arthur Wellesley I have the satisfaction to learn the 5th Battalion
of the 60th Regiment did its duty, and in consequence the appre-
hensions respecting the bravery of the corps are replaced by confidence and admiration.   Experience has taught me to ascribe,
with few exceptions, to the Commanding Officer of a Regiment
its perfections and imperfections. . . .

                                ' I am, dear Sir,
                                    ' Yours faithfully,
                                        ' GEORGE PREVOST.'

Major Davy received the gold medal habitually conferred on the C.O. of a unit after a victory.   In the
present case a special honour was paid to the Battalions
by the award of another medal to Captain Andrews,
senior company officer present with the H.Q. wing during
the crisis of the 27th.

## 7

Hardly had Sir Arthur had time to clear the battlefield
of Talavera, when a fresh danger presented itself.   News
arrived that Soult, marching from the north, had struck
the British line of communications at Plasencia.   On
August 3, Wellesley marched to encounter the new foe.
Next day he realised that he was in the presence of over
30,000 men, while others were not far behind; and that
the French force included not only the Army Corps of
Soult but those of Ney and Mortier in addition.   The
gravity of the situation was increased by the fact that
General Cuesta, who had undertaken to hold the position
at Talavera, abandoned it as soon as the British troops
were out of sight.   The consequence of this act was that
King Joseph at once advanced and occupied Talavera,
capturing the British wounded who had been left there in

hospital, and, among others, five Rifle officers, viz. Captain Wolff, Lieutenants Zuhleke, Friess, and Mitchell, and Ensign Altenstein.'

Without a moment's delay the British Commander despatched the 3rd Division to secure the bridge at Almaraz, and with his main body crossed the Tagus and took up a position at Jaraceio on the high ground overlooking the left bank. A fortnight later various considerations, of which the difficulty of supply was the most important, compelled him to retire, first on Merida, afterwards to the Portuguese frontier near Badajoz, whence he was in a position to hang on the enemy's flank, and prevent him crossing the Guadiana unless in great force. The position was admirably adapted for the defence of the south of Spain and Portugal. It was at this period that Sir Arthur was raised to the Peerage under the title of Viscount Wellington.

Meanwhile Major Davy was making every effort to replace the losses in his battalion, which was considerably under strength, notwithstanding the fact that its establishment had been reduced to 4 Staff Sergeants, 40 Sergeants, 40 Corporals, 20 Buglers, and 750 Riflemen.

Writing to the Secretary at War on October 17, he gives reasons for requesting the maintenance of the old establishment of N.C.Os. :—

' It is evident,' he says, ' that a Rifle Corps, acting as this has done (having constantly been employed during both the campaigns in Spain and Portugal by detachments not only of Companies but of Subalterns' and Sergeants' parties, and covering an extended chain of Outposts—one I recollect of more than 10 miles—as well as performing the duties of light troops) cannot fairly admit of a parallel with the situation of regiments of the Line, nor of an application of those regulations which are more particularly referrable to the latter : and . . . obviously demonstrates the necessity of having a greater number of officers and N.C.Os. to preserve that control which is requisite. . . .'

Major Davy goes on to remark that though by the creation of a sergeant a regiment of the line loses a marksman, a Rifle Corps suffers no diminution in this respect in its strength ; that on the contrary, sergeants are considered as the best marksmen, and by their example and instructions to others make themselves conspicuously useful. He adds that he would willingly give up ten buglemen for a similar augmentation of this class of N.C.O., the latter being by far too few for the important duties they have to discharge.

'It was a maxim of the late Sir John Moore, and which he several times repeated to me,' continues Davy, ' that " one rifleman should be deemed an equivalent, when posted, to three sentries of the line, and that the parties must in all cases be commanded by an officer or N.C.O." It would be impossible to adhere to it unless an adequate proportion of the latter is admitted. . . .'

Davy then emphasises his argument by saying how crippled he had been by the loss of so many officers at Talavera. Several of his Companies had only one officer, and one (apparently No. 10) was commanded by an Ensign (Barbaz). Under these circumstances a diminution of the establishment of N.C.Os. was particularly to be deprecated.

When this letter reached the Horse Guards the Commander-in-Chief seems to have felt that a good case had been made out, and the Battalion was allowed to retain the old number of N.C.Os.

The Secretary at War—in 1809 there was no S. of S. for War—addressee of Davy's letter, was Lord Palmerston, at that time little more than a boy, but in later life the revered Prime Minister of England.

During the month of November the distribution and state of the Battalion was as follows :—

| Place. | Division. | Brigade. | No. of Companies. | Present and fit for duty. | | | | | | | Sick. | | On Command. | | Missing. | |
|---|---|---|---|---|---|---|---|---|---|---|---|---|---|---|---|---|
| | | | | Majors. | Captains. | Lieuts. | Ensigns. | Staff. | Sergeants. | O. R. | Present. | Absent. | Sergeants. | O. R. | Sergeants. | O. R. |
| Badajoz | 1st | Guards. | 1, Capt. Hames, Lieut. Joyce | — | 1 | — | 1 | — | 4 | 43 | 5 | 7 | 1 | 1 | — | — |
| Lobon | — | Brig.-General Cameron | 1; Lieut. Furst | — | — | 1 | — | — | 5 | 47 | 3 | 4 | — | 4 | — | 1 |
| Montijo | 2nd | Major-General Tilson | 1, Capt. Wend | — | — | — | 1 | — | 2 | 46 | — | 4 | — | 2 | — | 6 |
| Campo Mayor | 3rd | Brig.-General MacKinnon | 5 | 2 | 4 | 7 | 6 | 6 | 20 | 252 | 17 | 24 | 3 | 7 | 1 | 19 |
| Olivença | 4th | Brig.-General Campbell | 1; Brevet-Major Galiffe, Lieut. Du Châtelet | — | 1 | 1 | — | — | 5 | 54 | 2 | 12 | — | — | — | 1 |
| | | Col. Kemmis | 1 Lieut. Muller | — | — | 1 | — | — | 4 | 37 | 6 | 3 | — | 2 | 1 | 8 |
| Total | . | . | . | 2 | 6 | 10 | 8 | 6 | 40 | 479 | 33 | 54 | 4 | 16 | 2 | 35 |

Now that active operations were for the time being at an end, Major Davy was anxious and obtained permission to concentrate the Battalion under his own command, and early in December it marched to Monte Mor o Novo on the Mourinho River, some twenty miles west of Evora, on the road to Lisbon.

MEMORANDUM BY MAJOR DAVY

' 3rd Dec., 1809 :—The Battalion (*i.e.* its Companies separately detached) etc. commenced to march to Montemor o Novo.

' 4th Dec.    Major Woodgate's detachment of 5 Companies marched to Estremoz. I directed Major Woodgate upon his arrival at Montemor o Novo on the 7th to send the Quartermaster to Lisbon for the clothing.

' 10th Dec.    The Battalion was this day united at Montomor o Nuove, having been separated in detachments since the 4th May.'

According to Rigaud (' Celer et Audax,' p. 106), the Companies of the Battalion were re-numbered as follows :—

No.  1 Brevet-Major Galiffe
     2 Captain  MacMahon
     3    ,,     Andrews
     4    ,,     Blassière
     5    ,,     Wend (or de Wendt)
     6    ,,     Hames
     7    ,,     Prevost
     8    ,,     Wolff
     9    ,,     Schöedde
    10    ,,     de Salaberry [1]

Brigadier-General Cameron, on giving up the Company of Riflemen which had been attached to his Brigade,

[1] N.B.—The Company numeration appears to have been changed more than once during the course of the Peninsular campaigns.

wrote of it to the Commanding Officer in complimentary terms :—

'My little Company of the 60th,' said he, 'marched from hence yesterday morning agreeable to Route. I was sorry to part with it as consisting of the best behaved men I have ever had attached to me in a military situation. There has been no instance of a complaint against any one of them since under my Command. And should a similar distribution take place on any future occasion I hope you will favour me with the same Company back again.'

Military operations being almost at a standstill, Major Davy obtained leave to go to England, handing over the Command to Major Woodgate ; but before quitting Portugal early in 1810, a recommendation on his behalf, made by Lord Wellington, had been carried out, and Davy was gazetted Lieut.-Colonel of the 7th Garrison Battalion. He never again went on active service. In 1830 he became a Major-General. In 1842 he was appointed Colonel Commandant of our 1st Battalion, and held the appointment until his death in 1856, being then General and K.C.H. His residence was Tracey Park, near Bath ; and in 1879 General Rigaud, author of 'Celer et Audax,' had access to a number of regimental letters and documents which were stored there. Tracey Park is still owned by Sir W. Davy's descendants, but they have quitted it ; and the present writer has never been able to get into direct communication with any of the family, or to ascertain whether the documents in question are still in existence.

Before his departure Davy had successfully recommended for promotion Lieut. G. H. Zuhleke, who, having been wounded and left in hospital at Talavera, had consequently been made prisoner by the French when they reoccupied that town on Lord Wellington's retirement across the Tagus. From Talavera Zuhleke was taken to

Madrid, whence, writing on December 20, 1809, to Major Davy, he says :—

' Of the Regiment under your command nobody but Captain Wolf and me are here now. . . . Captain Wolf desires to be most respectfully remembered to you ; he is recovering ; one of his wounds healed up, but he has lost his martial attitude. My wounds have a good appearance, but there are still rags or bones in them which retard the healing.

' We inhabit the cells of the Convent St. Francis converted into a hospital, and are well treated.'

On April 26, 1810, Zuhleke contrived to escape, and on arrival in Lisbon applied to be posted to the Portuguese Service. He had started his military career in 1794 in the Hesse Cassel Guards ; saw service under the King of Prussia in Germany and the Netherlands, and subsequently with the British Army in the Low Countries. When the 5th Battalion of the 60th was raised he received an Ensigncy therein, and served with the Battalion during the Irish Rebellion and afterwards in the West Indies, being present at the capture of Surinam. On October 22, 1810, he was appointed Major in the 2nd Caçadores. With them he was present at Fuentes d'Oñoro, Ciudad Rodrigo, Badajoz, and Salamanca. During the subsequent retirement from Burgos in 1812 the Major defended with a detachment the Bridge of Valladolid, and on November 17 was in action near the village of Muñhoz. During that same month the command of the battalion fell to him. In the following year he was present at the battle of Vitoria. On July 30 Zuhleke received the Brevet, and on August 27 the substantive rank of Lieut.-Colonel. He served at the battles of the Pyrenees; the Nivelle, the Nive, and Orthez. His name appears in the list of the 60th for the last time in 1814.

8

At Badajoz the British troops suffered terribly from Guadiana fever, and this circumstance, coupled with the fact of several reverses sustained by the Spanish Generals acting independently of Wellington, determined the latter to carry the bulk of his army to the Mondego for the better defence of Portugal, leaving one Division only on the Tagus.

During this year matters on the Continent had gone badly for British interests. In the spring war broke out between France and Austria, and on May 22 Napoleon sustained a serious check at Essling. This, however, was retrieved by a victory at Wagram on July 6, which was followed a few days later by an armistice, and ultimately led to the Peace of Vienna in the following October.

Great Britain attempted a diversion by sending a fine army of 40,000 men in July to capture Antwerp. The operation was mis-handled. Large numbers of men perished of disease, and the remnant, who were brought back to England in October, were so inoculated with fever contracted in the Walcheren marshes, that years elapsed before the soldiers engaged were fit for further active service.

The fault lay entirely with the Government. Instead of devoting its energies to the war, an attack was made in Parliament upon the Duke of York's private life. He consequently resigned his office as C.-in-C. Time was thus wasted, and the expedition, which had it landed at the time of the battle of Essling might have had important results, did not start until the Armistice between France and Austria had been signed. Once more the forces of Great Britain had been squandered to no purpose, and during the next two years the reinforcements needed by Wellington in Portugal were not forthcoming.

## CHAPTER VII

THE eastern frontier of Portugal from the northern-most point to the mouth of the Guadiana has a length of about 400 miles, but is not clearly defined by any mountain range or river barrier.

Communications were bad. Roads in the ordinary sense of the term were practically non-existent, nor were the rivers means of communication, for they flowed in deep beds banked by precipices, and were almost entirely unnavigable. Although communications were by degrees improved by the British Army, the roads were never good enough for wheeled transport on a large scale; and the Army was followed by interminable strings of mules, which carried its baggage and were led by Spanish muleteers, whose aversion to the Portuguese was even greater than it was to the French!

The fortresses of Almeida and Elvas guarded the respective approaches to Lisbon from Salamanca and Madrid. Opposite them on Spanish territory were the fortresses of Ciudad Rodrigo and Badajoz. As the two latter were still in Spanish hands a French invading army from the south-east would find itself stopped by Badajoz and Elvas; and one from Salamanca, by Ciudad Rodrigo and Almeida.

For the defence of this line Wellington had nominally at his disposal some 35,000 British and K.G.L. soldiers, nearly the same number of Portuguese regular troops, and

25,000 Portuguese Militia.  By this time the Portuguese
Regular Army was in moderately good condition ; a
result largely due to the efforts of the British officers
attached thereto.  The Militia was ill-armed and ill-
equipped.

There was in addition a species of local militia called
the Ordanança ;  formidable in numbers and eventually
reaching a nominal total of 300,000 men, but ill-armed,
nearly untrained and unreliable.

The British Army was organised in two Army Corps.
One upon the Tagus, under Lieutenant-General Hill,
comprised a British Cavalry Brigade and the 2nd Infantry
Division, two Brigades of Portuguese Cavalry and two of
Portuguese Infantry.  The other Corps under Wellington's
own command included two Cavalry Brigades posted in
the valley of the Mondego, the 3rd Infantry Division at
Pinhel, the 4th at Guarda ;  while the 1st Division, fifty
miles in rear at Viseu, formed a reserve for both.[1]  The
main body of the Portuguese Regular Army was concentrated about Thomar, a few miles north-west of Abrantes,
whence in the event of retreat it could join either Hill or
Wellington.  The Portuguese Militia prolonged the line
northward on Wellington's left and southward on Hill's
right.

Lisbon was the base of operations for the whole army
of defence.  Depots for the supply of Hill's Corps were
formed at Belem and Abrantes ;  for that of Wellington,
at the mouth of the Mondego and Pena Cova ten miles
above Coimbra.  Provisions had to be imported from
England.

Lengthened defence of the frontier would be impossible

---

[1] The 3rd Division, under Brigadier-General Craufurd, now consisted of the
Light Brigade (43rd, 52nd, 95th) and the 3rd Brigade, 45th, 88th, 60th.  The
87th, 24th, and 31st had been taken out of the Division shortly after Talavera,
and the 45th transferred from the 2nd to the 3rd Brigade.

in the event of the fall of the northern or southern fortresses ; in which case the only course for Hill and Wellington would be to fall back, converging upon one another and eventually concentrating upon Lisbon, in the neighbourhood of which the celebrated lines of Torres Vedras had been secretly constructed.

2

Early in February 1810, Major-General Thomas Picton reported his arrival and was given command of the 3rd Division. Craufurd consequently resumed command of his Brigade ; but Lord Wellington, unwilling to place him in a position inferior to that which he had been occupying, formed a new Division for his benefit, made up of the Light Brigade and two Portuguese Caçadore battalions. In order to fill the place of the Light Brigade in the 3rd Division, a Brigade was transferred to that Division from the 4th. The Brigades of the 3rd Division were thus commanded respectively by Generals MacKinnon and Lightburne.

Soon after the new position had been taken up a General Order of February 22 for the third time assigned a company of Riflemen to each Brigade of the Army, in accordance with the plan directed in the Orders of May 4 the previous year.

The General Order continued :

' The Commanding Officer, 5th Battalion, 60th Regiment, will detach three Companies to Lieut.-General Hill's Division. . . . And as the General Officers Commanding Brigades have invariably expressed the highest satisfaction at the uniform good conduct of this valuable body of men which has always continued effective under many trying circumstances, the Commander of the Forces desires that as far as possible the same Companies may be attached as formerly to the same Brigades.

' Two Companies to be attached to Sir J. Sherbrooke's Division ; one to the Brigade of Guards, and one to Brigadier General Cameron's Brigade. Two Companies to the 4th Division, being one for Major-General Cole's, and one for Brigadier General A. Campbell's Brigades ; and three Companies, with the Head Quarters of the Regiment with the 3rd Division.'

In a letter dated Coimbra, March 6, 1810, Major Woodgate, now in command of the Battalion, writes to Davy as follows :—

' Thus far into the bowels of the land have we marched without impediment, and until the last two days without rain ; but now it pours down in torrents. Galiffe and I dined yesterday with General Fane. A Brigade of Cavalry is stationed here. . . . The 3rd Division is Commanded by your friend General Picton, and contains our old Brigade of Colonel McKinnon and another of General Lightburne. Craufurd has a Division called the Light Division, and is out of our way. I have complied with his Lordship's directions about sending the same Companies, although against my wish ; but after the compliments the Generals have made I could not do otherwise. I much wish to keep Wend with me for many reasons, but as General Hill also asked for him I sent him there again. . . .
' I sent Wend's, McMahon's and Blassière's Companies to Portalégre, where General Hill is, from Rio Mayor. The other 7 Companies march with me to-morrow to Viseu where I shall crave admittance to his Lordship and thank him for his kindness.'

The British regiments belonging to the famous Light Division were the 43rd, 52nd, and 95th, which had been trained as Light Infantry on Colonel de Rothenburg's system by Sir John Moore.

Major Galiffe writing to his late commanding officer from Cerdeiva near Guarda on April 2 remarks that McMahon, Wend, and Blassière's companies

' are with General Hill at Portalégre ; Wolf's with General Cameron at Mengualda [Mangoalde] ; Hames' with the Guards at Viseu ;

Prevost's and mine, both under my Command, with General Cole at Guarda and neighbourhood; and Andrews, Schoedde, and Salaberry's with the Staff, are attached to General Picton's Division at Trancoso. . . . On our march from Coimbra to Viseu . . . the men in general behaved very well, and they are the spoilt children of the Division; in fact we are much more liked than the 95th.

'The Headquarters are going to be removed from Viseu to Cea, half way to Coimbra on the road from Guarda to this place. Everybody here says that we are to retire; in a little time we will be under the necessity of doing so even without an enemy before us, for we begin to feel the scarcity of provisions, and already there is almost no forage for our horses.

'The French have retired also to Salamanca and Astorga; Ney has left that Army to go to Paris, and they say that as there is to be war with Russia no reinforcements will come from France, and for that reason they are going to concentrate their forces near Madrid, and will act upon the defensive; and as we are in the same situation we may remain so the whole of the campaign.'

The rumour of war between France and Russia more than two years before it actually occurred is worthy of remark; but Major Galiffe's information was not entirely justified by the event. The French had no idea whatever of taking up a defensive position. On the contrary, the Emperor Napoleon was determined to complete his work by the expulsion of the British force from Portugal. To this end he placed the following army under command of Marshal Masséna, Prince of Essling, the very best of his lieutenants, with orders to drive Wellington into the sea :—

2nd Army Corps—General Reynier.[1]
6th Army Corps—Marshal Ney.
8th Army Corps—General Junot.

Had it been possible the Emperor would no doubt have taken command in person; but he was detained in France

---

[1] General Reynier had served against the British in Egypt, when the French Army evacuated the country under the cover of a Convention (1801). In 1806 he commanded the French Division which was defeated by Sir John Stuart at Maida.

by matters of State, not the least of which was his pursuit of the Continental System, *i.e.* the exclusion of British trade from continental ports in the hope of bringing England to her knees by the ruin of her commerce.

The Corps of Ney was pushed forward to invest Ciudad Rodrigo ; and even before the arrival of the 2nd Corps, the French outnumbered Wellington's force, being about 57,000 against 32,000 men.

### 3

G.O. of April 13, 1810, directed Captain Gustavus Brown of the 60th to report himself for duty with the Portuguese Army. Brown (originally Braun) was a soldier of experience who had served with Löwenstein's regiment during the campaign of 1794 in Holland, and took part in the gallant defence of Grave, which was reduced by famine on December 29 of that year. In 1796 he was with his regiment in the West Indies and transferred to our Regiment on the raising of the 5th Battalion. He served at the capture of Ste. Lucia and Grenada, attack on Porto Rico, and the submission of Surinam. He was now posted as a Major to the 9th Caçadores, which later on was attached to the 6th Division, and his distinguished career is elsewhere described. Suffice it to say here that in August 1812 he received the rank of Lieutenant-Colonel, and commanded his battalion at Salamanca, the Pyrenees, Nivelle, and the Nive, where he was badly wounded. During the war Brown received the Gold Medal with two clasps, and Gold Cross.

In September Captain Zuhleke also joined the Portuguese Army and was posted to the 2nd Caçadores as a Major. In that rank he was present in many actions, and commanded his battalion in the greater number. As Lieutenant-Colonel he was engaged at the affair of Hastingues on February 23, 1814, when his battalion greatly distinguished itself, and Zuhleke was mentioned in dispatches.

Among other officers of the Regiment who entered the Portuguese service Lieutenants Brunig (or De Breunay), killed at the storming of Badajoz, and William von Linstow, a Prussian who served in the Loyal Lusitanian Legion for five months during the winter 1808–9, and distinguished himself in the fighting at Braga and Oporto. He was then posted, as Captain, to the 1st Infantry Regiment; subsequently to the 6th Caçadores and became later on Staff Officer to Colonel—afterwards Sir Nicholas—Trant, a most enterprising and distinguished commander of Portuguese Militia. On quitting the Portuguese service in 1820 von Linstow is described in the Official Records at Lisbon as being ' discharged with honours.'

About this time Lieut.-Colonel William Williams, who had been appointed to the Command in place of Baron de Rothenburg, joined the battalion. Williams, born in 1776, got his first commission in 1792, had been in the 54th Regiment. During the following years he served under the Duke of York in the Low Countries. His next service was in the West Indies, and he was wounded in the attack on St. Vincent. Returning home Williams, now a captain, fought in the Irish Rebellion of 1798. He served under Sir R. Abercromby, and was said to be the first man to land in Egypt in 1801. In 1802 he got a brevet majority and two years later the substantive rank in the 81st, with which he served at the battle of Corunna, where he was again wounded and his distinguished conduct gained him promotion into our Regiment.

Other regimental news is given by Captain Andrews in a letter to Colonel Davy, dated Pinhel, May 19, 1820 :

' Shortly after the date of my former letter we moved up here to protect the front from Almeida on the left to Ciudad Rodrigo on the right. General Craufurd's left connects with us, extending his right to Gallegos and the villages perpendicular to Guarda.

' About the 28th ultimo the enemy moved a column from Salamanca to near Ciudad Rodrigo which caused our second line to close up upon the first when Headquarters removed at an hour's notice from Viseu to Celorico ; but tranquillity soon prevailed again, and the advanced posts of both armies are on speaking terms and only separated in one place by a small river.

' All the late accounts induce a belief that the enemy will not be in a condition to make a forward movement till very considerable reinforcements arrive which are not yet on the road ; this promises a lingering existence to us.

' Within the last few days we have had five recruits join us : Colonel Williams, Dr. Drumgold, and Ensigns Kruger, Larbusch, and Burghaagen, of whom I shall remark in rotation.

' Colonel Williams is a short man ; looking in profile at a short distance like Dumoulin (Assistant Surgeon) . . . his manner is quick, peremptory and at the same time courteous ; he speaks with rather a studied tone of voice, something between the agreeable and the commanding ; he bows like a Statesman entering into office, and from all I have observed I doubt not that General de Rothenburg has entrusted him with the talisman so essential to command this corps.

' Our Doctor Drumgold, is a being with a circular head fixed on between his shoulders without any neck. The deficiency is so well recompensed by the rotundity and size of his body that without any violence of language we may apply the old adage " things are as broad as they are long." . . . De Bree is still in Lisbon, very unwell ; I believe he is to be sent home to recruit. . . . Schoedde has been confined for 3 weeks with fever ; an application has been this day made to send him to Lisbon ; his constitution will not stand this kind of life any longer.'

Dr. Drumgold at a later date went to the 97th Regiment, and may possibly have been the original of the celebrated Dr. Slammer in ' Pickwick.'

On July 10 Ciudad Rodrigo fell.   Almeida still blocked the way, but an unfortunate or treacherous explosion rendered the fortress untenable ;   and on August 27 Almeida surrendered.   The two northern barriers against a French invasion of Portugal being thus demolished, Wellington retired behind the Mondego, and Masséna prepared to invade Portugal with a force of 65,000 men and sixty guns.   Captain Andrews, writing to Colonel Davy from La Rosa, September 12, describes the situation :—

' About the 29th July the Army moved to the rear of Celorico, except the Light Brigade and Cavalry.

' Our station was in a village half way up the Sierra de Estrella where although the climate was uncommonly fine, we got more sickly than ever I knew, as out of our three Companies we had near 50 in hospital, and I believe even bad as this appears we were more fortunate than the rest of the Army.

' The sick of the whole Army are immediately sent off to Coimbra and Lisbon, and everything kept in a state to facilitate a movement in that direction.

' Our Headquarters at present is at Gavio [? Càbra], a considerable town at the foot of the Estrella, four leagues behind Celorico. Cavalry at the latter place ;   Light Brigade at Minho ;   Cole's Division at Pinhanços ;   Picton's at Galizes, La Rosa, etc. ;   and the 1st Division at the Ponte de Marcello, about four leagues from Coimbra.   We are all in houses of some sort. . . .

' Our Corps has been going on in the old way for a long time ; the only change is that Prevost has arrived from America, and now commands his Company.

' Schoedde has returned quite well ; and we have only had one Officer recruit for some time, a Lieutenant O'Heir from Brunswick, a genteel looking man, an Austrian of Irish parents.'

Captain Schöedde had been sent down to Belem suffering from ague, where a Portuguese doctor prescribed a tempting regime, viz. ' the best of everything and not less than two bottles of Madeira a day ! '   This treatment proved so successful that he was one of the very, very few

who served in the Peninsula in 1808–14 without inter-
ruption.

As Masséna advanced Wellington fell back. He called
up Hill's Corps, including a newly formed 5th Division
under General Leith, from the Tagus ; and concentrated
his whole Army, about 50,000 strong, half of whom were
British, upon the ridge of the Serra de Bussaco, where he
resolved to give battle. General Hill joined him on
September 21, but even on the 25th only half his troops
were in line, and there were many gaps in his position, which
was indeed far too large for the numbers of the defending
force. Marshal Ney, commanding the leading Corps of
the French, wished to attack on the 26th ; but Masséna
would not give his consent until the next day, by which
time the British position had been fully prepared and
occupied. The Marshal failed to realise that it was liable
to be turned by the left.

The French General, Baron de Marbot, remarks :—

' The French troops were drawn up on stony ground sloping
steeply down to a great ravine which separated us from the Alcoba,
which was lofty, steep, and occupied by the enemy. From their
main position they could see all our movements, while we saw only
their outposts half way up the hill between the Convent of Bussaco
and the ravine, which at this point was so deep that the naked eye
could hardly make out the movement of troops which were marching
through it, and so narrow that the bullets of the British Riflemen
carried right across it. It might be regarded as an immense natural
ditch serving as the first line of defence to the natural fortress formed
by great rocks cut into an almost vertical wall. Besides this our
artillery, engaged on very bad roads, and obliged to fire upwards,
could render very little service, while the infantry had to contend
not only against a mass of obstacles and the roughest possible ascent
but also against the best marksmen in Europe.'

The Serra de Bussaco is an offshoot of the Alcoba
Mountain and runs from it for about nine miles in a

south-easterly direction.   The crest line varies in breadth
from a couple of hundred yards to half a mile.   Along this
ridge the Anglo-Portuguese force, fronting about north-
east, was drawn up as follows :—

On the extreme right the Corps of General Hill occupied
the highest ground overlooking the valley of the Mondego.
On its left was posted the 5th Division under General
Leith ;  and then in succession the 3rd Division, General
Picton ;  the 1st Division, Sir Brent Spencer, the Light
Division, Craufurd ;  while the 4th Division, Cole, formed
the left of the line.

The Serra is crossed by two roads, of which the more
northerly leading to Coimbra was guarded by the Light
Division, and the more southerly—three miles distant,
and occupied by the 3rd Division—ascends the mountain
along an embankment and runs to Palheiras.   Half-way
down the western slope on this road stands the church
of San Antonio de Cantaro.   The Serra is rocky in places
and covered with heather.   The exterior slope could
perhaps be taken at an average of 1 in 10.   The crest
line is from 300 to 500 feet above the plain.

In the 3rd Division the Light Companies of the 45th,
74th, 88th, and 94th, belonging to Mackinnon's Brigade
were formed into a Light Battalion with the three headquarter companies of the 60th,[1] which formed part of
Lightburne's Brigade, the whole being placed under the
command of Colonel Williams, and forming a line of
outposts on the low ground almost at the bottom of the
San Antonio ravine on the Palheirus Road.

On the opposite side of the valley was drawn up the
French 2nd Corps, commanded by General Reynier.   To
Reynier's right, opposite the road leading to Coimbra, was

---

[1] *I.e.* those of Andrews, Schoedde and De Salaberry.

posted Ney's (6th) Corps, while the 8th (Junot's) formed a general reserve in rear.

Reynier's Corps was drawn up in line of Divisional Columns, that of Merle being on the right and of Heudelet on the left.  The latter Division, with which we are chiefly concerned, was made up of the Brigades of Arnaud and Foy. The four battalions of the 31st Light Infantry in Arnaud's Brigade were in front.  In rear of them came the whole of Foy's Brigade, while the 47th, Arnaud's second regiment, formed the Divisional Reserve.

Captain Lemonnier-Delafosse, of the 31st, describes the situation as it presented itself to him.  After speaking of the British position he goes on to say :

' In front of these lines black specks at equal intervals indicated the sharp-shooter outposts driven in on the evening of the 26th— Riflemen—whose short rifles were so murderous.  Their bugle horns, a kind of short cornet lined with brass, could be heard emitting a sound so discordant and piercing that one might suppose they were sounding the call to awaken the dead at the Judgment Day.' [1]

Before daylight on the 27th firing commenced at the outposts ; and with the first rays of dawn General Merle with the eleven battalions which composed the Brigades of Generals Sarrut and Graind'orge launched his attack, driving back the Light Companies of the 74th, 88th, and 94th, who were on the north side of the San Antonio Road, and separating them from that of the 45th and the three Rifle Companies under Major Woodgate with whom Colonel Williams remained in person.

It was not long before Heudelet followed suit ; and the

[1] Lemonnier-Delafosse, p. 67.  ' En front de ces lignes des points noirs à égale distance indiquaient les tirailleurs d'avant-postes repoussès le soir du 26—Riflemen —dont les carabines etaient si meutrières.  On entendait sonner leurs buccins, sorte de cornet court en cuivre, donnant un son barbare et déclivant-on aurait dit de cris d'Alarme jetés au jugement dernier pour évoquer les morts.'

four battalions of the 31st Light Infantry of Arnaud, descending the opposite slope in overwhelming numbers, crossed the stream at the bottom of the valley and began to ascend the British side. To an assault in such overwhelming forces Williams's three companies, 150 strong, could offer no strong resistance ; but the Colonel, in accordance with the practice recommended in De Rothenburg's book, threw back his extended line and formed it on a knoll running parallel to the San Antonio Road, and on the left flank of the attacking column, which suffered heavy loss from its fire.[1] Nevertheless both French Divisions gradually scaled the hillside, reached the crest and gained a transient success, soon, however, nullified by a British counter-attack, which drove them back in bloody repulse. Further attempts, reinforced by the seven battalions of Foy's Brigade had no better fortune, and as Heudelet retired down the hill some companies of the 74th Regiment following him up joined the left of the Riflemen, who gradually changed front again to the east and occupied the position originally held during the previous night.

A little later Marshal Ney made an attack with the 6th Corps along the Coimbra Road, but met with no better fortune. His men were charged and driven across the valley by the Light Division aided by some units of the 1st.

By the middle of the day the action was over. The casualties of our Regiment included five officers wounded, viz. Colonel Williams, Captain Andrews, Lieutenants Joyce, Franchini, slightly ; and Baron Eberstein, severely ; three Riflemen killed, sixteen wounded, and five missing. Colonel Williams was hit twice. Our losses were apparently

---

[1] In his subsequent Dispatch General Picton remarks : ' The Light Corps of the Division, unable to resist such superiority of numbers in front, was most judiciously thrown in upon the flank of the advancing column by Lieut.-Colonel Williams.'

confined to the three companies under his personal command. Of the other units to which Rifle Companies were attached, the 2nd and 4th Divisions had no casualties, and the two Brigades of the 1st only nine.

Later in the day, when both sides were gathering their wounded close to one another, a German officer in the French service came into our lines in search of his brother who was in our Regiment. The latter proved to have been one of the three Riflemen killed.

The casualties of the Anglo-Portuguese Army as a whole amounted to about 1250. Those of the French were not far short of 5000. Wellington's Army had not been in action since the Battle of Talavera, fourteen months previously; and at Talavera probably not more than a quarter of the men present on this occasion had taken part. The chief value of the victory at Bussaco lay in the confidence gained thereby by the British and Portuguese troops.

On the day following the battle Masséna discovered a means of turning the left of Wellington's position; and the British General consequently retired within the lines of Torres Vedras, which he had prepared in the previous year. Masséna followed, but did not attack; and in the middle of November retired once more and took up a position in the neighbourhood of Santarem. Wellington also advanced and took up a position opposite, leaving the 3rd Division within the Lines.

In due course Wellington's dispatch on Bussaco arrived, and great was the regimental disappointment on finding no mention either of Colonel Williams or his Battalion. The Colonel ventured to write on the subject to General Picton, who replied on October 30 in the handsomest terms, explaining that he had sent in no written report to Lord Wellington, since the latter being at hand

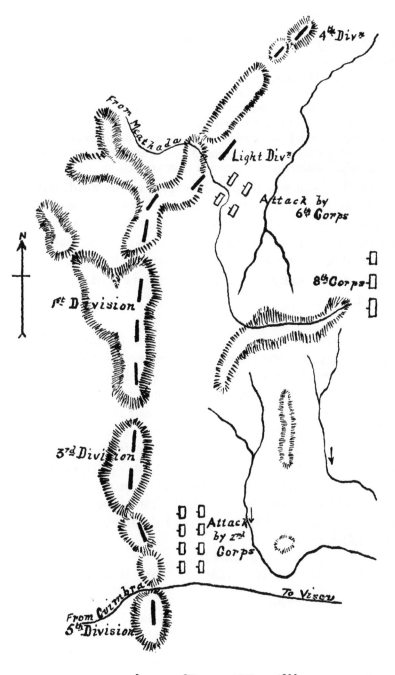

N

4th Divn

From Meathada

Light Divn

Attack by
6th Corps

8th Corps

1st Division

3rd Division

Attack
by 2nd
Corps

From Coimbra
5th Division

To Viseu

0    500    1000    1500
Scale of Yards

Allies
French

II.            SKETCH OF THE ATTACK ON THE POSITION AT BUSSACO.            K had personally witnessed the French attack on the 3rd Division,

' but,' added the General, ' on reflection I find the post you defended with so much gallantry for so many hours was situated so low in the ravine of St. Antonio d'Alcantara that he (Lord W.) would probably not have seen your situation or witnessed your exertions ; but you may be assured that I will take an early opportunity of mentioning to His Lordship that no Commanding Officer of any Corps had more claim to public notice on that occasion than yourself.'

General Picton was no doubt as good as his word, but Lord Wellington probably did not wish to re-open the question. Anyhow, the fact remains that nearly seventy years elapsed before ' Bussaco ' was added to the Battle Honours of our Regiment, and to the day of his death Col. Williams never received the medal clasp he had so well deserved.

## 4

In November Lord Wellington issued another G.O. complimentary to our Regiment, and added that the detached companies when not forming part of the battalion composed of the Light Companies of the several Divisions, were to be kept at the Headquarters of their Brigade.

Winter passed away, and the spring of 1811 appeared. Near the end of January a draft from the Isle of Wight, consisting of one ensign, one sergeant, and thirty-seven other ranks had joined the Battalion. The distribution of the Rifle Companies varied a little from time to time, but at the opening of the campaign of 1811 appears to have been as follows :—

### 1st Division

Stopford's Bde. (Guards)   No. 6 Rifle Coy.   Capt. Howard
                                              Lieut. Du Châtelet

Nightingall's Bde. (24th,
   42nd, 79th)   .   .   ,, 8 ,,   ,,   Lieut. J. Joyce [1]

[1] Hames late Captain of No. 6 Company was no longer in Portugal ; Captain Wolff was a prisoner of war.

### 2ND DIVISION
*M.-G. Hon. W. Stewart*

Colborne's Bde. No. 2 Rifle Coy. Capt. J. McMahon, Lieut. Ingersleben

Hoghton's　,,　,, 4　,,　,,　,,　I. Franchini, Lieut. Muller
Lumley's　,,　,, 5　,,　,,　,,　F. Blassière, Lieut. Sprecher
de Bernegg, Lieut. Barbaz

### 3RD DIVISION
*M.-G. Picton*

McKinnon Batt. H.Q. No. 3 Rifle Coy. Capt. G. Purdon,[1] Lieut.
Koch
,, 9　,,　,,　,,　J. Schöedde, Lieut.
Kent, Lieut. O'Hehir
,, 10　,,　,,　,,　A. Livingstone,[1] Ensign A. Wynne, En-
sign M. Furst

### 4TH DIVISION
*M.-G. Hon. G. d. Cole*

No. 7 Rifle Coy. Capt. Prevost

### 6TH DIVISION
*M.-G. Alexander Campbell*

No. 1 Rifle Coy. Bt.-Major J. Galiffe, Ensigns Ritter and Franchiosi

The country had been devastated, partly in accordance
with Wellington's orders, partly by the French, and by the
end of February Marshal Masséna found it impossible to
feed his army any longer.　He therefore decided to retire
behind the Mondego, where he hoped to be joined by the
9th Corps and other reinforcements, which would raise
his army to a strength of 70,000 men, and enable him to
resume his attack against Lisbon by both banks of the
Tagus.　The Marshal withdrew his troops with skill;
but on March 9 Wellington's advanced troops discovered
him in position in front of Pombal.　Two days elapsed

---

[1] Purdon and Livingstone joined early in May.

before the British Commander could bring a force large enough to attack him with advantage ; and this delay had allowed Masséna on the previous night to send his baggage and the main part of his army over the Soure River, covering his retirement by the 6th Corps. The Light Division attacked his rear-guard ; but Marshal Ney, its commander, made good his retreat, and on the following day was again found in position in advance of the village of Redinha, his left resting on the buildings of Saint Bernardo, his centre commanding the road from Pombal, and his right posted on the high ground beyond. On the side of the British the Light Battalion of the 3rd Division, commanded as usual by Colonel Williams, outflanked and threatened the enemy's left ; Pack's Portuguese Brigade and the 4th Division moved up in the centre, and the Light Division on the French right. Williams, advancing under cover of a wood which concealed his movements, completely turned the French left and reached a point actually nearer the village of Redinha than were the French. Ney's position was imperilled ; but having made the most of his army corps he observed that Wellington, uncertain of the numbers opposed to him, hesitated to attack. The French Marshal determined to hold his ground a little longer. He counter-attacked Williams' battalion, which, together with the remainder of the British force, was halted by the Commander-in-Chief. An hour elapsed before Wellington deemed himself strong enough to attack in force.[1] This hour was well employed by Ney in withdrawing his troops ; and when our advance was eventually made there was no enemy left. But

[1] Colonel Jenkinson, R.A., in a letter quoted in Wellington's 'Supplementary Dispatches,' vol. vii. p. 85, writes : 'Figure to yourself 14,000 men with their colours unfurled advancing in line supported by solid columns of infantry and cavalry on their flanks, and a second line in rear of the centre. The advance of the whole army in line which was formed as if by magic was majestic beyond description.'

Williams and his men, to use Napier's expression, ' chasing like heated bloodhounds,' forded the river at the same time as the French rearguard retired across the bridge. Unfortunately the road by which Ney, following Masséna, retired upon Coindeixa, ran in a westerly direction after passing through Redinha, that is to say, directly away from Colonel Williams. The main body of the 3rd Division could ford the river but slowly ; and for a time the Light Division on the left was unable to cross it. The enemy therefore got away without being much molested. In this action both sides lost about 200 men ; the casualties of the 60th amounting to ten Riflemen wounded and four missing.

Masséna's object was now to occupy Coimbra ; and General Montbrun, commander of his Reserve Cavalry, was directed to seize it ; but Colonel Trant, a British officer in the Portuguese service, made so stout a defence with a body of militia, that Montbrun not only failed to execute his commission, but sent such a doleful report to Masséna that the latter, relinquishing his matured plan of passing the Mondego, decided to retire eastward through the defiles of Miranda de Corvo. Marshal Ney established in a strong position at Condeixa covered the retreat. Wellington despatched the 3rd Division by a mountain path to turn his left, whereupon Ney abandoned his position and retired in the direction of Casal Novo. Colonel Williams once more distinguished himself and, in conjunction with a party of our cavalry, cut off the French from Fuente-Coberta with such speed that the French Commander-in-Chief himself narrowly escaped capture.

On the morning of the 14th General Erskine, in temporary command of the Light Division, advancing in a thick mist without military precautions, found himself involved in a frontal attack upon the Corps of Ney and

Junot posted at Casal Novo on almost inaccessible hills leading to the Pass of Miranda de Corvo through which the guns and baggage of Masséna were slowly defiling. Wellington's intention had, as usual, been to turn the French out of their position and threaten their flanks; but by Erskine's folly the combination was spoilt. Even when deployed in a single thin line the Light Division was unable to show a front equal to that of the enemy; and though protected by stone walls on the side of the mountain, was barely able to hold its ground. The situation was relieved by Picton, who with his Riflemen prolonged the line of the Light Division to the right, and with the remainder of his Division turned the enemy's left. Then the 1st, 5th, and 6th Divisions arrived in support of the centre; and the 4th appearing on Picton's outer flank still further enveloped the enemy's left. Grasping the situation General Picton outflanked the French so rapidly that they were obliged to abandon the hill with precipitation; and their guns, which had been playing upon our centre, were suddenly perforce withdrawn in order to open fire on the 3rd Division, which by this time was almost in their rear. For the moment Ney retired in confusion upon Miranda de Corvo, but did not fail to contest every valuable position on the way; 100 of his men were captured, and many others killed or wounded, but he gave time to Masséna to pass the defile and after dark followed him in person. The loss of our Regiment on this day was Lieutenant Wynne and three Riflemen wounded, and one Rifleman missing. In his dispatch Wellington stated that the Light Division and the Light Infantry of the 3rd Division under Colonel Williams had 'particularly distinguished themselves.'

'The result of these operations,' writes Lord Wellington to Lord Liverpool, 'has been that we have saved Coimbra and upper

Beira from the enemy's ravages ; we have opened communications with the northern provinces ; and we have obliged the enemy to take for their retreat the road by Ponte de Murcella on which they must be annoyed by the militia acting in security upon their flank, while the Allied Army will press upon their rear.' [1]

On this day the British Commander heard that Badajoz, the Spanish fortress on the Guadiana opposite the southern frontier of Portugal, had surrendered. Its fall was a heavy blow to him. But for one or two petty works and the two Divisions under Beresford, the road to Lisbon now lay open to Marshal Soult. Wellington therefore detached the 4th Division to reinforce Beresford, although by doing so he reduced the force under his own command to a strength appreciably below that of the enemy he was pursuing.

Masséna had now retired across the Ceira, but Ney, in disobedience to orders, retained his rear-guard consisting of a Brigade of cavalry and two Divisions of infantry on the near bank of the river, with his centre resting on the village of Foz d'Arouce.

Wellington, coming up at 4 P.M., held the enemy's right and centre with the Light Division, and turned his left

---

[1] Baron de Marbot in his Memoirs mentions an adventure on this day with a British officer who was possibly Lieutenant Sawatzky of our Regiment. He says : ' I was about to return to Masséna [to whom he was A.D.C.] when a young English Rifle officer trotted up on his pony crying, " Stop, Mr. Frenchman, I should like to have a little fight with you." I saw no need to reply to this bluster, and was making my way towards our outposts, 500 yards in rear, when the Englishman followed, heaping insults on me. At first I took no notice, but presently he called out, " I can see by your uniform that you are on the Staff of the Marshal ; I shall put in the London papers that the sight of me was enough to frighten away one of Ney's or Masséna's cowardly A.D.C.'s." I admit it was a serious error on my part, but I could no longer endure this impudent challenge coolly ; so, drawing my sword, I dashed furiously at my adversary. . . . We met ; he gave me a slash across the face ; I ran my sword through his throat ; his blood spurted over me, and he fell from his horse to the ground which he bit in his rage.' Sawatzky was returned as killed in the casualty list of the following day, but may have lived through the night ; and as no officer of the 95th appears to have been killed on the 14th or 15th, it seems possible that Marbot's victim was Sawatzky. He was the first officer of the regiment killed in the Peninsular War.

with the 3rd. The latter drove the enemy on to the bridge in confusion. Ney blew up the bridge and forded the river with the remainder of his rear-guard.

The British loss on this day amounted to only seventy killed and wounded, twelve of whom belonged to the 60th. The French admitted a loss of two hundred; Napier estimated it at five hundred, including a large number drowned.

On the 16th want of provisions compelled Wellington to halt. Next day he crossed the Ceira, and found Masséna in position on the further bank of the Alva, but hemmed in between that river and the Mondego. Once more he turned the Marshal out of his position by a flank movement; and performed the same feat on the day following. On the 21st the French Marshal reached Celorico with the 6th and 8th Corps, while Reynier occupied Guarda ten miles south-east of him.

Guarda, situated on the top of the mountain, bore the reputation of being the highest inhabited town in Europe. On the 28th Wellington determined to make a concentric attack upon Reynier's Corps. Picton, although he had the most difficult march, was the only Divisional General to be at his post at the appointed hour, when he cleverly placed his Division within a quarter of a mile of the enemy's left flank and rear. Shortly afterwards the Light and 6th Divisions approached from the other side. Reynier abandoned the mountain without firing a shot and retired across the Coa. Thus ended Masséna's retreat, during the course of which Lord Wellington had repeatedly mentioned Colonel Williams and his Light Battalion in terms of approbation.

On April 1 Wellington reached the Coa. The French occupied a triangle whose sides were about ten miles long, and at each angle had posted an army corps, the 2nd being

at Sabugal which formed the apex.   Wellington attacked
on the following day ;  but his orders were badly carried
out, and the Light Division exposed to attack by overwhelming numbers.   It fought with its usual distinction ;
and at the critical moment the approach of the 3rd and
5th Divisions, headed by Williams and his Light Battalion,
decided the contest.   Reynier fell back to Rendo.   Wellington remarked in his dispatch that this was one of the
most glorious actions in which the British troops had ever
been engaged ;  and that the glory had been gained by the
Light Division.   The result of the action—in which the
British loss was only 20 killed and 147 wounded, and
that of the enemy 760—was the abandonment of Portugal
by Masséna, who at once crossed the frontier into Spain
by Ciudad Rodrigo.

Writing in April to Colonel Davy, Major Woodgate
says :—

'I can date this letter from no particular place as I am not
aware that the few houses we occupy are sufficient to create a village,
or be called by any name ;  but we are near Almadilla, having followed
the enemy thus far "without impediment."   I shall not attempt to
give you a history of their retreat and our pursuit, as the despatch
will tell you all, and with more accuracy than I can.   I shall only say
that your friend General Picton's Division (and of them entirely
the light companies, and our three companies, brigaded as before
under Williams) together with the Light Division, were almost the
only part of the infantry of the Army engaged.   We lost a few men
and Sawatzky, whom, as an attentive and useful officer, I shall
much lament.

'There are about 1500 men in Almeida who are destined, by
our arrangements, to fall into our hands ;  not by besieging that
place, but we purpose starving them out, and for that purpose have
surrounded the fortress ;  but the opinion is that the French, after
getting themselves a little refreshed, will march a considerable
force to their relief, which may, if we choose to stand, bring on what
has been so long postponed, something like a general engagement.

'On the retreat we could never bring them to stand.   Their

rear-guard, and that of great strength, 20,000 strong, under Reynier,[1] was the only part of their army we could engage with, and they only fought because they were overtaken.

'Every town on the road, Leira, Pombal, Condeixa, etc., etc., has been burned to the ground, and the road strewed with dead bodies of all ages, horses, mules and baggage of every description.

'We had two fine views of the enemy. We saw about 30,000 men march out from Condeixa, and at Guarda about 16,000 marched out, while our division was within 600 paces of them. The whole army was intended to have met there, and, co-operating, to have attacked them; but it so occurred that we, who had marched over the mountains, were the only people who arrived at the appointed time, and the enemy retiring on our first appearance, we had a fine opportunity of seeing them march off. As our force consisted of not more than 4000, we could do no more than quietly to march into the town when they marched out of it, which we did that evening. . . . You must have seen Andrews' promotion. He leaves us in a few days, having just come up to settle his Paymastership accounts to March.

'De Bree has not left England yet to join us as Andrews hears from him last mail. He has most officiously (as usual) been sending recruits to us without the least authority or leave from Colonel Williams; we received 64 of his selection lately, the greater part of whom were taken six months ago in this country. This is not giving the men or the regiment a fair chance. . . .

'You will doubtless see Austin in London;[2] tell him I received his friendly letter from Martinique, and now that I know he is in England I shall write to him. . . .

'Steitz has been removed to the 1st Battalion, and Gilse has resigned the Adjutantcy, and Broetz is appointed.'

Major Woodgate's remarks about the devastation of Portugal is a reminder that Napier (Bk. xiv. Ch. 4) says that—

"at Figueras, where 12,000 people had taken refuge . . . the whole would have perished but for the active benevolence of Major von Linstow (of the 60th) an officer of Trant's Staff.'

---

[1] Ney it was who, as we have seen, commanded the rear-guard during the most part of the retreat.

[2] Lieut.-Colonel of our 4th Battalion.

5

Wellington had driven Masséna out of Portugal by a fine feat of arms ; yet the conduct of the retreat covered his opponent with glory.  The French commander retired to Salamanca, and put his army in quarters between that town and Ciudad Rodrigo to gain rest and be refitted.  The fortress of Almeida within Portuguese territory was, however, still held by the French, but was quickly blockaded by Wellington and placed in a critical position.  Towards the end of the month, Masséna with a view to its relief advanced on Salamanca, and on the 26th concentrated two cavalry divisions, the 2nd, 6th, 8th, and 9th Corps—a force of over 47,000 men and 38 guns at Ciudad Rodrigo. On May 1, Lord Wellington took up a position on a table-land behind the Duas Casas River, from Fort Conception on the left to the high ground behind Fuentes de Oñoro on the right—a length of about eight miles, which he could only lightly hold with 37,000 men and 48 guns. The river flowed through a deep ravine, but was fordable at several points, and bridged at Fuentes and Alameda.

The position was occupied on the left by the 5th Division.  Then—with an interval of a mile—came the 6th Division.  A gap of over two miles was only partially filled by the Cavalry Division ; while the 3rd and 1st Divisions took the right supported by the 7th and Light in second line.  Two Rifle Companies, mustering respectively 44 and 36 rifles, were attached to the 1st Division, and one (48 rifles) to the 6th.  The other three present (183 rifles) were with the 3rd Division.

On the 3rd the enemy was descried approaching ; his 6th Corps advancing on Fuentes de Oñoro by the road from Espeja : the 2nd and a Division of the 8th

Corps from Gallegos upon Alameda. The 9th Corps was in general reserve. The weight of the attack seemed to be aimed against the left, and the Light Division was sent to support the 6th. The attack was in reality intended against the British right, and General Loison hurled the 6th Corps against Fuentes de Oñoro.

The defence of this village had been entrusted to Colonel Williams of our Regiment, and that such a post should have been given him is a proof of his reputation. It straggled over three-quarters of a mile of ground, and in addition to the houses the gardens bordering the river, fenced with stone walls, lent strength to the defence. On high ground in rear a chapel formed a Réduit to the whole position which was occupied by twenty-eight Light Companies including the three (183 rifles) under Major Woodgate forming the H.Q. of our Battalion. The houses skirting the stream were captured by a French Brigade, but on the higher ground by the chapel a desperate fight was maintained in the face of overwhelming numbers. The result of the contest seemed doubtful, when a Brigade of the 1st Division came up, and a charge with fixed swords drove the assailants back over the river. But Williams who, to use the expression of the historian of the British Army, ' had so ably commanded the light troops at Bussaco,' fell, dangerously wounded, and was succeeded by Colonel Cameron of the 79th. The fight dwindled gradually to a passive defence. In Robinson's ' Life of Picton ' it is remarked that

' the spirited conduct of Lieut.-Colonel Williams of the 60th . . . who in the most gallant manner defended the village against the repeated attacks of the enemy called forth the warmest praise from General Picton and the whole army. It was not until after he was wounded that the light troops were arrested in their successful operation.'

The British casualties amounted to 259, those of the French to 652, including 160 prisoners.

The following day passed quietly except for another attack on Fuentes in the early morning; but on the 5th Masséna made another determined assault against the British right, which if successful would have commanded its line of retreat running through Frenada and Aldea Ribiera. He quickly gained an initial success, and Wellington's flank being completely turned, his position was little less than desperate. Fortunately for the British the success proved momentary. A want of harmony existed between some of the French commanders; the attack was not pressed home, and Wellington gained time to make fresh dispositions. With great coolness he threw back his right and with the 7th and Light Divisions re-established his line, albeit at the cost of lengthening it by five miles. Then another furious attack was made by a Division of the 6th Corps and one of the 9th against Fuentes de Oñoro, the greater part of which was captured. The defenders were, however, reinforced by the supports consisting of the 60th with the Light Companies of the 3rd Division under Major Woodgate; and Loison failed to dislodge them from the remaining houses. Fighting ceased at sunset. During the 5th, the British casualties amounted to 1522, those of the enemy to 2192. The losses of our Regiment included—on the 3rd—3 riflemen killed, 2 officers, namely Colonel Williams, Lieutenant Du Châtelet and 9 riflemen wounded, and 8 missing. On the 5th our casualty list showed Major Woodgate, Lieutenant Wynne, 1 sergeant, and 11 riflemen wounded, and 1 missing. The total of our regimental losses at Fuentes de Oñoro amounted therefore to 3 killed, 25 wounded, and 9 missing; about 10 per cent. of the strength.

In his dispatch Lord Wellington says that on the 3rd

the village of Fuentes 'was defended in a most gallant manner by Lieut.-Colonel Williams,' and mentions Major Woodgate in command of the battalion on the 5th. A few days later, he successfully recommended the latter for the brevet rank of Lieut.-Colonel.

Masséna having failed in both his objects, namely that of relieving Almeida and of driving Wellington from his line of retreat, retired to Ciudad Rodrigo. But had he been properly seconded by his subordinates and received from those nominally under his command the co-operation which he had a right to expect, Wellington would in all probability have been dealt a serious blow.

The fate of the garrison of Almeida now seemed to be sealed ; but the commander blew up the works, and with great courage and determination broke out, burst through the line of investment, and although not without loss, made good the escape of his main body.

Just at this time Masséna was ordered to hand over command of his army to Marmont, Duke of Ragusa—a Marshal at the age of thirty-seven.

### 6

It will be remembered that to General Hill had been allotted a small army corps consisting of the 2nd British and a Portuguese division with a proportion of cavalry, whose duty it was to guard the approaches to Lisbon by the Tagus, but that when Wellington retired down the Mondego he had called Hill to him, and the two forces were concentrated on the ridge at Bussaco.

When Masséna began his retreat in March, Hill's corps, commanded in his absence by Marshal Beresford,[1] was

---

[1] Beresford was a Marshal in the Portuguese Army only.

directed to march to the relief of the Spanish frontier fortress of Badajoz which had been invested by Marshal Soult during the last days of January. Before Beresford could arrive it surrendered, and excepting for one or two petty fortresses the road to Lisbon lay open to the French. The 4th Division was therefore despatched to reinforce Beresford. But just then Soult heard that the 1st French Army Corps under Marshal Victor had been defeated at the battle of Barrosa near Cadiz, and felt it essential to hasten southward to his colleague's assistance. Nevertheless the 5th Corps under Marshal Mortier besieged, and on the 21st occupied, Campo Mayor; but the place was regained four days later by Beresford, whose force by this time amounted in all to 25,000 officers and men with 18 guns ; including 4 rifle companies ; one being attached to Kemmis's Brigade of the 4th Division, and each of the others to a Brigade of the 2nd Division.

Mortier threw 3000 men into Badajoz, and had barely 10,000 for active operations. Beresford's instructions were to attack the fortress before the breach had been repaired ; and it is thought that had he carried out his orders with promptitude and intelligence he would have recovered it ; but delay on his part enabled the Governor to make good the damage and revictual the garrison.

On April 16 Olivenza, a Spanish fortress some twenty miles south-west of Badajoz, surrendered to the 4th Division. In Major-General Lowry Cole's dispatch to Sir W. Beresford he says, ' to the fire kept up by the British light companies and the rifle companies of the 60th and Brunswick regiments . . . I principally attribute the trifling loss we sustained.'

Beresford then invested Badajoz. On hearing the news Marshal Soult lost no time in marching to the relief of the fortress. Upon his approach Beresford, in accordance with previous arrangements, took post at Albuera, some ten miles south-east of Badajoz.[1]  Although the French were marching from the south Beresford drew up his infantry in a line fronting due east.  The 2nd Division was posted in the centre about half a mile behind the village and stream of Albuera, with the Portuguese Division on its left, and a Spanish Division under General Blake on the right.  The 4th Division was still in front of Badajoz; but a weak brigade consisting of two light battalions K.G.L. had been recently detached by Wellington to join Beresford, and occupied the village of Albuera.  Wellington himself was marching to reinforce the Marshal with the 3rd and 7th Divisions, and as it turned out Beresford would have been wise in retiring across the Guadiana pending his Chief's arrival.  Still—inclusive of 14,000 Spaniards—he had a force of 35,000 and 38 guns, and resolved to give battle.

Marshal Soult coming up on May 16 quickly realised the defective manner in which Beresford had occupied his position, and taking advantage of a wooded hill between the Feria and Albuera streams, massed behind it unknown to Beresford the whole of his reserve cavalry, the 5th Corps, and all but ten of his guns with a view to overwhelm and roll up the Allies' right flank.  His whole available force consisted only of 24,000 men, but his manœuvre went far to make up for his numerical inferiority.  His two remaining infantry brigades, supported by ten guns, and two brigades of cavalry, made a demonstration against the bridge and village of Albuera.

At the last moment Beresford realised his dire peril, and attempted to change front to the right with the

---

[1] In the trenches before Badajoz, between May 8 and 15, a Rifleman had been killed ; and during the repulse of a sortie on the 10th a Rifleman was killed.  Captain Prevost and seven Riflemen were wounded.

Spaniards, to reinforce them with the 2nd Division and to hold the bulk of the Portuguese in general reserve. He had already sent to Badajoz for the 4th Division. Before the change of front could be completed the attack burst like a storm upon the Spaniards. They fought well, but being outflanked by the enemy's cavalry and unable to manœuvre, fell back in confusion. At the critical moment up came General Stewart with the leading brigade of the 2nd Division. Stewart had commanded a battalion of the 95th, and it would appear that in this moment of excitement he gave a word of command familiar to riflemen but which was not understood by the troops of his division, and consequently disregarded. The result was that the Brigade was still in mass formation when it reached the top of the incline where, under heavy fire, it attempted to deploy ; but, attacked in flank and rear by a division of cavalry which up to the last moment had been concealed by thick mist, two battalions were overwhelmed and almost cut to pieces. The other two brigades of the division came up properly deployed and in good order, and formed in succession, one on the right the other on the left, of the first. Desperate fighting ensued. The slaughter on both sides was terrific, but the French columns were not deployed, and were torn through and through by the allied artillery.

Nevertheless Soult outflanked the British right ; our ammunition failed, and Beresford decided on a retreat. But by this time two brigades from the 4th Division at Badajoz were at hand ; and the brigade of the 2nd Division on the left wheeling inwards attacked the French right. Beresford countermanded his orders for retreat ; and the Fusilier Brigade of the 4th Division attacked the enemy's left with such success that Soult eventually gave up the contest and retired.

Never in its history had the magnificent fighting power

of the British infantry been more amazingly displayed. In this shockingly mismanaged battle the English, out of 10,000 present, lost about 4000 killed and wounded, and perhaps a couple of hundred prisoners.[1] The Spanish contingent, 14,000 strong, had about 1300 casualties. The Portuguese brigades were but slightly engaged. The greatest loss, in proportion, was sustained by the British infantry less than 7,000 strong, of which more than half were killed or wounded. In his subsequent dispatch Sir William Beresford wrote :

' It is impossible to do justice to the distinguished gallantry of the troops ; but every individual nobly did his duty ; and it was observed that our dead—particularly of the 57th Regiment—were lying as they had fought in ranks, and every wound in front.'

The French had a loss which probably exceeded 7000.

The 3 Rifle companies attached to the 2nd Division, viz. those of Macmahon, Franchini, and Blassière, went into action with 2 captains, 3 lieutenants, 13 sergeants, 4 buglers, and 112 men. Its casualties were 1 sergeant, and 1 rifleman killed ; Lieutenant Ingersleben, 2 sergeants, 16 riflemen wounded. Kemmis's Brigade of the 4th Division, to which Prevost's Rifle Company was attached, was unable to cross the Guadiana above Badajoz, and being obliged to march by Jerumenha did not come up until the following day.

It is of interest to note that immediately after the battle of Fuentes de Oñoro, a small detachment consisting for the most part of artillerymen but including Major Galiffe of our Regiment and a Rifleman named Daniel Loochstadt [2]

---

[1] That they contained many Scotch and Irishmen may be taken for granted, nevertheless it is remarkable that the regiments engaged in this supreme trial of ortitude were without exception English.

[2] On the issue of the silver War Medal in 1847, Loochstadt received it with no less than fifteen clasps. In the Peninsula he served partly in No. 4 (Blassière's) Company, partly in No. 6 (Howard's).

were despatched from the field of battle to join Beresford's Corps; and this detachment—in which Major Galiffe was the sole combatant officer—arrived in time to gain the

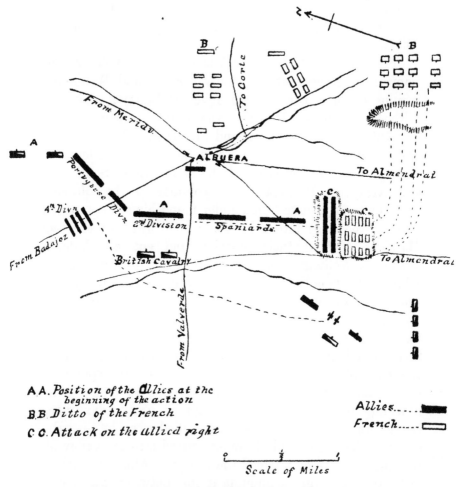

A A. *Position of the Allies at the beginning of the action*
B B. *Ditto of the French*
C C. *Attack on the allied right*

Allies......
French......

Scale of Miles

SKETCH OF THE BATTLE OF ALBUERA.

unique distinction of being present both at Fuentes de Oñoro and Albuera.

So terrible had been the casualties of the 2nd Division,

that after the action the 1st and 2nd Brigades were re-
organised in one provisional battalion, and the whole
division in one brigade which included our three Rifle
Companies and one of the 95th.   The division was then
strengthened by the transfer of a brigade from the 1st
Division.

Shortly after the battle of Albuera General Hill resumed
command of the southern Army Corps ;  and Wellington
coming down in person again invested Badajoz, this time
with the 3rd and 7th Divisions, the operations being covered
by Hill with the 2nd, 4th, and Spaniards.   For the siege
Wellington's means were quite inadequate ;  and learning
the approach of the Army of the South under Soult, and
the Army of Portugal, now commanded by Marshal
Marmont, the British Commander found himself compelled
to turn the siege into a blockade.   On June 18 the two
Marshals effected their junction and entered Badajoz
next day, Wellington having retired three days previously
behind the Guadiana.   He had been reinforced by the
remainder of his Northern Army Corps ;  but even so,
with something under 58,000 officers and men, and 60 guns,
he was confronted by 70,000 men and 90 guns of the enemy.
Soult and Marmont were in a position to strike a great
strategic blow ;  but both had had sufficient experience
of the fighting powers of British troops to make them
unwilling to try conclusions except at great advantage.
In the first days of July Soult was called away by disturbances in the south, and Marmont retired into the valley of
the Tagus where he fortified the bridge of Almaraz, with a
view to improving communications between his army
and that of Soult.

Resumption of the siege of Badajoz was for the moment
impracticable, and Wellington turned his mind to the
capture of the northern frontier fortress, Ciudad Rodrigo.

Leaving Hill again on the Guadiana, Wellington marched northwards. On August 12 he established his headquarters at Fuente Guinaldo and blockaded Ciudad Rodrigo. By the middle of September the fortress began to feel the pinch of famine; but Wellington, threatened by the approach of Marmont and the Army of the North under General Dorsenne, was not able to stop the provisionment of the place. He took up a position astride the Agueda and the Azava. The 3rd Division, in the centre of the line, was posted at El Bodon, with its light battalion and the 74th Regiment, both under command of Colonel Williams, thrown forward as outposts on the heights of Pastores within three miles of Ciudad Rodrigo. At daybreak on September 25 Marmont crossed the Agueda; eight of his squadrons forded the Azava and drove in our cavalry piquets; but were checked and driven back over the river by the fire of the 6th Division posted to the left of the 3rd. Meanwhile the latter, which was watching a considerable tract of country, was attacked by General Montbrun with fourteen battalions, thirty squadrons, and twelve guns. Montbrun fell upon Picton's left brigade, commanded by General Colville, and drove the whole division back to Fuento Guinaldo, five miles distant. The 60th and 74th were left to their fate; but Williams with ready wit forded the Agueda, and moving up the right bank, not only reached Fuente Guinaldo in safety during the night but captured a French cavalry patrol *en route*. On the 26th Wellington fell back to the village of Aldea-da-Ponte where some fighting took place, during the course of which Captain Prevost of the 60th received another wound from which he died the same evening. Aldea-da-Ponte remained in the enemy's hands; and Wellington fell back four miles to the Coa, where he took up a strong position which Marmont did not venture to attack. In fact he retired to the valley of the Tagus, and Dorsenne to Salamanca ; for difficulties of supply made it impossible to keep large forces concentrated many weeks.

Wellington then resumed the blockade of Ciudad Rodrigo, and his infantry was cantoned as follows :— 1st Division about Lagiosa ; 3rd, Albergueria ; 4th, Gallegos ; 5th, Guarda ; 6th, Freixada ; 7th, Penamacor ; Light Division Fuente Guinaldo. In the month of November the siege train was brought up and placed inside Almeida.

During the year 1811 four drafts, comprising in all 171 rank and file, sailed from the Isle of Wight and Lymington to join our battalion.

## 7

We must now turn again to the three companies of our Regiment attached to the 2nd Division. These were Nos. 2, 4, and 5, commanded respectively by Captains Macmahon, Blassière, and Franchini. After their losses at Albuera their whole number, present and fit for duty, on May 25 was only 4 officers, 14 sergeants, 3 buglers, and 117 riflemen.

General Hill with a Cavalry Division, the 2nd Infantry and the Portuguese Division, had been posted at Portalégre and the neighbourhood as a *vis à vis* to the French 9th Corps, commanded by Count D'Erlon. Of this Corps one division, commanded by General Girard, was watching the whole country between the Tagus and the Guadiana, a distance of nearly fifty miles. Its scattered position invited attack ; and in the middle of October, Hill obtained Lord Wellington's consent to strike a blow against Girard. Taking with him a Cavalry Brigade, the Brigades of Howard and Wilson (to each of which a company of the 60th was attached), and a Portuguese Brigade, Hill, who had secured the co-operation of the Spanish forces between two and three thousand strong, started on his march.  On the 23rd, he reached Albuquerque, thirty miles from Portalégre. Next day he came in touch with the Spaniards at Aliseda. In the night of the 25th and 26th, he marched to Malpartida, eight miles from Caceres.  During the march he found that Girard, who had been marching northwards, had left Caceres for Torremocha on the road to Merida.  Hill followed in hot pursuit ; and on the evening of the 27th, after a forced march of twenty-eight miles, reached Alcuescar, five miles south-west of Arroyo Molinos, the halting place of Girard, whose line of retreat was now gravely menaced.

Captain Blassière of the 60th, acting as Intelligence Officer, was sent to find his way into Arroyo and bring back information of the enemy's position, a mission which he executed with success.  At 2 A.M. on the 28th, Hill despatched Howard's Brigade under Colonel Stewart to march directly on the village.  Wilson's and the Portuguese Brigade, both commanded by General Howard, moved to the right with a view to intercepting Girard's retreat upon Truxillo.  Amid a storm of wind and rain—fortunately from the west—Stewart advanced, surprised the piquet, and at dawn marched up the main street of Arroyo. Remond's Brigade of Girard's Division had already started for Merida, but Dumbrowski and all the cavalry regiments under General Briche were surprised in the street.  Girard himself hastily mounted his horse and succeeded in rallying his men outside the village.

Their plight was, however, desperate ;  hemmed in against a perpendicular cliff by Stewart on one side and Wilson on the other, and charged by Hill's cavalry, the greater number surrendered on the spot. Girard finding himself at a point where the mountain was less inaccessible scaled the height with some hundreds of his men ; and although our troops followed rapidly in support, effected his escape. Still Hill's success was brilliant. With a loss to himself of only seven men killed and sixty-four wounded, he effected the capture of 1300 of the enemy, out of a total of about 2000 who were in the village at the time of its attack. Many others were cut down, and it was a mere remnant which eventually reached the Guadiana in safety.

General Hill, in his dispatch, noticed in terms of the highest approbation the conduct of Captain Blassière for ' his close recognisance of the enemy's position on the night preceding the action.'

Having effected his object, General Hill returned to Portalégre.

## CHAPTER VIII

By the end of 1811 Wellington's mind was set on resuming the offensive. The approaching war between France and Russia cast its shadow in advance ; French troops were withdrawn in large numbers from the Peninsula, and the position of affairs entitled the British Commander to prepare for an advance upon Madrid. It was, however, essential as a preliminary to capture the two frontier fortresses of Ciudad Rodrigo and Badajoz, the possession of which would open the gates into Spain.

Ciudad Rodrigo was the first to be taken in hand. The enemy—no longer commanded by a General of the highest class—was unwatchful and unsuspecting. His troops were not concentrated and there was a chance that the fortress might be taken almost by a *coup de main* before Marmont could march to their relief. To mystify, surprise, and mislead the enemy was the practice of Lord Wellington no less than of Stonewall Jackson. Even with his most trusted lieutenants Wellington kept his own counsel ; and as if to show that he intended to do nothing during the winter he allowed Colonel Murray, his Quarter-master-General, the Staff officer who might be regarded as indispensable, to go home on leave.[1]

By the end of the year the preparations of the British Commander were complete, and on the first days of 1812, he suddenly appeared in force before Ciudad Rodrigo, and rapidly invested it. The fortress, garrisoned by 2000

[1] The duties performed by Murray were more akin to those of Chief of the Staff than of a modern Q.M.G.

men under command of General Barrie, was according to
the reckoning of those days considered to be one of strength,
and surrounded by an enceinte wall, 32 feet in height.
On the north and on the south-west the fortified convents
of San Francisco and Santa Cruz formed redoubtable
outworks ; and still further westward two heights, termed
respectively the Greater and the Little Teson, were outer
barriers which would have to be captured before operations
could be undertaken for the possession of the town itself.

On January 5, Colonel Colborne of the 52nd—afterwards Field Marshal Lord Seaton—handling his men with
consummate skill, captured the redoubt on the Greater
Teson, and within twenty minutes reduced an outwork which
the French had reckoned would hold out for five days.

On the 15th, 300 volunteers of the K.G.L. and a Rifle
Company attached to the 1st Division captured the fortified
convent of Santa Cruz after sharp fighting, and at the cost
of 37 casualties.

Next morning the 1st Division was relieved by the 4th ;
and in accordance with the usual practice—an extremely
bad one—the trench guards were removed before the
arrival of the relieving party. The French seized the
opportunity to counter-attack ; and but for the timely
appearance of the relieving division would have recaptured
the convent. The enemy having been repulsed batteries
were erected and an assault ordered on the night of the
19th, the two Divisions detailed for the purpose being the
3rd under Picton, ordered to attack the north-western
salient of the main rampart, and the Light Division, destined
to assault a very small breach on the north side.

Mackinnon's Brigade of the 3rd Division, after momentary loss of direction, scaled the main breach ; but an
explosion disordered its ranks, and the body of troops
which entered the town was little more than a confused

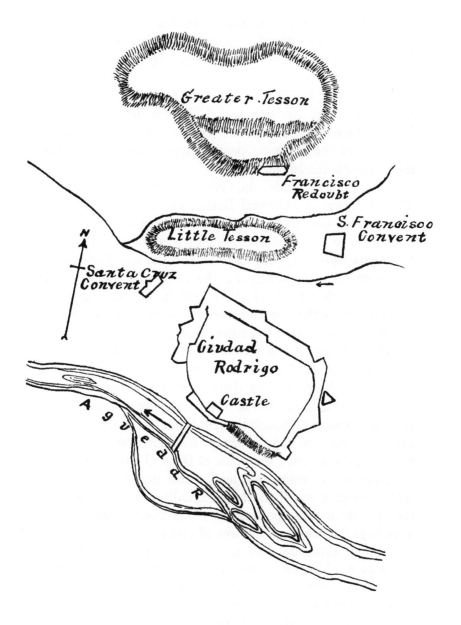

Greater Tesson

Francisco Redoubt

Little Tesson

S. Francisco Convent

N

Santa Cruz Convent

Ciudad Rodrigo

Castle

Agueda R

0   125   250   375   500

Scale of Yards

CIUDAD RODRIGO.

mob. At the critical moment the French fired a mine. Great loss of life ensued. General Mackinnon was among the killed and his body was subsequently found at a considerable distance.

Hardly had the 3rd Division got a footing in the place when the Light Division scaled the smaller breach. Its losses were considerable ; and its General, Craufurd, while cheering on the storming party, exposed himself and was shot through the lungs.

The French made, however, no further resistance. Their commander, General Barrie, gave up his sword and surrendered the fortress with the garrison, some 1800 strong. The loss of the assailants was 500 ; that of our Regiment being one sergeant killed : Captain Livingstone and four riflemen wounded ; the former so severely that he was unable to take further part in the war, and was transferred to the 3rd Garrison Battalion. Two riflemen had also been wounded the previous day.

The capture of the fortress was followed by a loss of discipline and a disgraceful orgy of drunkenness. Men were found drowned with their heads immersed in barrels of wine. The Light Division dispersed to plunder ; and the 3rd Division, although the fact is not as clearly recorded, may have been as bad. Perhaps not, however ; for above all the din was heard the voice of General Picton, in tones of thunder, expressing his unalterable conviction that all within earshot would be lost eternally. The energy and tone in which these pessimistic views were expressed may have kept his men in some sort of order.

General Craufurd died a few days afterwards, but his connection with the Regiment had been so slight that it deserves no more than a passing mention. He was perhaps ' felix opportunitate mortis ' ; for although a man of ability and education, his insubordination and lack of self-control were on more than one occasion a peril to the whole Army.

At this period the Rifle officers were distributed with their companies among the several Divisions of the Army as follows :—

1st Division : Captains Schaw and Franchini ; Lieut. Joyce.
2nd    ,,    Captains Macmahon and Blassière; Lieut. Barbaz.
3rd    ,,    Major Woodgate ;  Captain Schöedde ;  Lieuts. Wynne, Broetz, Du Châtelet, and O'Hehir.
4th    ,,    Captain Holmes.
6th    ,,    Brevet-Major Galiffe ; Lieut. Muller.

## 2

Lord Wellington did not allow the grass to grow under his feet.  As soon as he had found time to place Ciudad Rodrigo in a state of defence and had handed it over to the Spaniards, he marched his troops southward with a view to the capture of the more formidable fortress of Badajoz.  Since its siege of the previous year Badajoz had been considerably strengthened.  It was garrisoned by 5000 men, commanded by the celebrated General Philippon.

Badajoz stands at the confluence of the Guadiana River —500 yards broad—and the Rivillas, a small stream. The confluence is crowned by a work 100 feet high, on which stands an old castle.  The town was fortified by eight curtains and bastions and the following outworks— Lunette of San Roque covering a dam on the Rivillas ; the Picurina, a redoubt in advance of the Rivillas south of San Roque ;  and the Pardaleras, 200 yards beyond the south-westernmost bastion.  The problem was stiffened by the fact that in the British army there was no trained unit of Sappers and Miners, that Wellington's siege train was by no means equal to the occasion ;  and that for the purpose of gaining time the principles laid down for the conduct of a siege could not be carried out in their entirety. General Hill, with his Corps, watched the approaches from the east, and General Graham, with the 1st, 6th, and 7th Divisions was posted near Llerena, some eighty miles to the south-east with a view to guard against an attack from Marshal Soult, at present engaged in the blockade of Cadiz.

The force detailed for the siege of Badajoz included the well-tried 3rd, 4th, 5th, and Light Divisions. It was not strong enough for a total investment, but the siege was begun on March 17. On the night of the 25th–26th the outlying fort of Picurina was captured by the 3rd Division after a stout resistance. On April 6 the batteries had effected a breach in the Santa Maria and Trinidad Bastions, and although the counterscarp had not been blown in, Wellington resolved to capture the fortress by assault that night. The 4th and Light Divisions were detailed for the assault of the two bastions. The 5th Division was ordered to make a feint at the Bastion of St. Vincente. The 3rd Division was directed to make a false attack upon the Castle.

Nothing turned out as Lord Wellington intended. The main breach was found to have been so strongly retrenched that the Light and 4th Divisions, which by mistake failed to assault the exact spot intended, were, despite magnificent heroism, time after time repulsed.

The 3rd Division was at first commanded by General Kempt, for Picton had been warned too late that the hour of the assault had been changed from 10 to 9 P.M., and did not appear until the attack was well under way.

To Colonel Williams was assigned the duty of covering the advance, with the seven light companies of the Division and the three headquarters of the 60th ; the latter being under the immediate command of Lieut.-Colonel Fitz-Gerald. Williams, taking a circuitous route, pressed on

quietly till he reached the Rivillas stream. This had
been inundated ; but by crossing a mill-dam one by one,
and wading the water, which was knee deep, the men
reached the further bank. In rear of the Riflemen followed
the Light Companies. Behind them came in succession
the Brigades of Kempt, Campbell, and Champalimaud.
Just as the 60th had crossed the Rivillas a French sentry
in the ' covered way ' discharged his musket and the
Riflemen, believing themselves discovered, opened fire.
The garrison very quickly returned it with interest. Some
of the assailants were shot, some fell into the inundation
and were drowned ; but the ladder party, arriving at the
Castle, planted its ladders. They were found to be too
short and the assault failed. Then search resulted in the
discovery of a spot where the wall, having been breached
in the siege of the previous year, was lower—not more
than 20 feet high, and against this the ladders were
once more erected. Two officers of the 5th Fusiliers were
among the first to ascend. Their men rushed up behind
them, and though some were killed scaling the wall and
others hurled back by the defenders on to the bayonets of
their own support, a number of men held their ground and
the Castle was won. The French appear to have walled
in all the gates excepting one. Through that one they
retreated and then locked it from the outside. Partly,
therefore, from this cause, and partly from the fact that
both Generals Picton and Kempt were wounded ; partly
also perhaps from the feeling that the Division, having
converted a false attack into a real one and made it a
success, was not intended to do anything more, word was
merely sent to Lord Wellington of the capture of the
Castle ; but nothing was done to assist the 4th and Light
Divisions in their desperate struggle at the breaches.
Then the French suddenly threw open the gate already

mentioned and charged into the Castle yard; but a short fight resulted in their expulsion.

While these scenes were being enacted, Lord Wellington was sitting at a short distance from the scene of strife. Message after message of bad import reached him; but, excepting for the deathly pallor of his face, he made no sign. At last came a message from the breaches that the assault of the 3rd and 4th Divisions had definitely failed, that nearly all the officers had fallen and that the casualties among the men had been overwhelming. He turned to an officer at hand and said quickly: 'Go and tell Picton that he must try and succeed at the Castle.' But just at that moment came a message from that General to say that he was in possession of it. The revulsion of feeling must have been indescribable.

Previous, however, to this, sounds of musketry and of cheering had been heard from the north-western corner of the city. The 5th Division, as already mentioned, had been ordered to make a demonstration at the Bastion of St. Vincente. The General Officer in command had been, rather reluctantly, allowed to take with him a few scaling ladders, with which he escaladed the wall, and as it happened with little loss. His leading regiments advanced along the ramparts but were afterwards driven back; but then the Divisional General succeeded in getting his Division into the town and marched through the streets towards the breaches. His bugles, playing 'the Advance,' were answered by Picton's bugles from the Castle; and the French, finding themselves in danger of attack in flank and rear, realised that their defence of the breaches had been in vain and abandoned them. The 3rd and Light Divisions marched up the breaches unopposed, and the place was won. General Philippon, crossing the Guadiana, reached the fort of St. Christoval, whence he sent a message

to Marshal Soult in the hope of maintaining himself for a
few days.   Next morning he found it impossible to do so,
and surrendered.

Through the siege, from start to finish, Wellington's
casualties amounted to nearly 5,000.   Those which occurred
at the storming were 3,713.   During the siege and assault
our 5th Battalion had four companies engaged ; three with
the 3rd and one with the 4th Division.   Our losses were
as follows : From  March  18–22,  3  riflemen  killed  and
3 wounded ;  23rd–26th, 1 wounded ; March 31 to April 2,
1 killed, and 4 wounded ;  and during the storming on the
6th and 7th, Lieut. Sterne and 6 riflemen killed, Lieut.-
Colonels Williams and Fitzgerald—both of whom were
specially mentioned by General Picton in his dispatch—
Lieut. De Gilse and Lieut. and Adjutant Broetz with
2 sergeants and 24 riflemen wounded.   Our total losses
before Badajoz amounted, therefore, to 10 killed and 38
wounded, out of about 250 officers and men engaged.[1]

The casualties in the British battalions of the Divisions
principally engaged were as follows : —

| | |
|---|---|
| 3rd Division    .    .    .    . | 521 |
| 4th     ,,       .    .    .    . | 925 |
| Light   ,,       .    .    .    . | 919 |
| 5th     ,,       .    .    .    . | 536 |

Our men had not been engaged in so terrible a contest
since Ticonderoga.   An eyewitness, describes the inundation as appearing the next morning to be a lake of blood.

The horrors perpetrated by the assailants on the
inhabitants of Badajoz after it had been taken have often
been mentioned and need not be touched upon here.

---

[1] Lieutenant A. F. de Bruney (or Bruning) recently transferred from the 5th
Battalion, was also killed in the assault, while serving with the 3rd Caçadores.

The capture of Ciudad Rodrigo and Badajoz, almost under the noses of Marshals Marmont and Soult was a fine feat of arms and added greatly to the lustre of Wellington, who received the reward of an earldom.

As after Talavera, the compliment of two gold medals was paid to the Battalion: Colonels Williams and Fitz-Gerald being the recipients.

### 3

Through the winter General—now Sir Rowland—Hill had executed his duties on the frontier of Spain and Portugal; and, as already noticed, during the siege of Badajoz commanded the covering force. After the capture of the fortress, Wellington entrusted him with an important strategical operation, the object of which was to sever the principal line of communication between the French Army of Soult in Andalusia and that of Marmont which was quartered in the neighbourhood of Salamanca. This communication was formed by a bridge of boats across the Tagus at Almaraz. The task assigned to Hill was a difficult one, for the approaches were difficult and the bridge was securely guarded by permanent forts.

Taking with him the troops with which he had achieved the surprise at Arroyo Molinos, Sir Rowland effected his object on May 19, with a loss of only 33 killed and 144 wounded. The operation, although in a sense a minor one, was of such difficulty that Napier remarks that had it failed, its conception would have been attributed to madness. As it turned out it was entirely successful. The Rifle Companies of Blassière and Macmahon accompanied the force but had no casualties. One was with Howard's Brigade; the other formed a reserve with the 6th Portuguese Regiment, and in his subsequent dispatch

General Hill pays this body a compliment for its discipline
and steadiness.

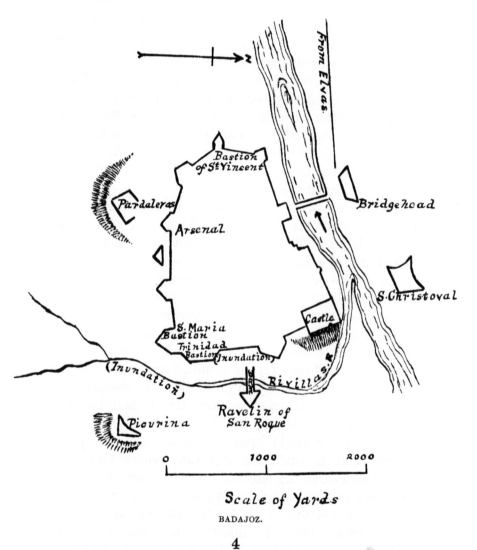

Scale of Yards

BADAJOZ.

4

The situation in Spain and Portugal was now as follows.
Marshal Soult, Duke of Dalmatia, with the Army of the

South, was in occupation of Andalusia, and besieging Cadiz ; a task which the British command of the sea made toilsome and difficult.

Marshal Marmont, Duke of Regusa, had distributed the Army of Portugal in the neighbourhood of Valladolid.

General Cafarelli commanding the Army of the North, guarded the lines of communication between Madrid and Bayonne and occupied the Biscayan Provinces.

In the Eastern Provinces of Spain Marshal Suchet kept in check such Spanish forces as operated against him.

King Joseph with his own Guards and the Army of the Centre, occupied the neighbourhood of Madrid and formed a central reserve to the whole.

The capture of Ciudad Rodrigo and Badajoz had opened for Wellington the gates into Spain. Communication between Hill and Wellington was materially shortened by the repair of the Alcantara bridge, whereas the destruction of the bridge at Almaraz had greatly lengthened that between Soult and Marmont, who could no longer hope to act in close co-operation. The moment was favourable for an Anglo-Portuguese advance ; the question was the direction that it should take. Lord Wellington had the choice of three lines. (1) He could advance on Valladolid against Marmont, and by so doing threaten the main line of French communication between Madrid and Bayonne. (2) He could advance by the line of the Tagus against Madrid. (3) He could bring his main force to operate against Soult in Andalusia and thus relieve Cadiz. Of these three courses he resolved upon the first, despite the risk of being overwhelmed by superior numbers in the event of the Armies of the North, Centre, and Portugal concentrating against him. To minimise this risk Wellington stirred up a hornet's nest in the Northern Provinces by requesting the Spanish forces in Galicia to operate against

Astorga, and the British Admiral off the Biscayan coast
to harass the northern sea ports.  Further, he arranged
with Lord William Bentinck, who commanded the British
forces in Sicily, to land at Alicante in Murcia such force
as he could spare.

The innumerable difficulties caused by want of specie,
transport, and supplies, must be studied in a larger history
of the campaign : suffice to say that they were enough to
break the heart of an ordinary man.

## 5

Our 5th Battalion was still commanded by Colonel
Williams, but Brevet Lieut.-Colonel Woodgate had gone
home soon after the capture of Ciudad Rodrigo and served
no more in the Peninsula.  He got command of the 4th
Battalion in 1814, but quitted the Regiment three years
later on reduction.  Woodgate died in Paris in January,
1861, being then 80 years of age.

Woodgate was succeeded as senior Major by John
Forster FitzGerald—who in later life was destined to reach
the rank of Field Marshal.  FitzGerald's career had been
somewhat remarkable.  We all know the story of bitter
cries proceeding from a nursery, which were explained
as due to the fact that ' the Major had fallen out of his
cradle ' !  Of this story FitzGerald might almost have
been the hero.  Born either in 1784 or 1786, he was given
a commission as ensign in 1793, when only seven, or at
most nine years old.  By May 9 in the following year
he was a Captain in the 79th, and (in accordance with the
rule in that period) was by virtue of seniority promoted
in September 1803 to the Brevet rank of Major !  In
November 1809 he was given a regimental majority in
the 60th, and next year—again by virtue of seniority—

a Brevet Lieutenant-Colonelcy. In the spring of 1812 he was transferred to the 5th Battalion.

The rifle companies were no doubt distributed as in the year previous. Captain Shaw probably commanded No. 6 and was attached to the 1st Division; Captain Holmes to the 4th; Captain A. W. Mackenzie only joined the Battalion on July 22, and his company is uncertain; Captain Schoedde was still with the 3rd Division.

6

During the first days of June, Wellington concentrated behind the Agueda the whole force under his immediate command, leaving Hill's Army Corps about Villa Franca and Los Santos to watch Soult.

Seven companies of our Regiment, *i.e.* the whole Battalion with the exception of the three companies attached to the 2nd Division, were present with Wellington; viz. two companies with the 1st Division; Headquarters and three companies with the 3rd; one with the 4th and one with the 6th. The strength of the Battalion as a whole was about 620 rifles.

It is worthy of note that the 7th Division was at this period commanded by John Hope [1] who, although a Major-General, was still shown in the Army List as a Lieut.-Colonel of our 2nd Battalion, to which he had been posted in 1804.

On June 9 Wellington crossed the Agueda in three columns; that on the left including the 3rd Division with our three Headquarter Companies under Colonel Williams. North of the Duero the Portuguese force commanded by General Silvieira and General D'Urban marched parallel

---

[1] Not to be confused with his distinguished namesake Sir J. Hope, who took command at Corunna when Moore was mortally wounded.

to the Anglo-Portuguese Army, whose total strength was about 53,000 men with 60 guns.

On the 13th Wellington entered Salamanca amid transports of joy on the part of the population and lost no time in attacking the three forts which guarded the town. All fell on the 27th after a stubborn resistance, which cost him 380 casualties, those of our own Regiment being 1 rifleman killed and 2 wounded.

Major Gustavus Brown of the 5th Battalion—at this time commanding the 9th Caçadores attached to the 6th Division—did good service in the assault of Fort La Merced by occupying the neighbouring houses and keeping down the fire from the fort. This officer had made the campaign of 1794 in Holland, and was present at the gallant defence of Grave, which was reduced by famine at the end of that year. In 1796 he served in the West Indies with his regiment (Löwenstein's) which had been taken into the British service and had, as we have seen, been merged into the 60th in the year following. Brown took part in the capture of St. Lucia, Grenada, and Surinam. In 1809 he joined the Portuguese Army and was present at Bussaco.

## 7

On hearing ot Wellington's advance, Marmont had concentrated about 25,000 men and marched to the relief of the Salamanca Forts. Being told of their capture he retired northward and took up a position behind the Duero, with his right on the Hornijas River and his left at Simancas. Wellington followed him and on July 3 attempted to turn his right flank. At the instance of the Commander-in-Chief the Riflemen of the 3rd Division successfully forded the Duero at Pollos, but the river was so high that Wellington feared to follow them up with the main body of the

Division and abandoned the operation. The enemy, eight battalions strong, was posted about a mile from the river but made no attempt to molest the Riflemen, who were left isolated. Marmont—whose army, made up of the Divisions of Generals Maucune, Thomières, Sarrut, Brennier, Bonet, Clausel, Foy, and Ferey, had been reinforced to a strength of about 50,000 men—then in his turn crossed the river by his left and advanced westward, threatening the British right flank. Wellington retired, and some days of manœuvring followed. A hot fight took place on the 18th at Castrejon where a rifleman was killed and two were missing. On the 21st, Marmont was on and partly across the Tormes, while Wellington facing east was posted with his left a little in advance of Salamanca and his right about half a mile north of the village of Arapiles. That night a terrible storm, which caused a stampede among the cavalry horses, proved, as was not infrequently the case in the Peninsula, the precursor of a battle on the following day.

At dawn on the 22nd Colonel Waters, the acting Adjutant-General, drew the Commander-in-Chief's attention to the importance of two hills respectively fourteen and sixteen hundred yards east of the village of Arapiles after which they were named. The French skirmishers were already almost in possession of both, but the northern hill was taken from them and successfully occupied.

Wellington had had an extraordinary piece of good luck. A letter written by him to the Spanish General, Castaños, stating that he could no longer hold his ground, had been intercepted by the French; and Marmont thereupon resolved to fight a general action without awaiting, as he had intended, the arrival of the Armies of the South and Centre. Another circumstance tended to illusion him; for Lord Wellington had amassed his baggage upon the

road to Ciudad Rodrigo, and the dust caused thereby
led Marmont to the conclusion that his opponent had
already begun his retirement. Unfortunately for the
Marshal, he was unable to gauge the strength of the British
Army massed near the village of Arapiles. Intent on
striking a decisive blow, he ordered General Maucune with
his own Division and that of Thomières—the latter leading
—to move to the west and strike the Ciudad Rodrigo road.
General Bonet's Division occupied the southern Arapile
hill. The Divisions of Foy and Ferey were on the Tormes
about Calvarissa. That of Clausel was near Bonet, but
the rest of the French army was still entangled in the
forest of Babila Fuente, and the execution of the movement made of course a large and increasing gap between
Maucune on the left and Bonet in the centre. Wellington
was not slow to seize upon the fatal error. Drawing up
the 5th and 4th Divisions supported by the 7th and 6th,
near the village of Arapiles, fronting due south, and
leaving the 1st and Light Divisions on his left to make
head against Foy and Ferey, the British Commander
galloped off to the 3rd Division, which had been posted at
Aldea Tejada, partly for the purpose of covering the retreat
of the baggage, and ordered its Commander, who happened
to be Edward Pakenham, his brother-in-law, to attack the
head of Marmont's army, while with his centre he struck
it in flank. The space between Wellington's centre and
the 3rd Division, was filled by the British cavalry under
General Stapleton Cotton.

It was now nearly 5 P.M. The sudden appearance of
the 4th and 5th Divisions, advancing in line of battle to
the attack, filled Marmont with amazement. In an instant
he realised the peril of his position. The gap between the
left of Ferey and the rear of Maucune's column was already
between three and four miles. In vain did the French

Commander send messengers at full gallop to accelerate
the arrival of his remaining Divisions still entangled in
the forest, and other officers to halt Maucune. The dazz-
ling line of advancing bayonets threatened to separate
irrevocably the centre of his army from the right.

Just at this moment the Marshal fell, severely wounded.
General Bonet had been already seriously hurt, but General
Clausel, one of the very ablest of Napoleon's soldiers, took
command. It was too late to recall Maucune. Before
he could be halted, the head of Thomières' column, travers-
ing a pine wood, was suddenly attacked by the full force
of the famous 'fighting 3rd Division' in column of Brigades
—headed by that of Wallace (late Mackinnon's) which
included the three Headquarters Rifle Companies. In vain
did the head of the French column sacrifice itself to give
time to the rear; in vain did the artillerymen die at their
guns; nothing could withstand the onset of Pakenham and
his men. Wallace and his Brigade, headed by the Riflemen,
fought their way through Thomières' Division, only to
find themselves isolated in the midst of that of General
Maucune, for the supporting Brigades had been unable to
keep pace with them; but at the critical moment a Brigade
of British Heavy Cavalry, under General Le Marchant,
charging down, completed the débâcle; and the Divisions
of Maucune and Thomières no longer existed as an organised
unit.

But with the coolness of a born commander of men,
General Clausel made a grand effort to snatch victory out
of defeat. With his own Division he not only filled up
the gap between Bonet and Thomières, but had something
in hand for the reinforcement of the left wing. The
Divisions of Sarrut and Brennier were brought up from the
forest and connected the French centre with its right.
Maucune succeeded at length in rallying his column. The

centre fought a desperate battle against the British 4th

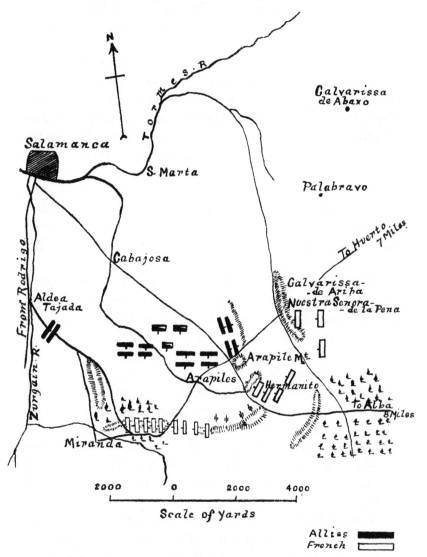

SKETCH OF THE SALAMANCA POSITION AT 4 P.M., JULY 22, 1812.

and 5th Divisions.  Fortune favoured first one side then

the other. The 6th Division, brought up in support, suffered terrible losses, but by sunset Wellington was victorious along the whole line. Although an error in the direction of pursuit saved the French from still worse disaster, their casualties reached a figure of 11,000 ; more than double those of Wellington exclusive of several thousands dispersed over the country side. Our own Regiment, whose parade strength of July 15 had shown 17 officers and 472 N.C.Os. and men present, lost in the action 6 riflemen killed, Colonel Williams, Major Galiffe, Ensign Lerche, and 24 riflemen wounded and 6 missing.

The casualties among the General officers on both sides were considerable. Four French Generals were killed, two wounded, one made prisoner. In the British Army General Le Marchant was killed, Beresford, Alten, Leith, Cole, and Cotton were wounded.

The French also lost 11 or 12 guns and 2 eagles, but it was the ulterior effects of the victory that were far reaching. So scattered was his army that General Clausel could, for several days, collect only 20,000 men, with whom he crossed the Duero in retreat. The defence of Madrid was for the moment impossible. On August 11 King Joseph abandoned his capital, which was entered on the following day by Wellington's victorious army. At the same time came news of the landing of the Anglo-Sicilian force at Alicante on the east coast. Marshal Soult was ordered to abandon the siege of Cadiz and join the King. The whole of the south of Spain had been torn from the hands of the enemy.

Colonel Williams, wounded for the sixth time, now bids farewell to our Regiment. He had commanded the battalion with distinction at Bussaco, Fuentes de Oñoro, Ciudad Rodrigo, Badajoz, and Salamanca. There are few who deserve a higher place in our annals as a commander of light troops ; and the 13th—now known as the Somerset Light Infantry, into which he exchanged—is equally indebted to him.

8

Lord Wellington's stay at Madrid was but brief. General Clausel had barely collected his 20,000 men when he moved down the Duero and threatened the British communications. On September 1 Wellington quitted Madrid for Arevalo taking with him about 25,000 men, principally made up of the 1st, 5th, 6th, and 7th Divisions, leaving behind him at Madrid the 3rd, 4th, and Light Divisions. With Wellington marched the two companies of our Regiment attached to the 1st Division and that attached to the 6th. These three companies mustered about 220 of all ranks.

On Lord Wellington's approach, Clausel retired before him as far as Briviesca, a few miles south of the Ebro ; and Wellington, following him up the Arlanzan River, decided to take the castle of Burgos with the 1st and 6th Divisions, while pushing the remainder of his army beyond it as a covering force against Clausel.

Burgos was not a very strong fortress, and was garrisoned by only 1800 men under the command of General Du Breton ; but its strategical importance was great, for it was situated on the Royal road, forming the main line of communication between Madrid and Bayonne. It contained Clausel's reserve artillery, and for more reasons than one its capture was an object of importance.

On September 19, the 1st Division captured the outwork called Fort St. Michael. But its capture was dearly won, for the affair was mismanaged ; our losses amounting to 430, of whom one Rifle sergeant and two men were killed; while Lieutenant O'Hehir was severely wounded, one rifleman was missing and seven others wounded.

Lord Wellington, remembering the services of Colonel Gustavus Brown (of our Regiment) at the attack on the Salamanca forts, gave orders that he was to be placed in charge of the detachments in the town of Burgos belonging to the 6th Division, with instructions to occupy with his battalion of Caçadores the houses close to the ramparts. During the remainder of the siege this officer, who was commanding the 9th Portuguese Caçadores, was constantly employed in work demanding skill and courage, and the references to him in Wellington's Dispatches give evidence of the Commander-in-Chief's good opinion and confidence.

On the 23rd General Hope—still nominally Lieut.-Colonel of our 2nd Battalion—was compelled by ill health to quit the army. Wellington in his Dispatches speaks of him as being very attentive to his duties and regrets his departure. Hope, born in 1765, began his career in 1778 as a cadet in a Scottish regiment in the service of Holland. Rapidly passing through the grades of N.C.O. and subaltern, he became a captain in 1782. Transferred then to the British service, he was appointed to the 60th in 1788, but transferred next year to the 13th Dragoons. Having served in the campaigns 1793–5, and at the Cape of Good Hope, he was made Lieut.-Colonel of the 2nd Battalion of the 60th. In 1810 he was promoted to the rank of Major-General. In 1821 Hope was knighted; and later on received the G.C.H. He died in 1836.

The siege of the Castle continued, but no great progress was made, for the two Divisions had little experience in siege operations; and Wellington soon had cause to regret the loss of so many of his experienced senior Generals of Division at Salamanca, and that he had left behind at Madrid the Divisions which had captured Ciudad Rodrigo and Badajoz.[1]  The British Commander was also ill provided with a siege train, and things began to go from bad to worse.  General Clausel was succeeded in the command of the Army of Portugal by General Souham, who having formed junction with the Army of the North, came down during the third week in October in an attempt to relieve Burgos.  At his approach, Wellington left his trenches in charge of a Portuguese Brigade, and with the 5th and 6th Divisions joined Sir Edward Paget, who having recovered from his wound received at the passage of the Douro, had just rejoined the army and been placed in command of the covering force which consisted of the 1st and 7th Divisions.  Nevertheless, the Allies with 25,000 British soldiers, were threatened by 44,000 of the enemy ;  and not only that, but were exposed to attack which if successful would force them to retire by a road under the very guns of the Castle of Burgos.  Happily at the moment of imminent peril, General Souham received an order from King Joseph to delay his offensive.  Just at the same time the British Commander received news that King Joseph and Marshal Soult were advancing upon Madrid, and that Sir Rowland Hill, who had arrived with the 2nd Division and had taken command of the Army Corps in the neighbourhood of the capital, was unable to hold his ground.

It was evident to Lord Wellington that he had no alternative but an immediate retreat.  During the night of October 21 and 22, he skilfully withdrew his army to the south of Burgos, the siege of which had cost Du Breton something over 600 men and Lord Wellington nearly 2000.

---

[1] Graham and Picton had been invalided just prior to Salamanca.  At that battle Cole, Leith, Alten, and Cotton had been wounded ; and at the present moment every Division in the Army was commanded by a ' locum tenens.'

Wellington retired to Cabezon on the Pisuerga, whence on the 26th, he wrote to Sir Rowland Hill at Madrid informing him of the retirement from Burgos and asking that officer to join him on November 4, if possible, at Arevalo on the Adaja.

Continuing his retreat to the Duero, Lord Wellington found his communications threatened by General Souham. Having no alternative but to continue the retirement, he crossed to the left bank of the Duero, but had hardly done so when he heard that Souham had seized the bridge at Tordesillas in his rear. Wellington with difficulty regained his line of retreat and being in force insufficient to hold the long line of the river was compelled to relinquish his plan of forming a junction with Hill on the Adaja, so fell back to the Tormes, reaching the heights of San Christoval on November 8.

Meanwhile, immediately after the battle of Salamanca, Marshal Soult had been ordered to raise the blockade of Cadiz and join King Joseph. General Hill had then advanced from the Spanish frontier to Madrid, where he arrived at the end of September ; and having taken command also of the 3rd, 4th, and Light Divisions, found himself at the head of a force of 40,000 men with 30 guns. He took up a line on the right bank of the Tagus from Toledo to Aranjuez, covering Madrid from the south. Spanish levies to the amount of 20,000 men, with 12 guns, acted to some extent in co-operation with him by holding the road from Cuenca to Valencia.

On October 3, Soult formed his junction with King Joseph at Almanza. On the 19th Joseph advanced against Madrid with the Army of the Centre—12,000 men and 12 guns—commanded by Count d'Erlon, and that of the South under Soult, comprising 46,000 men with 72 guns. The former moved upon Cuenca and Tarancon ; the latter upon Aranjuez. By this time the Tagus was fordable, and the Spaniards being unable to stop the French, General Hill found it impossible to hold the line of the river. On the 20th he fell back through Madrid to join Wellington. On November 3, Madrid was re-occupied by King Joseph.

On the 8th Hill was behind the Tormes about Alba, and in touch with Wellington at San Christoval. On the same day King Joseph formed a junction with General Souham ; and the united French forces amounted to 100,000 men and 130 guns, against only 70,000 of the Allies with 68 guns. Moreover, of the latter, 16,000 men and 12 guns were Spaniards of inferior discipline. On the 10th and 11th, some fighting took place about Alba de Tormes, during the course of which 8 riflemen were wounded. On the 11th the King deprived himself of 10,000 men and 10 guns by sending the Army of the North back to Burgos.

On the 15th, Marshal Soult, to whom the King had given command in chief, successfully manœuvred Lord Wellington out of his position, and the latter found it necessary to resume his retreat. The Army of Portugal was in his rear ; Soult was attempting to intercept him at Tamames. Lord Wellington barely escaped the trap on the 18th, but Soult, finding himself foiled, desisted from further pursuit. On the 19th, the British Army reached Ciudad Rodrigo. Both wings had retreated for nearly 300 miles, the troops had undergone great hardships ; they were dying of hunger, and the discipline of the whole Army was, chiefly through the failure of the Commissariat, much affected. From the time of the first attack on Burgos till the end of the retreat, the British losses had been about 9,000 men, 3,500 of whom were prisoners. In our Battalion 115 were ' missing ' ; a large proportion being undoubtedly killed, starved to death, or wounded.

II.                                                                 N

On reaching the Portuguese frontier, Lord Wellington sent Sir Rowland Hill to take up a position with his Army Corps by Coria and Placensia. The Light Division held the line of the Agueda; and the Army went into winter quarters, the cavalry being billeted along the valley of the Mondego and the infantry near the Douro.

As regards the detail of our Regiment, the company attached to Sterling's Brigade of the 1st Division was ordered to remain with that Division when, on November 11, the Brigade was transferred to the 6th. Sterling then received the company hitherto attached to Hulse's Brigade of the 6th Division.

A return given in Wellington's Supplementary Dispatches, vol. vii. p. 523, gives the following figures regarding the 5th Battalion, dated October 25, 1812: Effective 729; sick 111; wanting to complete establishment 170; in the depôt at Lymington, 63. The proportion of sick to effective is 15 per cent. Only four battalions show a small ratio; the average of the whole Army in the Peninsula being at this time no less than 58 per cent. of sick to effectives.

The results of the campaign appeared at first sight disappointing; but, as pointed out by the British Commander, nearly 20,000 French prisoners had been taken; Astorga, Guadalexara, and Consuegra had been captured; the siege of Cadiz had been raised; and the south-west of Spain delivered from the enemy.

During this time events on the Continent had followed one another with startling rapidity. Napoleon's ' Continental System ' was recoiling upon himself. It alienated his ally the Emperor of Russia, who refused to exclude British commerce from his ports. Napoleon's remonstrances were disregarded, and he decided that no course remained but to compel Russia by force of arms to accede

to his wishes.   On June 24 he crossed the Niemen, at
the head of an army of a strength hitherto unapproached
in civilised warfare.   On September 14 he occupied
Moscow.   Still the Russians refused to make peace.   The
condition of the French army, at such an enormous distance
from its base, became daily more and more critical, and
almost on the same day that Wellington retired from
Burgos, the Emperor Napoleon began that disastrous
retreat from Moscow from the effects of which his power
never recovered.

## CHAPTER IX

THE year 1813 opened with bright prospects for the Allied Powers. The Grand Army of the Emperor Napoleon had been practically destroyed in Russia. The remnants were retiring through Germany followed by the Russians, with whom the Prussian Government, which under the cunning system of Scharnhorst had secretly trained a large army, was making common cause. The veteran armies of France in Spain had been drawn upon in the past year for the purpose of the Russian expedition. They were now still further depleted in order to stiffen up the battalions of young soldiers who were by this time almost the sole resource of the French Empire.

In Spain the French armies held the line from the Galician mountains to Valencia distributed as follows :—

In the Provinces of Biscay and Guypuscoa, General Caffarelli, with the Army of the North, 48,500 strong, guarded the main line of communication with France : a duty made excessively difficult by the incessant harrassing of the mobile guerilla bands termed " Partidas," which attacked convoys, cut off stragglers, and almost destroyed postal arrangements.

The Army of Portugal, 42,700 in all, commanded by General Reille, held the line of the Esla and Tormes with supports at Valladolid and the neighbourhood. On

Reille's left was posted the Army of the South—46,000—with its Headquarters at Toledo. Its commander was Count Gazan, for Marshal Soult had been made Chief of the Staff, but being too blunt and downright for King Joseph was in February transferred to Germany, and succeeded by Marshal Jourdan.

Further eastward, Marshal Suchet with 2,000 cavalry, 16,000 infantry, and 30 guns, held the line of the Zucar against the Anglo-Sicilian contingent at Alicante, nearly 15,000 strong and aided by Spanish troops to the amount of 35,000 men and 50 guns.

The Army of the Centre, almost 20,000 strong under Count d'Erlon, formed a central reserve, and was posted from the Guadarama mountains to the district of Madrid.

At Madrid, King Joseph remained with his Court till March, although warned by Napoleon that he should hold the capital merely as a post of observation and maintain his G.H.Q. at Valladolid.

The nominal strength of the French armies in Spain still exceeded 230,000 men ; but of this number 30,000 were in hospital.

2

As the spring of 1813 wore on and the grass began to grow, Lord Wellington, who had been rejoined by nearly all his old Divisional Commanders, completed his arrangements for resuming the offensive.[1] Sir Thomas Graham with five Cavalry Brigades, the 1st, 3rd, 4th, 5th, 6th, and

---

[1] Tin camp kettles were substituted for the heavy iron ones hitherto carried ; and for the first time tents were provided for the officers and three per company for N.C.Os. and men.

7th Infantry Divisions and two Portuguese Brigades was directed to cross the Douro, move up the right bank of that river and occupy the line of the Esla with a view to forming a junction with the Galicians in the north. Wellington himself, retaining the Light Division in his own hands, called up Hill and the 4th Spanish Army, 40,000 strong, under General Castaños, in order to cross the Agueda and force the line of the Tormes.

The Count de la Bispal with the 1st Army of Reserve, 15,000 strong, was directed to cross the Tagus at Almaraz and co-operate with Lord Wellington by advancing upon Valladolid.

In the eastern provinces the Anglo-Sicilian Army, under Sir John Murray, was instructed to take the fortress of Taragona and to prevent Marshal Suchet despatching reinforcements to King Joseph. The 1st Spanish Army under General Copons, 10,000 strong, was to operate in Catalonia; the 2nd under General Elio, 20,000 men, in Murcia; and the 3rd, 12,000 men under del Parque, in the Morena.

The State of the Anglo-Portuguese Army was as follows :—

### STAFF

Commander-in-Chief of the Allied Forces in the Peninsula : Genl. the Marquis of Wellington, K.B.[1]

Second in Command of the British Army : Lieut.-General Sir Thomas Graham, K.B.

Adjutant-General : Colonel Lord Aylmer (acting).

Quartermaster-General : Brig.-General G. Murray.

Commander of Portuguese Army : Lieut.-General Sir Wm. Beresford, K.B.[1]

---

[1] Wellington had the rank of Marshal General, and Beresford that of Marshal in the Portuguese Army.

## Cavalry Division

### *Lieut.-General Sir Stapleton Cotton, K.B.*

|  | Brigades. | Effective sabres. |
|---|---|---|
| Household Cavalry . . | Major-Gen. O'Loghlan . | 725 |
| 5th Dragoon Guards . .<br>3rd & 4th Dragoons . . | } Br.-Gen. Hon. W. Ponsonby | 1085 |
| 3rd D.G. & Royal Dragoons | Major-Gen. J. Slade . . | 696 |
| 13th Light Dragoons . | Major-Gen. Long . . | 320 |
| 12th & 16th L.D. . . | Major-Gen. G. Anson . | 725 |
| 14th L.D.. . . .<br>1st & 2nd Hussars K.G.L. . | } Major Gen. V. Alten . . | 998 |
| 1st & 2nd Dragoons K.G.L. | Major-Gen. Bock . . | 512 |
| 10th, 15th, 18th Hussars . | Col. Grant . . . | 1530 |
| 4th & 6th Portuguese Cavy. | Col. Campbell . . . | 655 |
| 1st, 11th, 12th ,, ,, | Br.-Gen. D'Urban . . | 746 |
|  | Total . . | 7992 |

## 1st Infantry Division

### *Major-General Howard*

|  |  | Bayonets. | Rifles. |
|---|---|---|---|
| 1st & 3rd Bns. 1st Guards<br>1 coy. 60th. . . | } Maj.-Gen. Howard . | 785 | 44 |
| 1st. Bn. Coldstream Guards<br>,, 3rd Guards<br>1 coy. 60th . . . | } Maj.-Gen. Hon. E.<br>Stopford . . . | 1320 | 43 |
| 1st, 2nd, 5th Bns. K.G.L. | Maj.-Gen. Baron Lowe | 1519 |  |
| 1st & 2nd Light Bns. ,, | Col. Halkett . . | 1222 |  |

N.B.—A Light Battalion made up of the Riflemen and the Light Companies of the Guards served during the whole campaign of 1813, and was present at the battles of Vitoria and the Nive.

## 2ND DIVISION

### *Lieut.-General Sir W. Stewart, K.B.*

|  |  | Effective Bayonets. | Rifles. |
|---|---|---|---|
| 1st Bn. 50th, 71st, & 92nd 1 coy. 60th | Col. O'Callaghan . | 2324 | 65 |
| 1st Bn. The Buffs, 57th, 31st, 66th . . . 1 Coy. 60th . . . | Br.-Gen. Byng . | 2183 | 56 |
| 1st Bn. 28th, 39th . . 2nd Bn. 34th . . 1 Coy. 60th . . . |  | 2212 | 70 |
| 6th & 18th Portuguese . 6th Caçadores . . |  | 2671 |  |

## 3RD DIVISION

### *Lieut.-General Sir T. Picton, K.B.*

| 1st Bn. 45th, 74th, 88th . 3 coys. 60th. . . | Br.-Gen. Brisbane . | 1612 | 194 |
|---|---|---|---|
| 1st Bn. 5th ; 2nd Bn. 83rd, 87th, 94th . . . | Maj.-Gen. Hon. C. Colville . . | 1782 |  |
| 9th & 21st Portuguese . 11th Caçadores . . | Br.-Gen. Power . | 2128 |  |

## 4TH DIVISION

### *Lieut.-General Sir G. L. Cole, K.B.*

| 3rd Bn. 27th, 1st Bn. 40th, 48th, 2nd 53rd . . 1 coy. 60th . . . | Maj.-Gen. W. Anson | 2273 | 56 |
|---|---|---|---|
| 1st Bn. 7th, 20th, 23rd, . 1 coy. Brunswick Oels. | Col. Skerrett . . | 1622 |  |
| 11th, 23rd Portuguese . 7th Caçadores . . | | 2537 |  |

## 5TH DIVISION
### *Major-General Oswald*

| | | Effective Bayonets. | Rifles. |
|---|---|---|---|
| 3rd Bn. 1st, 1st Bn. 9th, 38th . . . . 1 coy. Brunswick Oels. | Major-Gen. Hay | 1694 | |
| 1st Bn. 4th, 30th, 44th 2nd Bn. 47th, 59th 1 coy. Brunswick Oels. | Br.-Gen. Robinson | 2183 | |
| 3rd & 15th Portuguese 8th Caçadores . | Br.-Gen. Spry . | 2082 | |

## 6TH DIVISION

| | | Effective Bayonets. | Rifles. |
|---|---|---|---|
| 1st Bn. 11th, 42nd, 61st, 79th, 91st . . . 1 coy. 60th . | Col. Stirling . . | 3096 | 51 |
| 1st Bn. 32nd, 36th . . | Col. Hinde . . | 900 | |
| 8th & 12th Portuguese 9th Caçadores . | Br.-Gen. Madden . | 2469 | |

## 7TH DIVISION
### *Lieut.-General Lord Dalhousie*

| | | Effective Bayonets. | Rifles. |
|---|---|---|---|
| 6th, 24th, 58th . 9 coys. Brunswick Oels. | Br.-Gen. Barnes . | 1860 | |
| 51st, 68th, 82nd Chasseurs Britanniques . . | | 2031 | |
| 7th, 19th Portuguese 2nd Caçadores. . | Br.-Gen. Le Cor . | 2102 | |

## LIGHT DIVISION
### *Major-General C. Alten*

| | | Effective Bayonets. | Rifles. |
|---|---|---|---|
| 1st Bn. 43rd, 1st & 3rd Bns. 95th . . | Major-Gen. Kempt | 895 | 881 |
| 1st Bn. 52nd, 2nd Bn. 95th . . . | Major-Gen. Vandeleur | 819 | 398 |
| 17th Portuguese . 1st & 3rd Caçadores . | | 1983 | |

PORTUGUESE DIVISION

*Lieut.-General Silvieira*

| | | Bayonets. |
|---|---|---|
| 2nd & 14th Portuguese. | | 2101 |
| 4th & 10th Portuguese | } | 2559 |
| 10th Caçadores | } | |
| 1st & 16th Portuguese | } Br.-General Pack . | 1979 |
| 4th Caçadores. | } | |
| 5th, 13th, 24th Portuguese | Br.-Gen. Bradford . | 2329 |

Total of Sabres, Bayonets, and Rifles  .  . 67,122 : Guns 90

3

It may here be noticed that the Battalion States which after Colonel Williams' departure had been signed by Major J. F. FitzGerald, commanding 5th Battalion from August 1812 to March 1813, were signed in April 1813, by Colonel John Keane, successor to Wm. Williams. Born in 1781, Keane was fortunate enough to join the army at the age of 13 with the rank of Captain. In 1802—being then barely 21—he got a majority in our Regiment, and in the following year a Lieutenant-Colonelcy in the 13th : which in 1809 he commanded at the capture of Martinique. On January 1, 1812, Keane received the Brevet rank of Colonel and exchanged with Colonel Williams. It would, however, appear that he was almost immediately after-wards appointed to the Staff, and FitzGerald to the command of a light Battalion in the 2nd Division ; for the subsequent States until the end of the war were all signed by Major Galiffe, who no doubt remained with the Head-quarter Companies of the Battalion in the 3rd Division, which after the retreat from Burgos were quartered during the winter 1812–13 at Paradosa, just within the Spanish frontier.[1]

Jean Pierre Galiffe, who was destined to command the Battalion in no less than six general actions, was the scion of a very old French family which eventually settled at Geneva. Born in 1767, he was not only very much older than either Keane or FitzGerald but had much greater professional experience. At the age of 17, Galiffe had received a commission in the Swiss regiment of Château Vieux, one of those embodied in the old Royal Army of France. After the massacre of the Swiss Guard and the deposition of Louis XVI. on August 10, 1792, the regiment decided that it was absolved from its oath of allegiance, quitted the garrison of Bitsche and joined the army of Condé in Germany. The officers then returned to Geneva, which was threatened, and afterwards besieged, by the French. In 1794–5 Galiffe fought in the Netherlands as Captain in the Red Hussars of Timmerman. Thence he was in 1796 transferred to the York Rangers in the British Service, which in the year following was merged into our 5th Battalion.

The State of the Battalion on April 25 showed :—

| 1 Lieut.-Colonel. | 2 Majors. | 7 Captains. |
|---|---|---|
| J. Keane. | J. F. FitzGerald, | |
| | J. Galiffe. | |
| 15 Subalterns. | 5 Staff. | 61 Sergeants. |
| 15 Buglers. | 719 Riflemen. | |

Of the officers 18 were British and 12 foreign. With one exception the whole of the N.C.Os. and men were foreigners.

---

[1] Nevertheless, in a letter of July 23, Wellington speaks of Colonel Keane as commanding the 5th Battalion, and it was not until August 8 that the latter became permanent commander of a brigade in the 3rd Division.

The Rifle Companies seem to have been thus distributed and commanded.

*1st Division*

No. 5, Captain E. Purdon.
  ,,  9, Captain J. W. Harrison.

*2nd Division*

No. 6, Captain F. S. E. Erskine.
  ,,  8,   ,,    P. Blassière.
  ,,  4, Lieut. A. F. Evans or
            Ensign G. Lerche.

*3rd Division*

No. 2, Captain R. Kelly.
  ,,  3,   ,,    I. Franchini.
  ,,  7,   ,,    J. H. Schöedde.

*4th Division*

No. 10, Captain J. Stopford.

*6th Division*

No. 1, Lieut. H. Muller.
      ,,  C. J. de Franchiosi (attached to the Staff of the Division
          as Interpreter).

4

On March 15 King Joseph, in accordance with the Emperor's wish, quitted Madrid—as it turned out for the last time—and took up his Headquarters at Segovia and the neighbourhood. The French Army in Spain at this period showed, as already noticed, a nominal strength of about 230,000 men ; but the force available for active operations could hardly be reckoned at over 160,000, of whom about 50,000 were employed either in the Northern Provinces or on the east coast, leaving about 110,000 men to oppose Wellington.

Of the latter force, the Army of the Centre extended from Burgos to Arevalo. The Army of the South, under General Gazan, occupied the space between the Duero, Tormes, and Adaja, retaining a detachment at Madrid and posts on the Tagus. The Army of Portugal, under General

Reille, had one Division at Valladolid ; while the remaining five Divisions had been perforce detached to help General Clausel, who had succeeded Caffarelli in command of the Army of the North, to contend with the Partidas ; a wearying and toilsome business which was not completed until well into the spring of 1813, after which Clausel and his troops had no time for rest before Wellington was in the field threatening the very existence of the French in Spain.

5

In accordance with Lord Wellington's plan of operations adopted after mature consideration, Sir Thomas Graham, at the head of five Brigades of Cavalry, six Infantry Divisions and two Portuguese Brigades (45,000 sabres, bayonets, and rifles with seven batteries), crossed the Duero on May 24, and entered the Province of Tras-os-Montes.  Advancing in three columns he formed a line, on a front of forty miles facing east with his left at Braganza, his centre (including the 3rd Division with the Headquarter Companies of our 5th Battalion) at Vimioso, and his right at Miranda de Duero.  On the 29th he occupied Tabara, Losilla, and Carvalajes overlooking the river Eslar.

Wellington in person, quitting Portugal at the same time with three Cavalry Brigades and the Light Division, advanced on Salamanca ; where he had ordered Hill and the Spanish General Morillo to join him.  The combination was effected with admirable precision, and Wellington with 4 Cavalry Brigades, 2 British, a Portuguese and a Spanish Division, took command of a force of 30,000 sabres and bayonets, with 5 batteries.

Graham's line of advance had already turned the flank of General Villatte who held Salamanca.  The latter attempted, however, to make a stand, but on finding Wellington's cavalry threatening his rear made a hasty retreat. During the 27th Wellington advanced northward to the Duero, and on June 1 united the two wings of the army on that river.

Meanwhile General Gazan, Commander of the French Army of the South, was making desperate attempts to concentrate his army; but the concentration was slow, and Gazan was compelled to retire in succession behind the Trabanços, the Zapardiel, and the Duero.

Wellington, halting somewhat unfortunately on the 3rd, enabled the French armies to complete their concentration and extricate large convoys moving northward by the Pisuerga. But King Joseph and Marshal Jourdan, his Chief of the Staff, felt a further retirement essential, and Gazan accordingly retreated on Torrelobaton; D'Erlon with the Army of the Centre upon Dueñas; Reille on Palencia. Next day Lord Wellington resumed his advance in four columns of Army Corps, counting from the left as follows :—

1. Giron's Spaniards : 3 Divisions, by Herrera and Polientes on the Ebro.

2. Sir T. Graham, with the Cavalry Brigades of Bock and Anson, the 1st and 5th Infantry Divisions and the two Portuguese Infantry Brigades of Pack and Bradford, by Rio Seco, La Piedra and Villar cayo.

3. Headquarters : comprising the Household Cavalry Brigade and those of Ponsonby and D'Urban, the Light, 3rd and 4th Divisions, and two batteries of reserve artillery, by Palencia, and Villadiego.

4. Sir R. Hill, with the Cavalry Brigades of Long, Fane, and Victor Alten, the 2nd, 6th, 7th British, a Portuguese and a Spanish Division marched by Castrogeriz.

On the 6th the French were behind the Carrion ; on the 7th behind the Pisuerga. King Joseph, who had collected 53,000 men, hoped to fight a battle for his crown near Burgos ; but that fortress was now reported untenable and Joseph had no alternative but to retire behind the Ebro, where he expected to be joined by the Army of the North and the detached Divisions of the Army of Portugal. The Royal Road, his line of retreat, was, however, choked up with a mass of vehicles containing stores and baggage ; and Wellington, throwing forward his left, continuously outflanked the retreating enemy.

On the 13th Joseph blew up the castle of Burgos ; and Wellington—despite his long and rapid march over difficult country—turned a deaf ear to the suggestions of his Staff urging him to halt ; determining to make a great effort to drive the French from Spain before they could be reinforced from Germany, where an armistice had followed Napoleon's victories at Lutzen and Bautzen. An immediate British victory was essential to prevent the armistice leading to a general peace.

To this end Wellington changed the direction of his march slightly to the left, and then crossed the Ebro at Polientes with his left corps on the 14th and at San Martin, Quintana, and Puente Areñas with the centre and right on the following day.

This flank march severed the French from the sea coast, and the British Commander took advantage of the fact to organise a new base at Santander on the Bay of Biscay. By this action he cast off his old line of communications with Portugal and formed one infinitely shorter, with the additional advantage that it enabled him to descend almost perpendicularly on that of the French. Continuing to bring his left round, Wellington poured his troops through mountain pass and defile into the main road leading from

Frias to Bilbao whence he was in a position to roll up the enemy's left.

The author of ' The Bivouac' describes the march in glowing words.

> ' One while, the columns moved through luxurious valleys inter-sprinkled with hamlets, vineyards and flower gardens ; at another they struggled up mountain ridges or poured through Alpine passes overhung with topping cliffs. . . . If the eye turned downwards, there lay sparkling rivers and sunny dells ; above rose naked rocks and splintered precipices ; while moving masses of glittering soldiery, now lost, now seen, amid the windings of the route, gave a panoramic character to the whole that can never fade from the memory of him who saw it.'

And in his ' History of the Peninsular War' Napier emphasises the fact that

> ' neither the winter gullies nor the ravines, nor the precipitous passes among the rocks retarded even the march of the artillery ; where horses could not draw men hauled ; when the wheels would not roll the guns were let down or lifted with ropes ; and strongly did the rough veteran infantry work their way through these wild but beautiful regions. Six days they toiled unceasingly ; on the seventh, swelled by the junction of Longa's division and all the smaller bands which came trickling from the mountains, they burst like raging streams from every defile and went foaming into the basin of Vitoria.'

During this forced march the troops lived upon the abundant corn of the country ; the huge strings of commissariat mules being far in rear. But the French were extended for convenience of supply, and the extension appeared to the King warranted by the position taken up ; for by this time they were behind the rocks and defile of Pancorbo. The entrance to the defile was guarded by outposts, the idea being to resume the offensive on the arrival of reinforcements. Gazan occupied the centre; D'Erlon the left, with flank guards extended to the Burgos-Logrono road; and Reille the right; one of his Divisions —that of Maucune—being posted at Frias in touch with Gazan, and those of La Martinière and Sarrut in rear and still further to the right at Espejo and Osma respectively.

Marshal Jourdan, Chief of the Staff, warned Joseph that Wellington was probably manœuvring to turn his right; and Reille was accordingly directed to re-occupy Valmaceda. But on the 18th his two Divisions were attacked at Osma by Sir Thomas Graham with the 1st, 3rd, and 5th British Divisions, while Maucune was simultaneously assailed and defeated by the Light Division from Wellington's column.

This blow made it clear to the French Staff that a large force of the Allies was already in their rear. Immediate retreat was inevitable; and the French army fell back behind the Ebro, narrowly escaping disaster during the operation. King Joseph then took up a position in front of Vitoria, on a tract of low-lying land about eight miles square which lay on the left bank of the Zadorra, forming part of the ' basin of Vitoria '; the position being occupied not so much for the purpose of a general action as of gaining time for the arrival of Generals Clausel and Foy, who had been sent for in hot haste. It was bordered on the north and west by the Zadorra; on the east and south by mountains. Two miles north of the city of Vitoria, General Reille was posted to defend the bridges of Gamara Major and Arriaga, on the roads leading respectively from Durango and Orduña. Four miles thence down stream the Zadorra makes a bend southward, and here the Army of the South under Gazan was drawn up astride the Royal Road from Madrid to Bayonne: its right resting on the river, and its left posted behind the village of Subjana de

II.                                                            O

Alava. Bridges spanned the Zadorra at Mendoza, Tres Puentes, Villodas, and Nanclares and were commanded by 50 guns. D'Erlon's Army of the Centre formed a second line in rear of Gazan ; and the bulk of the cavalry and the King's Guards were held in reserve at various points between Vitoria and Ariñez.

Wellington's Anglo-Portuguese Army as concentrated for battle consisted of about 72,000 officers and men with 90 guns, and had the aid of 20,000 Spaniards. The 4,000 composing the Division of Carlos d'España had been left at Miranda de Ebro ; and De La Bispal with 3,000 men was still a few marches distant.

The French mustered less than 70,000 of all arms, but were provided with 153 guns.

The gap of four miles separating the left of Reille from the right of Gazan was an unpardonable error, and the peril of King Joseph's situation was aggravated by an enormous mass of baggage crowding the streets of Vitoria.

On the 19th Sir Thomas Graham reached Zuazo. Hill occupied the Puente Lara on the Ebro. The Centre was about Espejo and Subjana de Morillas : the 3rd and 7th Divisions, combined under command of Lord Dalhousie, being at Carcamo and Berberana.

On the 20th Graham reached Murguia ; Dalhousie, the Bayas River ; and Hill, Pobes. On this day Lord Wellington found that, contrary to his expectation, the enemy intended to fight on the line of the Zadorra.

The memorable June 21 dawned in mist and rain. The battle was begun on our right by Sir Rowland Hill, who sent forward a brigade of Morille's Spanish Division to capture the heights of La Puebla on which the French left rested. Morille was successful ; but was very shortly afterwards counter-attacked in such force that Hill sent up in haste support consisting of the 71st and his Light

Battalion under Lieut.-Colonel J. F. FitzGerald of the
60th, made up of the three rifle companies attached to the
2nd Division and the Light Companies of its 1st and 3rd
Brigades. The two battalions were commanded by Lieut.-
Colonel Cadogan of the 71st, who on falling mortally wounded
was succeeded by FitzGerald. Thus reinforced Morille
not only held his ground but captured a second ridge which
placed him in rear of the French line ; but once more a
counter-attack drove back the allied force to the ridge
originally taken. For a long time the fighting was desperate ; and only subsided when, the French having been
defeated in the centre, a general retirement of their line
took place.

Meanwhile on Hill's left the 4th and Light Division
had reached the banks of the Zadorra and Kempt's Brigade
crossed it at Tres Puentes about 1 P.M., with the consequence that Gazan's flank was laid bare, and his army
retired in the direction of the town of Vitoria. The
4th Division crossed at Nanclares ; General Picton at
the same time crossed the river still further to the left
at Mendoza, and Vandeleur's Brigade of the Light Division
forded the river higher up and captured the village of
Margarita. Wellington observing the weakness of the
French centre, then directed Brisbane's Brigade to advance
diagonally at a run and seize the eminence in front of the
village of Ariñez known as the ' Hill of the English.' [1]
The battalions of the 95th Rifles, belonging to Kempt's
Brigade of the Light Division, joined in the advance, which
was so well and rapidly executed that the whole of the
French line south of the Royal Road was outflanked.

---

[1] Credit for this advance has been claimed for General Picton, and it is quite
likely that the idea struck him simultaneously. It may be added that Wellington
was not always accurate in his account of the part that he personally played in
events. Nevertheless the fact that he was on the spot at the moment is strong
evidence that he directed the movement.

The enemy still held on however to Ariñez, and Brisbane was directed to capture the village. His first line was formed by three companies of the 74th and the three Headquarter Companies of our Regiment under Major Galiffe. These dashed forward with so much élan that they charged right through the village. The French then retired upon the villages of Ali and Armentia, the last available position in front of Vitoria ; and here their massed artillery covered the retreat of Gazan's and D'Erlon's infantry on Salvatierra.

During this time Sir T. Graham, who had in his Corps the two Rifle Companies attached to the 1st Division, was hammering at General Reille and the Army of Portugal on the upper Zadorra. The Spaniards captured the village of Duraña on the Royal Road, and by so doing intercepted the enemy's main line of retreat. But Reille steadfastly held the bridge spanning the river, and it was only when the centre of the French army gave way that he retired ; effecting his retreat at nightfall with great skill along the Salvatierra road and forming a rearguard to the beaten armies of Gazan and D'Erlon.

The horror of the last scene is graphically described by Sir W. Napier :

' Behind the French was the plain on which the city stands ; and beyond the city thousands of carriages and animals and women and children were crowding together in all the madness of terror, and as the English shot went booming overhead the vast crowds started and swerved with a convulsive movement while a dull and horrid sound of distress arose ; but there was no hope, nor stay for army or multitude ; it was the wreck of a nation.'

Yet the French retreated in a compact body, and although of 153 guns they lost all but two, not many prisoners were taken ; and their casualties did not much

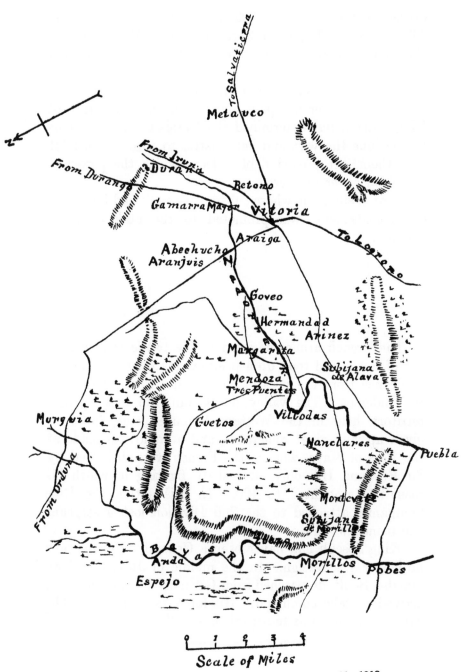

Scale of Miles

SKETCH OF THE POSITION AT VITORIA, JUNE 21, 1813.

exceed 7,000 : about 2,000 more than those of the Allies. In our Regiment 3 riflemen were killed, Captain Franchini, Lieut. Joyce, 4 sergeants and 43 riflemen wounded.

The battle of Vitoria not only struck a heavy blow at the French power in Spain and forced the abdication of King Joseph, but determined the Russians and Prussians to continue the war, and led Austria to throw in her lot with them. It proved a blow from which the power of Napoleon never recovered.

Lord Wellington, whose substantive rank had been that of Lieut.-General, was promoted to the rank of Field Marshal.

6

The French armies engaged in the battle were driven out of Spain and took up a position guarding the frontier of France from the pass of Roncevalles to the mouth of the Bidassoa. The valley of Baztan, the last Spanish territory held in the Western Pyrenees, was, however, given up with great reluctance, and a stand was made at the Pass of Maya, whence the French were dislodged by Sir Rowland Hill only after sharp fighting in which his casualties amounted to 120. In this action the Rifle Companies attached to the 2nd Division lost a sergeant and 5 men wounded.

King Joseph had never been a General. He was now no longer a sovereign. In the middle of July, therefore, Marshal Soult, despatched from Germany by the Emperor, arrived to take command. His force, now entitled ' The Army of Spain,' was re-organised as follows :—

### Chief of the Staff, Count Gazan
### Right Wing, Count Reille

|  | Present under arms. | Total strength |
|---|---|---|
| 1st Division, General Foy . . . . | 5,922 | 6,784 |
| 7th    ,,    ,,   Maucune . . . | 4,186 | 5,676 |
| 9th    ,,    ,,   La Martinière . . | 7,127 | 9,806 |
|  | 17,235 | 21,366 |

### Centre, Count D'Erlon

| | Present under arms. | Total strength |
|---|---|---|
| 2nd Division, General D'Armagnac . . | 6,961 | 8,580 |
| 3rd    ,,    ,,   Abbé . . . | 8,030 | 8,728 |
| 6th    ,,    ,,   Daricau . . | 5,966 | 6,627 |
|  | 20,957 | 23,935 |

### Left Wing, General Clausel

| | Present under arms. | Total strength |
|---|---|---|
| 4th Division, General Conroux . . . | 7,056 | 7,477 |
| 5th    ,,    ,,   Vandermasen . . | 4,181 | 5,201 |
| 8th    ,,    ,,   Taupin . . | 5,981 | 7,587 |
|  | 17,218 | 20,265 |

### Reserve, General Villatte

| | Present under arms. | Total strength |
|---|---|---|
| French troops . . . . | 14,959 | 17,929 |
| Foreign ,, General St. Pol . . . | — | — |
| ,, Casabianca . . | — | — |

### Cavalry

| | Present under arms. | Total strength |
|---|---|---|
| General Pierre Soult . . . . . | 4,723 | 5,098 |
| ,, Treillard . . . . | 2,358 | 2,523 |
|  | 7,081 | 7,621 |
| National Guards, etc. . . . . | 14,938 | 16,946 |
| Second Reserve . . . . . | 5,595 | 6,105 |
| Grand Total of Field Army . | 97,983 | 114,137 |

*Garrisons*

|  | Present under arms. | Total strength. |
|---|---|---|
| San Sebastian . . . . . . | 2,731 | 3,086 |
| Pamplona . . . . . . | 2,951 | 3,121 |
| Santona . . . . . . | 1,465 | 1,674 |

In the eastern provinces of Spain Marshal Suchet commanded about 70,000 men. The Peninsula thus occupied the attention of nearly 200,000 men : veterans, whose presence in Germany would have ensured victory.

7

Of the army under Marshal Soult, the Left Wing rested on St. Jean Pied de Port : the Centre occupied the heights about Espelette and Ainboa : the Right Wing, the Puente (or Pass) of Vera. The reserves held the line of the Bidassoa from Irun to the mouth of the river. Bayonne was the base of operations.

On the side of the Allies, who occupied the southern slopes of the Pyrenean mountains, Hill's Corps, fronting north-east, was posted between the Pass of Roncevalles and the valley of Baztan, both inclusive. The 4th Division at Viscarret, and the 3rd at Olagué formed respectively a support and reserves to Hill.

On Hill's left the 7th Division occupied Echallar and the Light Division Vera ; both on the upper Bidassoa. The 6th Division at St. Estevan formed a central reserve to any point of the line.

Spanish troops, supported by the 1st Division, were posted on the line of the lower Bidassoa. G.H.Q. were at Lesaca three miles south of Vera.

The 5th Division and two Portuguese brigades under Sir Thomas Graham besieged San Sebastian. The Spanish force under Count De La Bispal blockaded Pamplona. Partida bands severed the direct line of communication between Soult and Suchet.

After deducting the 14,000 men employed in the operations against the two fortresses, Wellington had 80,000 to oppose to Soult, whose available effective force may be reckoned at 90,000 men and 86 guns.

In this mountainous country communication was difficult; movement was confined to a few tracks, and rapid concentration was impossible. In this respect, however, Soult had the advantage, for a pretty good road ran from Bayonne to St. Jean Pied de Port.

Marshal Soult did not delay in taking the offensive, his object being the relief of the fortresses of Pamplona and San Sebastian which, with one or two smaller places, were still held by the French. At daylight on July 25th Clausel attacked Byng's Brigade of the 2nd Division about a mile north of Altobiscar, while Reille simultaneously assailed Morillo's Spaniards and drove them back through the Val Carlos, thus threatening to cut Byng from the passes of Ibaneta and Atalosti. Byng, whose Brigade included a Rifle Company, held his ground for nine hours, but at 3 P.M. his ammunition began to fail and he fell back on Altobiscar. A brigade of the 4th Division came up in support; but the Allies, outnumbered by three to one, retired at nightfall beyond Viscarret, abandoning the crest line of the Pyrenees. Byng, separated from his own—the 2nd—Division accompanied the 4th. Next morning General Picton arrived with the 3rd Division, but decided to continue the retreat till the following day, when after some hesitation he took up a position between Zabaldica and Sorauren, covering Pamplona, which was about four miles distant. During the retirement a rearguard action had taken place in which Ensign C. B. Martin, of the

Rifle Company attached to the 4th Division, was severely wounded.

Meanwhile on the 25th at 9 A.M. the main body of the 2nd Division had been attacked by General D'Erlon. In the temporary absence of General Stewart—an absence for which he was afterwards severely rebuked by Lord Wellington—command of the Division was assumed by Major-General Pringle, who occupied the passes of Aretesque, Zurella, and Maya, a line of about five miles from east to west. An outlying piquet on the rock of Aretesque on the right was driven in; but at the instance of the Brigade Major, Captain Peregrine Thorne, who as it happened was a few days later transferred to the 60th, the three Light Infantry Companies of the Brigade with the Rifle Company attached thereto were already moving up in support; and on their arrival the ground was held with the utmost gallantry. The camp of the Brigade was, however, three miles distant; and General Stewart having dismissed the men to cook their dinners, much time was lost before the defenders of the rock were reinforced by successive companies of the 34th Regiment. Then their right flank was completely turned by the advance of a French regiment, which prevented the further reinforcement of the remaining battalions of the brigade. A retreat by way of the left was the sole remaining resource, and this was effected—although with the loss of many men captured by the enemy—and a second position to the left rear taken up. Another stand was made, aided by the arrival of successive half-battalions from Cameron's Brigade on the left, to which had been entrusted the defence of the Zurella and Maya passes. The fighting was magnificent and the slaughter on both sides terrific. Cameron fell badly wounded and the command of his Brigade was taken by Lieut.-Colonel FitzGerald of our

Regiment who had doubtless been in charge of its Rifle and Light Companies. The defenders were, however, gradually pushed back by overwhelmng numbers. The Zurella pass was lost and the appearance of another division of the French army necessitated the abandonment of the Col de Maya, hitherto held by a half-battalion of the 71st. The unexpected arrival of a brigade from the 7th Division enabled the Col to be recovered ; but news of the retreat of the 4th Division arrived, and the flank of the 2nd being consequently laid bare the latter Division was withdrawn during the night, first to Irurita, fifteen miles southward ; afterwards to Lizasso. Nevertheless the defence of its position reflected the utmost glory on the 2nd Division. Two Brigades, aided only at the end of the fight by part of a third, had withstood the assault of three French Divisions. Their losses, including 500 (mostly wounded no doubt) captured, amounted to 1500, about 50 per cent. of their strength. But the casualties inflicted on the enemy were about 2500 and Count D'Erlon made no attempt to improve the advantage he had gained.

It was not until after nightfall that Lord Wellington got news of the French attacks ; even then twenty-four hours elapsed before he got definite reports on the situation. On the morning of the 27th, having ordered up the 6th and 7th Divisions in support of his right wing, he joined Picton on the heights of Zabaldica, and assumed command in person just as the enemy was about to launch his attack upon the British position, which was occupied by the 4th Division, two Spanish battalions and a Portuguese Brigade. The 3rd and the Spanish Divisions of Morillo and De La Bispal were drawn up two miles in rear immediately covering Pamplona.

The loud cheers which greeted Lord Wellington, were

heard by the Commander of the French Army, who—with some excess of caution—postponed his attack until the following day, when at 11 A.M. General Clausel assailed the front and flank of the 4th Division which held the left of our line. But Soult's delay had been fatal to his chances of success. Clausel's attempt to outflank the 4th Division was frustrated by the timely arrival of the 6th which assailed his right while two Brigades of the 4th Division, wheeling up, struck it on the left, and hurled the French back into the valley below. Reille, who had not been quite ready when his colleague advanced, then attacked the Portuguese and Spaniards on our right; but with no greater success and though the fighting was desperate the repulse of the enemy was complete.

The 29th passed without renewal of the fight; and though the Corps of General D'Erlon had now reached Ostiz, that of Hill posted between Lizasso and Arestequé and connected with Wellington by the Marcalain-Oricain road was in a position either to hold D'Erlon at bay, or to menace Soult's rear.

Marshal Soult then gave up all hope of relieving Pamplona, but resolved to place himself between the Allies and the valley of Baztan; and by so doing to regain his base by the Doña Maria pass, and concentrate to relieve San Sebastian. But Hill being in the way, Count D'Erlon with over 20,000 men and supported by La Martinière's Division of Reille's Corps and a Cavalry Division, was ordered to attack him in his position on the Buenza-Marcalain heights. Accordingly next day (the 30th) D'Erlon drove back Hill's two Divisions and recovered the Doña Maria pass. In this engagement Sir Rowland Hill alludes in terms of high praise to the services of his brigadier Lieut.-Colonel FitzGerald.

But at daylight on the 30th, Lord Wellington, seeing

that the enemy in his front had in part evacuated his position, directed the 3rd and 7th Divisions to encircle respectively Soult's right and left. Both were successful. Clausel was indeed retiring up the valley of the Lanz, but his hindermost Division was roughly handled; and Wellington observing the success of his flanking movements attacked Reille's Corps in front and flank. One of its Divisions broke in headlong rout; the other, under General Foy, was forced to retire upon a separate line to Aldiudes. Soult's casualties amounted to 5000, and to such an extent were the Corps of Clausel and Reille dispersed that between them they could number only 15,000 men.

The position of Marshal Soult was by this time precarious. He had, it is true, the 20,000 men of D'Erlon's Corps; but in his front was Sir Rowland Hill, reinforced; while the 3rd, 4th, 6th, and 7th Divisions threatened his flank and rear. In a few hours he would be surrounded by 50,000 men.

At midnight on the 30th–31st he began his retreat towards the Doña Maria pass. At 10 A.M. on the 31st Hill came up with the Corps of Count D'Erlon which forming Soult's rearguard was ascending the pass. It was veiled in a thick fog, but Sir Rowland at once attacked it with the Brigade of FitzGerald. This officer drove the enemy's covering skirmishers to the summit of the hill, where he found the main body in such force that his advance was held up and the Divisional Commander suspended the attack pending the arrival of the 7th Division on his right. FitzGerald was severely wounded and taken prisoner. On the arrival of the 7th Division D'Erlon was driven from the summit and forced half-way down the further side of the mountain.

Soult reached St. Estevan in safety. The only road available for his further retreat into France followed the right bank of the Bidassoa by Sumbilla to Vera ; and Sir Thomas Graham, whose Corps at Irun had formed the extreme left of Wellington's Army, was sending a Spanish Division to block the road at Yanci and Vera. The Marshal appeared to be unconscious of his imminent peril. The British army was close at hand. General surrender seemed inevitable, when in the early morning of August 1, the capture of a few British stragglers enabled Soult to grasp the situation and without a moment's hesitation to resume his retreat. The road ran through a narrow defile hemmed in on the left by the Bidassoa, at this point a narrow but swift and rocky stream, and on the right by steep sliffs.

The advance guard of the 4th Division assailed Clausel's rearguard. At Yanci D'Erlon quitted the defile by a track running eastward to Echalar ; but before Reille could follow him that officer was assailed by the Light Division which after a terrific march appeared upon the heights overlooking the left bank of the Bidassoa. Eventually the French army reached Echalar, but with terrible loss not only of men but of morale. Clerc states that during the week's operations 101 officers had been killed and 277 wounded. Of N.C.Os. and men 10,448 were killed or wounded, and 3,200 taken prisoners. Wellington's losses (exclusive of Spaniards) amounted to 7,100. Those of our Regiment on the 25th, Lieutenant John W. van Dahlmann killed, Lieutenants John L. Barbaz and John Joyce wounded and prisoners. The casualties of ' other ranks ' would seem to have been about 7 killed, 40 wounded, and 18 missing. All those officers were attached to the 2nd Division ; but to what extent the casualties were incurred at the respective points attacked is uncertain. In all probability, however, the large majority took place at Aretesque and Maya. Barbaz died of his wounds ; and the same report was made of Joyce, but erroneously.

On the 26th Ensign C. B. Martin (4th Division) was wounded. On the 31st Colonel FitzGerald was wounded and captured. Between July 30 and August 1 one rifleman was killed. Adjutant Kent and fourteen of other ranks were wounded.

Soult now disposed his army thus : Reille's Corps on the western flank occupied two lines of defence ; the first, from the mouth of the Bidassoa to the Mandale mountains ; the second, seven miles in rear, from St. Jean de Luz to Ascain on the Nivelle, where General Villatte connected Reille with Clausel, whose corps occupied the Mandale and Great Rhune mountains as far as Amotz. D'Erlon prolonged the line to Ainhoa ; and General Foy with the detached Division of Reille's corps and some National Guards was at St. Jean Pied de Port. On the extreme left a Division under General Paris held Oloron.

The British army returned to the positions it had occupied previous to the battles of the Pyrenees.

## 8

Just at this time we come across a remarkable testimonial of the coolness and accuracy of our Riflemen, written —of all people—by Marshal Soult, Commander-in-Chief of the French army in Spain to the Minister of War at Paris.

'St. Jean de Luz, 1st September, 1813.

' The loss in prominent and superior officers, sustained for some time past by the Army, is so disproportionate to that of the rank and file that I have been at pains to discover the reason ; and have acquired the following information, which of course explains the cause of so extraordinary a circumstance.

' There is in the English army a battalion of the 60th consisting of ten Companies—the regiment is composed of six battalions, the other five being in America or the West Indies. This battalion is never concentrated, but has a Company attached to each Infantry Division. *It is armed with a short rifle ; the men are selected for their marksmanship ;* they perform the duties of scouts, and in action are expressly ordered to pick off the officers, especially Field and General Officers. Thus it has been observed that whenever a superior officer goes to the front during an action, either for purposes of observation or to lead and encourage his men, he is usually hit.

' This mode of making war and of injuring the enemy is very detrimental to us ; our casualties in officers are so great that after a couple of actions the whole number are usually disabled. I saw yesterday battalions whose officers had been disabled in the ratio of one officer to eight men ; I also saw battalions which were reduced to two or three officers, although less than one-sixth of their men had been disabled. You can imagine that if these casualties should recur, it would be very difficult to provide for the replacement of officers even though the nominations were made beforehand.' [1]

Referring to this letter, Colonel Dumas (' Neuf mois de Campagnes à la suite du Maréchal Soult ') is responsible for the statement that ' " Les Riflemen " killed all our

---

[1] ' Les pertes en officiers supérieurs et particuliers que l'armée à éprouvées depuis quelques temps sont tellement hors de proportion avec les pertes en soldats, que j'ai dû considèrer quelle cause pouvait y donner lieu. Voici des renseignements que j'ai recueillis à ce sujet lesquels appliquent naturellement un effet aussi extraordinaire.

' Il existe à l'armée anglaise 1 battalion du 60me composé de 10 compagnies (le régiment a 6 bataillons, les 5 autres sont en Amérique on aux Indes). Ce bataillon n'est jamais réuni ; il fournit 1 compagnie à chaque division de l'infanterie de l'armée ; *il est armé des carabines ; les hommes sont choisis parmi les meilleurs tireurs ;* ils font la service d'eclaireurs et, dans les affaires, il leur est expressement recommandé de tirer de préference sur les officiers et particulièrement sur les chefs et les génèraux. Ainsi, il a été remarqué que dans une affaire, lorsqu'un officier supèrieur est dans le cas de se porter en tête, soit pour observer, soit pour diriger sa troupe, ou soit même pour exciter le combat, il est ordinairement atteint.

' Cette manière de faire la guerre est de nuire à son ennemi nous est très désavantageuse ; les pertes en officiers que nous éprouvons sont si considerables que dans deux affaires ils sont ordinairement tous hors de combat. Hier j'ai en vu des bataillons qui en eu des officiers hors de combat dans la proportion de 1 sur 8 hommes ; j'ai vu aussi des bataillons qui étaient réduits à 2 ou 3 officiers, quoiquils n'eussent pas le sixième de leurs hommes hors de combat. Vous concevrez que si ces pertes se renouvelaient, il serait très difficiles de pourvoir au remplacement des officiers ; eût-on même les nominations faites d'avance.'

officers between July 25 and August 31, viz. 500 officers and 8 Generals.'

In connection with this correspondence may be recalled the tradition in our Regiment that the French did not consider our men entitled to quarter for the reason that their fire was aimed, a practice considered unfair in those days. Whether this was or was not the theory of the French, they certainly did give the Riflemen quarter in their customary chivalrous manner.

## 9

San Sebastian fell on September 10, and on October 7 Lord Wellington, by a fine combination and splendid audacity, forced the passage of the Bidassoa and established his army on French soil; a feat, the moral courage of which can be the better appreciated that it was performed eleven days prior to the great defeat of Napoleon at Leipzig.

The allied army was now organised in three Army Corps. That on the right, commanded by Sir R. Hill, included the 2nd, 6th, a Portuguese and a Spanish Division. The central Corps was commanded by Sir W. Beresford, and comprised the 3rd, 4th, 7th, and Light Divisions as well as the Spanish units. That on the left, under Sir John Hope, who had just relieved Sir Thomas Graham, was composed of the 1st and 5th Divisions, a British and two Portuguese Brigades unattached to any Division, and the Galician army.

Wellington now had under his command—inclusive of troops detached—an effective force of about 44,000 British and K.G.L., 25,000 Portuguese, and 30,000 Spaniards; roughly 100,000 men. Opposed to him was a field army of about 52,000, posted as follows : Villatte

II.                                                            P at St. Jean de Luz ; south of the Nivelle, which was fordable at many points even in its lower reaches, were posted Reille's troops from the sea west of Ciboure to Ascain. Clausel held the hills from Ascain to Amotz with the river forming a re-entrant behind him. D'Erlon behind the Nivelle carried on the line to the Mondarrin mountain overlooking the Nive. General Foy was higher up that river at Biddarray ; Paris at St. Jean Pied de Port.

The position of the French army was strongly fortified, and supported by entrenched camps ; but the length of the line from the Nive to St. Jean de Luz was sixteen miles, and Wellington expressed the opinion that he would ' easily carry it.' The opposing armies had both received reinforcements, but Wellington had on the field a great superiority in numbers : 90,000 (74,000 being Anglo-Portuguese) and 95 guns against 67,000 of the enemy, although with probably a greater number of guns.

The plan of the British Commander was to contain the enemy's troops on the flanks, but to burst through the centre of his line in overwhelming numbers and to force the passage of the Nivelle south of the village of St. Pée. Thus with 19,000 men Sir John Hope near the mouth of the Nivelle occupied the attention of the 25,000 under Reille and Villatte ; and a division of 6000 indifferent Spanish soldiers contained double that number com-manded by Foy and Abbé. But Hill with 20,000 was opposed by only 5000 men of D'Armagnac ; Beresford with 24,000 men assailed two divisions under Clausel 10,000 strong, while the Light Division and a Spanish one under Longa—8000 men in all—were hurled against the remaining division of 3000 men under Taupin.

The columns of attack were brought up under cover of night ; and as the morning of November 10 dawned brilliantly, the Light Division sprang forward ; and General

Kempt's Brigade captured the Little La Rhune mountain although strongly fortified. The 4th and 7th Divisions advancing at the same time drove the remainder of Clausel's Corps across the Nivelle in disorder. On their right the 3rd Division under Major-General Colville preceded by its Light Battalion with the three Rifle Companies commanded by Major Galiffe, carried the bridge of Amotz at a run and entered the village of St. Pée severing the communication between Clausel and D'Erlon. This success paved the way for Sir R. Hill, who on the allied right was not less successful and carried the positions occupied by D'Erlon.

Galiffe's conduct was personally noticed by General Colville in his report to Sir William Beresford, the Corps Commander. After mentioning the names of several battalion commanders he continues : ' to these names I am happy from my own observations to add that of Major Galiffe of the 60th, who commanded the advance of the Division.' [1]

Meanwhile the Light Division pushing across the Sare-Ascain road drove the enemy opposed to it over the Nivelle about 2 P.M. Matters were critical for the French, but Marshal Soult now brought up all available reinforcements from his camp at Serres. To meet the new assailant Wellington wheeled the Light, 4th, and Giron's Division to the left, thus protecting the left of Beresford's Corps. Soult's attack was in consequence not pushed home ; but his demonstration had had a considerable measure of success, for Lord Wellington, having now only the 3rd and 7th Divisions remaining in hand, was forced to await the arrival of the 6th Division before driving Maransin's Division from its position and descending into the lower ground to threaten the retreat of Reille and Villatte.

---

[1] Galiffe's command included the Light Companies of the Division. The three Rifle Companies were commanded by Major J. H. Schoedde, who was awarded the Gold Medal in consequence.

The Division on the left of Hill's Corps was the 6th commanded by General Henry Clinton. Its advance guard, consisting of the 9th Caçadores and the Light Companies of Lambert's Brigade, the whole under command of Lieut.-Colonel Gustavus Brown, moved forward at 6.30 A.M.; and having with little difficulty driven the enemy's outposts over the Nivelle followed them by two fords the discovery of which by Brown enabled the Division to cross unobserved by the French. 'The active intelligence of Lieut.-Colonel Brown,' observed General Clinton in his subsequent report to Sir R. Hill, 'enabled me to place the troops destined to make the attack upon the right of the enemy at the very foot of the hill.' Brown, under cover of a wood which skirted the foot of the hill, having gained the enemy's right, was directed to dislodge the French skirmishers on the higher ground. The ascent was steep and rocky but the Colonel not only executed his task, but pursued the enemy down the reverse slope, forcing him —by the General's testimony—to abandon guns and wagons.

The 6th Division at length established itself on the right of the 7th, and on its arrival Wellington crossed the Nivelle and with the 3rd and 7th Divisions captured the opposite heights occupied by Maransin. He was now established in rear of the enemy's right, but the failure of daylight prevented the British commander completing his manœuvre; and at night Soult withdrew his right wing along the Royal Road to Bayonne and escaped the peril. According to the French military historian Clerc, his losses during the whole action amounted to 3,339 killed and wounded—including 447 officers—and 1,231 prisoners. He also lost 69 guns and the whole of his field magazines at St Jean de Luz and Espelette. In the course of an autumn day the French had been driven from a line of positions which had taken them three months to fortify.

The losses of the Allies amounted to 343 killed, 2,278 wounded, and 73 missing. In our Rifle Battalion Lieutenant Thomas Eccles—attached to the 2nd Division— 3 sergeants and 4 riflemen were killed; Captain James Stopford, Lieutenant John P. Passley, Ensign H. Shew-bridge, 3 sergeants, a bugler, and 54 riflemen were wounded. Two riflemen were missing. Shewbridge—6th Division—was slightly hurt; Passley—3rd Division—and Stopford—4th—were severely wounded.

The battle was fought in a region of romantic beauty with wooded mountain and bracken-covered glen, one in which the qualities of Riflemen would have the utmost scope, and which will amply repay a visit from Riflemen of the present generation: even from those at the present moment [1] victoriously engaged in a far more terrible contest in the opposite corner of France.

'The deadly strife,' observes an eye-witness, 'was surprisingly grand; yet the sublimity of the scene defied all attempts at description. . . . The fierce and continued charge of the British was irresistible; onward they bore, nor stopped to breathe, rushing forward through glen, dale, and forest. . . . Yet they seemed to chase only the startled red deer, prowling wolf or savage wild boar, until they arrived at the steel-bristling stronghold of the foe.'

Lieutenant Lerche (misprinted 'Levicke' in Well-ington's Supplementary Dispatches) commanding the Rifle Company acting with the Light Companies of Byng's Brigade of the 2nd Division, is mentioned in terms of appreciation by that general officer in his report to the Divisional Commander.[2] Marshal Beresford reported to Lord Wellington that the conduct of the Brigade of the

---

[1] October 25, 1918.

[2] Lerche had formerly been sergeant-major to the Battalion, but was promoted to commissioned rank in May 1812.

3rd Division commanded by Colonel Keane of our Regiment ' was highly distinguished in the attack.' Unfortunately the report did not reach G.H.Q. until after Lord Wellington's dispatch had been transmitted to England.

Sir Henry Clinton, commanding the 6th Division, was loud in his praises of Lieut.-Colonel Gustavus Brown, whose conduct he related at length in his Report dated November 11 to Sir Rowland Hill, characterising him as ' an officer of the greatest zeal and much professional experience and ability.' [1] *Vide* Wellington's ' Supplementary Dispatches,' vol. viii. p. 358.

## 10

After the battle of the Nivelle Marshal Soult retired to his entrenched camp in front of the fortress of Bayonne, his retrograde movement being facilitated by heavy rains which stopped Wellington from molesting it. His right was posted in the village of Anglet ; his centre carried the line up to the Nive ; and his left occupied an entrenched position on the right bank of that river as far forward as Lahoussoa.

On the side of the British, Sir John Hope in advance of Bidart held a ridge the continuation of which as far as Ustaritz on the Nive was occupied by Beresford. Hill was posted along the river bank to Itsazu, the whole position from left to right extending over fourteen miles. Comfortable cantonments were secured within the villages in rear and the piercing winds of the Pyrenees became a nightmare of the past. Nevertheless from the strategical

---

[1] This dispatch of Sir H. Clinton is given at length in Wellington's ' Supplementary Dispatches,' vol. viii. p. 358, and is largely a narrative of Colonel Brown's exploits. The latter must have been well versed in the theory and practice of the Rifleman's art.

point of view the position of the Allies on exterior lines and cramped up as it was between the ocean and the Nive, a navigable river, left much to be desired.

Incessant rain for a time prevented movement, but early in December the weather cleared and orders were issued to get elbow room by forcing the passage of the Nive at daylight on the 9th. Sir Rowland Hill, with two brigades of cavalry, two infantry divisions and fourteen guns forded the river above and below Cambo, without meeting much opposition, while Morillo on his right crossed simultaneously at Itsazu, and the 6th Division from Beresford's Corps on his left crossed on a pontoon bridge at Ustaritz. Hill then advanced on Bayonne and found Count D'Erlon posted on the heights of Lursinthoa within three miles of the city. The 6th Division drove the enemy out of Villefranque, but the main body of Hill's Corps being hindered by bad roads the fighting was broken off: the Allies having done all they had expected.

Marshal Soult at once realised the advantage gained by his opponents; but he also appreciated the fact that Wellington's army was now cut in two by the Nive. Making rapid use of his interior lines and bringing D'Erlon and his four divisions by the Bayonne bridge to the left bank of the river in support of Reille, Clausel, and Villatte, the Marshal concentrated 60,000 men and 40 guns against the allied centre containing only half that number of men and 24 guns. In the early morning of the 10th, the Light Division at Arcangues was attacked by Clausel, but the mass of the French army was pushed along the Royal Road against Sir John Hope's advanced position at Barrouilhet. His piquets were driven in, but the line of resistance was held while the main body of his corps gradually arrived on the scene. From the right also Wellington was approaching with the 3rd, 4th, 6th, and 7th Divisions and on their advance the French Commander fell back on Bayonne.

During the two following days sharp local combats took place ; but Soult had decided to make his principal effort against Sir R. Hill on the right bank of the Nive. On the evening of the 12th he concentrated seven divisions at Mousseroles in the hope of overwhelming Hill next morning.   But his troops had been put in motion too early in the day and were descried by the British in the act of crossing the Nive at Bayonne ;  whereupon Hill brought up from Urcuray Barnes' Brigade of the 2nd Division ; and posted it in rear of the Portuguese Brigade of that division at the village of St. Pierre astride the Bayonne-St. Jean Pied de Port road.   Byng's Brigade, in a valley, formed the right and Pringle's on the Villefranque ridge, covering the pontoon bridge, the left of the line.   The two wings being separated from the centre by a marsh were compelled to fight as independent units.   The Portuguese Division, commanded by General Le Cor, was held in reserve behind Barnes.

When morning broke on the 13th it was found that the pontoon bridge had been washed away ;  and Hill was completely isolated with less than 14,000 men to oppose 35,000 in his front, while the division of General Paris, opposed only by Vivian's Cavalry Brigade and 4000 Spaniards, threatened his rear.  There was nothing for it but to hold on till the bridge could be repaired.

At 8.30 A.M., as the morning mists cleared away, Count D'Erlon was seen advancing with a division of cavalry, three of infantry and twenty-two guns.   His Corps was supported by the division of Foy and Maransin ; and the two remaining were held in reserve.   Darricau's Division attacked Pringle ;  D'Armagnac's, Byng.  Abbé hurled

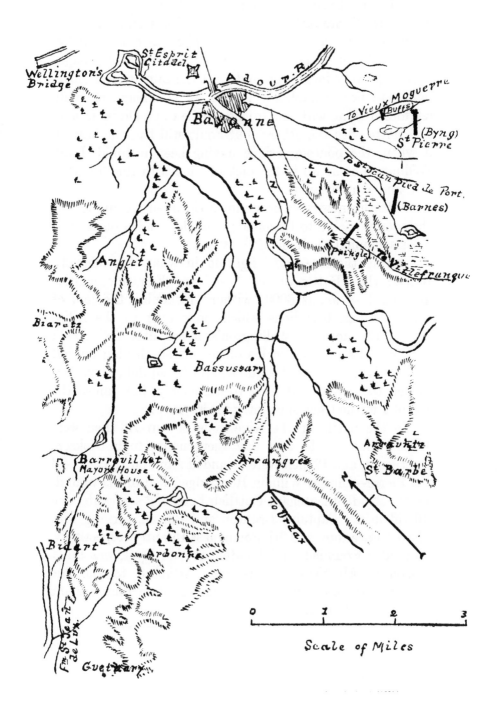

Scale of Miles

himself against the centre and gained the crest of his position. Sir W. Stewart, who commanded the 2nd Division, threw Barnes into the fight; but Abbé was reinforced by Foy and Marausin and the British centre forced back though disputing every inch of ground. At the same time Byng's right had been completely turned by D'Armagnac. Hill reinforced Byng with one brigade of Le Cor's Division and the centre with the other; but Soult still had the best of it when, the pontoon bridge having been restored, Lord Wellington came up in the nick of time with the 3rd, 4th, 6th, and 7th Divisions.

With this new force he assumed the offensive and forced the enemy back into Bayonne; and thus ended one of the most sanguinary actions of the war.

During the five days' fighting in what was officially termed the 'Battle of the Nive' the French casualties amounted to about 6000, without reckoning the loss of three German battalions which came over to the Allies during the action. Wellington's losses exceeded 5000; but he had succeeded in his object of establishing himself on the right bank of the Adour, and by so doing not only denied the use of that river to the enemy, but gained the use of a fertile tract of country for his cavalry.

In the battle of the Nive none of our riflemen were killed; but on the first day Lieutenant Hugh Dickson, 1 sergeant, and 12 men were wounded; and on the last—at St. Pierre—Ensign William Routledge (died of his wounds on the 22nd), Lieutenant Gottlieb Lerche, 2 sergeants, and 16 riflemen. Captains Brunton and Phelan, serving with the 6th Caçadores and 6th Portuguese Infantry respectively, were also wounded, and Lieutenant Van Dyck was taken prisoner. All these officers were attached to the 2nd Division, the only two unhurt being Captain Blassière and Lieutenant A. F. Evans.

On the 13th also Colonel Gustavus Brown was so
severely wounded that he took no further part in the war.
His name was mentioned in Sir R. Hill's dispatch.    At
the peace Brown was knighted—an honour exceptional
for one who had commanded no force greater than a
battalion.    He remained for some years in the Portuguese
Service, and lived long; but never rejoined our Regiment.
He had been mentioned in dispatches more often probably
than any regimental officer in the army.

## 11

Thus closed the memorable year 1813, begun in Portugal,
ended in France ;  and in the course of which two fortresses
had been captured and four general actions won.

The small  proportion  of  sick  to  effectives  in  our
battalion during October 1812 has already been mentioned.
A return given in Wellington's 'Supplementary Dispatches,'
vol. viii. p. 474, again gives evidence of the good health
of the Riflemen.    The average number of deaths in hospital
during the year ending December 20, 1813, was 100 per
battalion.    Those of the 60th were only 35 ; but it must
be regretfully stated that during the last six months 100
men had been returned as deserters.    The term desertion
at that period appears to have included short periods of
absence without leave.    There is, however, little doubt
that the battalion still contained men who had served in
the French army.

## CHAPTER X

DURING the year 1813 the value of the 60th was still further recognised by an Act of Parliament authorising the augmentation of the regiment by four new battalions. The 7th was accordingly raised on September 1 in Guernsey from Dutch, German, and other prisoners of war ; and the 8th at Gibraltar out of an existing provisional battalion also composed chiefly of prisoners of war. No further step seems to have been taken to carry out the authority to raise a 9th and 10th battalion.

The 7th battalion, dressed in green and including 640 Light Infantry and 200 Riflemen, had as its Colonel Commandant Major-General Sir George Murray, K.B., Wellington's celebrated Q.M. General, and as its Lieut.-Colonel Henry John. In January 1814 its effective strength was : 1 lieut.-colonel, 7 captains, 9 lieutenants, 8 ensigns, 5 staff, 42 sergeants, 8 buglers, 684 men ; wanting to complete establishment, 131.

Among the captains was Colin Campbell, whose previous zeal and gallantry in the Peninsula had earned promotion, and who in later years was known to fame as Field Marshal Lord Clyde, the subduer of the Mutiny in India.

The 8th was wholly a Rifle battalion. The appointment of Colonel Commandant was given to Major-General Sir James Kempt, who had been associated with light troops since the battle of Maida, where he commanded a light battalion, and was at the present time serving in France as head of a brigade of the celebrated Light Division.

The Lieut.-Colonel of the 8th Battalion was the gallant J. P. Hunt, formerly of the 52nd, who had led the storming party of the Light Division in the assault on San Sebastian.

The battalion was raised at Cadiz, and in January 1814 its parade state signed by Henry FitzGerald showed, 1 major, 3 captains, 2 lieutenants, 33 sergeants, 20 buglers, and 590 riflemen.

Among the officers of the newly raised battalions, were several of note in addition to those already mentioned.

Of the majors, one was Baron E. O. Tripp, who was serving in Wellington's army as D.A.A.G. to the Cavalry Division, and subsequently fought at Waterloo as an A.D.C. to the Prince of Orange—Lord John Somerset, also of the 60th, being another.

The captains included P. F. Thorne, who had served with distinction as a Staff officer in the 2nd Division, and is mentioned in Sir W. Napier's account of the fight at the Pass of Maya.

John Barton Pym had been in civil life a miniature painter and attracted the notice of General Sir John Doyle, who gave him a commission in his own regiment. In December 1812 Lord Wellington went to Cadiz to consult the Government ; and the Regency, wishing to make him a present, commissioned Captain Pym to paint him in miniature. The portrait was pronounced to be a most striking likeness.

Captain Pym, who had served in all but the first Peninsular campaign, died at Gibraltar of malarial fever in 1814.

Charles Leslie, who had served in the 29th Regiment,

wrote in later years a book on light drill,[1] an enlargement
and amplification of the work by Captain Thomas Mitchell,
Adjutant to our 1st Battalion, which had been based on
that published by De Rothenburg in 1798.

Leslie wrote a military journal of the period 1807–32,
which was not published till many years after his death,
but contains much of interest to our Regiment.

Captain R. Brunton had served with distinction in the
6th Caçadores, and was severely wounded at the battle
of the Nive. He subsequently fought at Waterloo as a
D.A.Q.M.G. and at a later stage became Colonel of the
13th Light Dragoons.

Captain (afterwards Sir) John Scott Lillie had served
throughout the Peninsular War in command of a Caçadore
battalion formed out of the Lusitanian Legion.


2

The following was the 'State' of the Anglo-Portuguese
Army on January 16, 1814 :—


CAVALRY DIVISION

*Lieut.-General Sir Stapleton Cotton, K.B.*

|  |  |  | Effective rank and file. |
|---|---|---|---|
| Maj.-Gen. O'Loghlin . | . | 1st Life Guards . | . 217 |
| | | 2nd ,, ,, . . | . 260 |
| | | Roy. Horse Guards, Blue . | 277 |
| ,, | Hon. W. Ponsonby | 5th Dragoon Guards . | . 336 |
| | | 3rd Dragoons . . | . 358 |
| | | 4th ,, . . | . 386 |

[1] 'A Treatise on the Employment of Light Troops on Actual Service,' by
Lieut.-Col. Charles Leslie, late 60th King's Royal Rifles. (W. Cloon & Sons, 1843.)
   'A System of Light Drill, as practised in the First Battalion 60th, or the King's
Royal Rifle Corps,' by Lieutenant and Adjutant Mitchell. (Manchester,
J. A. Robinson, Military Printer, Deansgate, 1842.)

|  |  | Effective rank and file. |
|---|---|---|
| Maj.-Gen. Vandeleur . . | 12th Light Dragoons . | . 387 |
|  | 16th ,, ,, . | . 415 |
| ,, Hon. H. Fane . | 13th ,, ,, . | . 348 |
|  | 14th ,, ,, . | . 417 |
| Colonel Vivian . . . | 18th Hussars . . | . 427 |
|  | 1st ,, K.G.L. . | . 426 |
| Maj.-Gen. Boek . . . | 1st Dragoons ,, . | . 339 |
|  | 2nd ,, ,, . | . 332 |
|  | 3rd Dragoon Guards . | . 350 |
|  | Royal Dragoons . | . 359 |
| ,, Lord Edward | 7th Hussars . . | . 513 |
| Somerset | 10th ,, . . | . 459 |
|  | 15th ,, . . | . 466 |
| Viscount Barbacena . . | 1st Portuguese Cavalry | . 255 |
|  | 6th ,, ,, | . 268 |
|  | 11th ,, ,, | . 172 |
|  | 12th ,, ,, | . 199 |
| Colonel Campbell . . . | 4th ,, ,, | . 26 |

8230

### INFANTRY DIVISIONS

### 1st Division

*Major-General Hon. E. Howard*

| | | |
|---|---|---|
| Maj.-Gen. Hon. E. Stopford . | 1st Batt. 1st Guards . | . 785 |
|  | 3rd ,, ,, | . 776 |
|  | 1st Batt. Coldstream Gds. . | 767 |
|  | 1st ,, 3rd Guards . | 864 |
|  | 1 Coy. 5th Batt. 60th Rifles | 50 |
|  | Capt. J. W. Harrison |  |
|  | Lieut. R. Pasley |  |
|  | ,, J. Hamilton |  |
| ,, Hinuber . . | 1st Line Batt. K.G.L. . | . 574 |
|  | 2nd ,, ,, ,, | . 532 |
|  | 5th ,, ,, ,, | . 482 |
|  | 1st Light ,, ,, | . 568 |
|  | 2nd ,, ,, ,, | . 585 |

### *Unattached Brigade*

| | | Effective rank and file. |
|---|---|---|
| Maj.-Gen. Lord Aylmer . | . 1st Batt. 62nd . . | . 427 |
| | 76th . . . | . 546 |
| | 85th . . . | . 430 |
| | 77th . . . | . 170 |
| | | 7826 |

### 2nd Division

#### *Lieut.-General Hon. Sir William Stewart, K.B.*

| | | | |
|---|---|---|---|
| Maj.-Gen. Barnes . . | . 1st Batt. 50th . . | . 345 |
| | 1st ,, 71st . . | . 498 |
| | 1st ,, 92nd . . | . 391 |
| | 1 Coy. 60th . . | . 49 |
| ,, Byng . . | . 1st Batt. the Buffs . | . 530 |
| | 1st ,, 57th . . | . 438 |
| | 1st Provisional Batt. . . | |
| | 2nd Batt. 31st . . | . 271 |
| | 1st ,, 66th . . | . 278 |
| | 1 Coy. 60th . . | . 45 |
| ,, Pringle . . | . 1st Batt. 28th . . | . 485 |
| | 2nd ,, 34th . . | . 410 |
| | 1st ,, 39th . . | . 565 |
| | 1 Coy. 60th . . | . 47 |
| Colonel Harding . . | . 6th Line Regt. Portuguese . | 715 |
| | 18th ,, ,, ,, | 901 |
| | 6th Caçadores ,, | 302 |
| | | 6270 |

The Rifle officers present with the Division appear to have been—

Capt. F. B. Blassière
Lieut. A. F. Evans
  ,,   G. Lerche
  ,,   S. O'Hehir
  ,,   J. J. Stewart

### 3rd Division

*Lieut.-General Sir Thomas Picton, K.B.*

|  |  | Effective rank and file. |
|---|---|---|
| Maj.-Gen. Brisbane . . | 1st Batt. 45th . . . | 496 |
|  | 74th . . . . | 438 |
|  | 1st Batt. 88th . . . | 738 |
|  | 4 Coys. 5th Batt. 60th . | 197 |
|  | Major J. Galiffe (commanding the Batt.) |  |
|  | ,, J. H. Schöedde |  |
|  | Capt. R. Kelly |  |
|  | ,, S. Purdon |  |
|  | ,, I. Franchini |  |
|  | Lieut. J. P. Passley |  |
|  | ,, C. A. Fourneret |  |
|  | Ensign H. Shewbridge |  |
|  | ,, E. J. Bruce |  |
|  | Lieut. & Adjt. J. Kent |  |
|  | Q.Mr. A. Reckney |  |
| Colonel Keane (5th Batt. 60th) | 1st Batt. 5th . . . | 640 |
|  | 2nd ,, 83rd . . . | 371 |
|  | 2nd ,, 87th . . . | 305 |
|  | 94th | 350 |
| Maj.-Gen. Power . . . | 9th Line Regt. Portuguese | 783 |
|  | 21st ,, ,, ,, | 784 |
|  | 11th Caçadores ,, | 215 |
|  |  | 5317 |

### 4th Division

*Lieut.-General Hon. Sir G. L. Cole, K.B.*

| Maj.-Gen. W. Anson . . | 3rd Batt. 27th . . . | 564 |
|---|---|---|
|  | 1st. ,, 40th . . . | 468 |
|  | 1st ,, 48th . . . | 413 |
|  | 2nd ,, 53rd (2nd Provisional Batt.) . | 480 |
|  | 1 Coy. 60th . . . | 45 |
|  | Major J. Stopford |  |
|  | Lieut. J. Moore |  |

|                       |                        | Effective rank and file. |
|-----------------------|------------------------|------:|
| Maj.-Gen. Ross        | 1st Batt. 7th Fusiliers | 604 |
|                       | 20th                   | 395 |
|                       | 1st Batt. 23rd Fusiliers | 420 |
|                       | 1 Coy. Brunswick Oels  | 42 |
| Colonel Vasconcellos  | 11th Line, Portuguese  | 794 |
|                       | 23rd  „        „       | 899 |
|                       | 7th Caçadores          | 265 |

5389

### 5th Division
*Major-General Hon. Charles Colville*

|                       |                        |      |
|-----------------------|------------------------|-----:|
| Maj.-Gen. Hay         | 3rd Batt. 1st          | 320 |
|                       | 1st  „   9th           | 482 |
|                       | 1st  „   38th          | 364 |
|                       | 2nd  „   47th          | 256 |
|                       | 1 Coy. Brunswick Oels  |     |
| „   Robinson          | 1st Batt. 4th          | 344 |
|                       | 2nd  „   59th          | 268 |
|                       | 2nd  „   84th          | 294 |
|                       | 1 Coy. Brunswick Oels  | 20 |
| Colonel de Regoa      | 3rd  Line, Portuguese  | 608 |
|                       | 15th  „        „       | 427 |
|                       | 8th Caçadores          | 189 |

3597

### 6th Division
*Lieut.-General Sir Henry Clinton, K.B.*

|                       |                        |      |
|-----------------------|------------------------|-----:|
| Maj.-Gen. Pack        | 1st Batt. 42nd         | 669 |
|                       | 1st  „   79th          | 594 |
|                       | 1st  „   91st          | 458 |
|                       | 1 Coy. 60th            | 37 |
|                       | Lieut. H. Muller       |     |
|                       | „   C. J. de Franchiosi |    |
|                       | „   J. Currie          |     |

|  |  | Effective rank and file. |
|---|---|---|
| Maj.-Gen. Lambert . . | 1st Batt. 11th . . | . 477 |
|  | 1st ,, 32nd . . | . 474 |
|  | 1st ,, 36th . . | . 365 |
|  | 1st ,, 61st . . | . 438 |
| Colonel Douglas . . . | 8th Line, Portuguese . | . 724 |
|  | 12th ,, ,, . | . 820 |
|  | 9th Caçadores . . | . 231 |
|  |  | 5287 |

### 7th Division
*Major-General Walker*

|  |  |  |
|---|---|---|
| Colonel Gardiner . . . | 1st Batt. 6th . . . | 709 |
|  | 24th ,, (3rd Provisional Batt.) |  |
|  | 2nd ,, 58th |  |
|  | 9 Coys. Brunswick Oels . | 250 |
| Maj.-Gen. Inglis . . . | 51st . . . | . 268 |
|  | 68th . . . | . 238 |
|  | 1st Batt. 82nd . . | . 489 |
|  | Chasseurs Britanniques | . 288 |
| Colonel Doyle . . . | 7th Line, Portuguese . | . 684 |
|  | 19th ,, ,, . | . 854 |
|  | 2nd Caçadores . . | . 374 |
|  |  | 4609 |

### LIGHT DIVISION
*Major-General Baron Charles Alten*

|  |  |  |
|---|---|---|
| Maj.-Gen. Kempt . . . | 1st Batt. 43rd . . | . 724 |
|  | 1st ,, 95th Rifles . | . 422 |
|  | 3rd ,, 95th ,, . | . 365 |
| Colonel Colborne . . . | 1st Batt. 52nd . | . 714 |
|  | 2nd ,, 95th . . | . 350 |
|  | 17th Line, Portuguese . | 611 |
|  | 1st Caçadores . . | . 421 |
|  | 3rd ,, . . | . 318 |
|  |  | 3925 |

PORTUGUESE DIVISION
*Major-General Le Cor*

| | | | Effective rank and file. |
|---|---|---|---|
| Brig.-Gen. Da Costa . . | 2nd Line, Portuguese | . | 971 |
| | 14th ,, ,, . | . | 831 |
| ,, Buchan . . | 4th ,, ,, | . | 888 |
| | 10th ,, ,, | . | 909 |
| | 10th Caçadores . . | . | 172 |
| | | | 3771 |

*Unattached Portuguese Brigades*

| | | | |
|---|---|---|---|
| Maj.-Gen. Bradford . . | 13th Portuguese | . | 547 |
| | 24th ,, . | . | 609 |
| | 5th Caçadores . . | . | 293 |
| ,, Campbell . . | 1st Portuguese | . | 563 |
| | 16th ,, . | . | 676 |
| | 14th Caçadores . . | . | 222 |
| | | | 2910 |

Total of Effective Rank and File, Cavalry and Infantry, 57,131.[1]

In addition to the combatant troops were the following :—

| | | | |
|---|---|---|---|
| Lieut.-Col. Dundas . . | Royal Staff Corps | . | 154 |
| Capt. Gibson . . . | Veteran Battalion | . | 871 |

The Spanish troops under Wellington's immediate command were these :—

| | | | |
|---|---|---|---|
| Army of Galicia . | Gen. Freyre (afterwards Giron) . | 12,000 men |
| Division . . | ,, Morillo . . . . | 4,000 ,, |
| ,, . . | ,, O'Donnell . . . | 6,000 ,, |
| ,, . . | Prince of Anglona (detached from 3rd Spanish Army) . . | 8,000 ,, |
| | Total . . | 30,000 ,, |

[1] If we add one-sixth, for officers, sergeants, etc., and for the *personnel* of the artillery, we have a gross total of about 66,500.

The Anglo-Portuguese Army Corps were organised as follows :—

*Sir John Hope's Corps :* 1st and 5th Divisions ; Lord Aylmer's Brigade, the Portuguese Brigades of Bradford and Campbell.

*Sir William Beresford's Corps :* 3rd, 4th, 6th, 7th, and Light Divisions.

*Sir Rowland Hill's Corps :* 2nd, and the Portuguese Division of Le Cor. At the outset of the campaign the 3rd Division was temporarily attached to this Corps, and the Headquarter Companies of the 60th were at Hasparren and Urcuray.

## 3

The year 1814 dawned brightly for the allied countries in arms against France. In 1813 fortune, excepting in the Peninsula, had favoured now one side, now the other, until the defeat of Napoleon at Leipzig in October had destroyed the greater part of his army and had driven the Emperor over the Rhine and into France, whither his Army of Spain had, as we have seen, already preceded him. It was difficult to believe that the war could continue much longer. Large bodies of French troops still remained indeed in Germany, but the fortresses which they occupied were blockaded by the enemy ; and elsewhere, Holland, Northern Italy, and the eastern provinces of Spain alone acknowledged the presence of that Grand Army which had so recently dominated the continent of Europe.

In the south of France the Allies were in overwhelming numerical superiority ; and although the moral difference was not as great, it was evident that Marshal Soult would be forced to remain on the defensive, at all events until he could effect a junction with the army under command of Marshal Suchet, which at present was still defending the north-eastern provinces of Spain against the Anglo-Sicilian contingent now under command of Lieut.-General William Clinton, brother to the commander of the 6th Division.

The force at Soult's command was now reduced to a field army of 40,000 men supported by a miscellaneous force of National Guards, Gens d'Armes, Conscripts, etc., to the number of about 14,000.

With such slender resources the Marshal found it impossible to retain his hold of Bayonne. Having, therefore, detached the veteran division of General Abbé with the National Guards, etc., to strengthen the garrison of the fortress, he took up a position with his remaining 35,000 men : his right behind the Adour, ten miles above Bayonne, his left and centre in occupation of the Urt—St. Jean Pied de Port road—a menace to Wellington's flank and communications.

Lord Wellington, on the other hand, had at his command a field force of about 66,000 men and 95 guns, in addition to 30,000 Spaniards with 5 guns ; but 28,000 men—Hope's Army Corps—were required for the investment of Bayonne ; and of the Spaniards, ill-fed and ill-disciplined, only 10,000 men were employed in France. After making, therefore, the necessary deductions for sick men, detachments, etc., Wellington could operate against Soult with not many more than 45,000 men. On January 25 the strength of our 5th Battalion—still commanded by Major Galiffe— was 747 of all ranks, the private Riflemen present being 470.

Early in February, a frost having dried the ground which had hitherto been little better than a quagmire, Lord Wellington began his campaign by directing Sir R. Hill with his Army Corps, which formed the right of the line, to turn the rivers beyond the Nive near their sources.

That officer having been relieved by the 6th and 7th Divisions, concentrated his Corps and the 3rd Division—20,000 men with 16 guns—at Urcuray and Hasparren on February 12 ; and on the 14th drove General Harispe from Helette to Meharin ; turned the Joyeuse River, and cut the direct communication between Soult and St. Jean Pied de Port.   In this action Captain F. P. Blassière of our regiment, who at least since 1810 had been attached to the 2nd Division, was mortally wounded and died on the 22nd.   This officer had been employed in intelligence work and had gained distinction, notably at Arroyo Molinos in 1811.   He had also fought at Roliça, Vimieiro, the Douro, Talavera, Bussaco, Badajoz, Albuera, Vitoria, Pyrenees, the Nivelle, and the Nive.[1]

On the 15th Hill occupied Meharin ; and after a march of eleven miles and a hard fight, drove Harispe with 4000 men from the Garris mountain and over the Bidouse to St. Palais with a loss of 450 men.   Hill's casualties were 163.   During this day Lieutenant Gottlieb Lerche of our 5th Battalion and Lieutenant Stepney of our 2nd were wounded.   Three other Riflemen were killed, and eight wounded.   Lerche lost an arm, and two years later was placed on the half-pay of the Royal Wagon Train.   He had got his first commission in 1812 ; having previously been sergeant-major to the Battalion.

On the 16th and 17th Hill crossed the Bidouse and the Saison.   During these operations his casualties amounted only to 31 killed, 189 wounded, and 12 missing ; but of the wounded no less than 22 were officers.

On the 23rd while Sir Rowland occupied the road running southward to Navarreux from Sauveterre on the

---

[1] Blassière—no doubt somewhat caricatured—is one of the characters in James Grant's ' The Romance of War,' where—by novelist's licence—he meets his death at the battle of the Nive.

Gave d'Oloron, Lord Wellington reinforced him with the 6th and Light Divisions, and forded the river.   The 4th and 7th successfully contained Soult's right and centre, but the extension of the allied army on a line of twenty-five miles, gave the enemy a chance, of which he did not avail himself, to interpose between Wellington and Sir John Hope whose Corps, consisting chiefly of the 1st and 5th Divisions, was now opposite Bayonne.

Such a stroke would have spoiled Wellington's plans for the passage of the Adour, an operation which with great audacity he had determined to execute below Bayonne, despite the breadth, tides, and rapidity of the river.   But as it turned out Wellington's operations had had the effect of diverting his opponent's attention from Bayonne and Sir John Hope seized his opportunity.   The 6th Division, which had marched on the 21st to join Wellington, was relieved at Mousseroles by the 5th, while two Spanish divisions were called up from the Bidassoa and posted about Arcangues.

During the night of the 22nd–23rd the 1st Division was brought up to the Adour at a point opposite the village of Le Boucau some two miles below Bayonne.   It had been intended to construct a floating bridge by means of the boats known as Chasse-marées, but stormy weather just now prevented them coming up the river and Hope boldly decided to throw a covering party across in pontoons notwithstanding the breadth of the river, nearly 300 yards, at the point selected, which was about three miles below Bayonne.   The Rifle Company commanded by Captain J. W. Harrison was naturally chosen for the purpose ; and soon after 10 A.M. on the 23rd was rowed over the river, each pontoon being able to convey only six or eight men.

The position of the company—perhaps forty-five strong

—was very much that of a forlorn hope at the assault of a fortress, with the additional probability of being upset in and drowned in the surf. Its destruction appeared a certainty, but as it happened, when the Riflemen landed the guard of a battery on the right bank withdrew— apparently in sheer astonishment at the hardihood of the enterprise ; and the green jackets having taken up a covering position, a raft pontoon attached to a hawser stretched across the river was constructed upon which the Light Companies of the Coldstream and 3rd Guards came across and were gradually followed by four other companies of the latter regiment. The crossing was happily unmolested ; but in the afternoon the detachment, some 500 strong, was attacked by two battalions of the Bayonne garrison, which, thanks in great measure to the success of a British rocket battery, were completely repulsed.

The following accounts relate to the operation, which it must be allowed was one of singular hardihood :—

### The Passage of the Adour, February 23, 1814

(*From Captain Batty's ' Campaign in the Western Pyrenees,*' 1823.)

' A cannonade between some English 18-pounders and a French corvette at anchor in the river under the walls of Bayonne, kept the enemy's attention on the alert opposite to the entrenched camp, whilst great exertions were making to ferry over the troops at the mouth of the river, on a pontoon raft which was soon constructed for that purpose : and before evening 4 companies of the 3rd Regiment of Guards, the two light companies of the Coldstream and 3rd Guards, together with two companies of the 60th Regiment,[1] attached to Major-General Stopford's Brigade, were conveyed to the right bank. The French sent two Battalions to drive this little band into the sea. The strength of these two battalions amounted to upwards of 1300 men : the troops who had been ferried over

---

[1] There is a mistake here. Although two Rifle Companies had until recently been attached to the 1st Division, one had by this time rejoined the Headquarter Companies in the 3rd Division.

amounted to barely 500. Major-General Stopford immediately prepared to receive the attack, and placed his little corps as favourably as circumstances would permit, resting its right flank on the Adour, and its left towards the sea, in an oblique direction across the sandy point, the prolongation of which forms a bar opposite the mouth of the river. A few rocket-men were hastily sent across the river, and posted on the sandhills to aid in repelling the enemy ; and 2 guns of the troop of horse artillery were so placed on the left bank of the river, as to be able to flank, by their fire, the troops coming on to attack the front of the guard.

'The enemy came on a little before dusk with drums beating the pas-de-charge, the Rifles gradually and steadily retiring. The Guards awaited the approach of the French columns till within a short distance of their front and then commenced a well-directed fire ; the guns on the left bank began to cannonade them, and the rockets on the sand-hills were discharged with terrific effect, piercing the enemy's column, killing several men, and blazing through it with the greatest violence. The result was the almost immediate rout of the French who, terror-struck at the unusual appearance and the effect of the rockets, and surprised at the immovable firmness of the little corps, made the best of their retreat back towards the citadel leaving a number of killed and wounded on the ground. This gallant little combat closed the events of the day.'

Extract of a letter dated March 20, 1814, from an officer in the 3rd Guards, to a friend :

'Since I wrote to you last a part of our Battalion has been in a dangerous predicament. On the 15th of last month, we hastily broke up from St. Jean de Luz and marched to Biaritz about 9 miles, and halted there, while preparations were being made for the passage of the river Adour at its mouth.

'The rifle company of the 60th Regiment passed first ; then our Light Infantry, and the Coldstream Lt. Infantry who advanced, beat up the ground, sent out piquets and reported "all well." Immediately the 8th, 7th, 6th and 5th companies of our battalion, with me in the latter company passed, and the 4th company following. On the next return of the raft owing to the very heavy surf and quick return of the tide, the raft became wholly unmanageable and was expected every moment to be sent to the bottom, or driven to Bayonne with the tide. During this dilemma, the enemy were seen advancing over the hills towards the beach, to the amount of two or three thousand men, who opposed to our little band (not five hundred men) with the sea in our rear and on our left ; a large impassable river on our right ; an immense superiority of number of the enemy in our front, and no assistance or road for retreat, had a very strange appearance I assure you.   However our advance posts were cautioned to fight deliberately and cool ; and the remaining four companies were formed from the river side covering the raft (which was loaded with troops and in a dangerous predicament) which they left on the sea ; and a brigade of field rockets of 50 men, each carrying 4 rockets, was passed over in a small cock boat, which formed in two divisions on the right and left of our little line.

' It was getting dark when the enemy advanced with loud cheering : our light companies opened a well-directed fire upon them ; the rockets appeared to set the firmament in a blaze ; and our field guns from the opposite bank of the river flanked their columns with shrapnel shell.   Thus after repeated efforts to take the sand-hills on which our advance was formed and seeing the determined bravery of our troops (our light infantry alone having turned the head of the enemy's columns three different times), and the destruction caused by the field rockets (which had not been used before) the enemy returned, leaving the ground strewed with their killed and wounded, and amidst the loud huzzas of our friends on the opposite bank where Sir J. Hope had been stationed and who, after remaining in perfect silence, during the whole affair anticipating our fate, was heard to say " Well done, Guards ! Bravely done, Guards ! " but I must certainly allow that had it not been for the field rockets, we must have had a duck and a swim for it.   Our Light Infantry had one sergeant and six men wounded ; the Coldstream, one man wounded ; the 60th none.   I should estimate the enemy's loss at about 300 in killed and wounded.'

On the 24th the main body of the 1st Division crossed the Adour, and next day the force on the right bank was further reinforced by the Portuguese brigades of Campbell and Bradford.   On the 26th a line was taken up facing south and extending over four miles from Le Boucau to a point a mile west of the Bayonne-Bordeaux road. Advancing southward on the 26th the village of St. Etienne was captured, and the line contracted to Mill of St. Bernard—

St. Etienne—Hayet. On the left bank Lord Aylmer's Brigade occupied Anglet, the 5th Division extended thence to the Nive, while a Spanish division took the space between the Nive and Adour. The investment of Bayonne was therefore completed.

The construction of a bridge over the Adour was next taken in hand. It was made by the mooring of twenty-six Chasse-marées head and stern at intervals of 40 feet. Two cables were then carried across their decks and planks were laid thereon. The bridge was completed by the 28th and protected by a boom. The constructing engineer, Major Todd, stated—contrary to the ideas of the world at large— that the soldiers were more full of resource, and being accustomed to more regular discipline, were better men for the purpose and appreciably quicker and more handy than the sailors.

Vandeleur's cavalry brigade succeeded in establishing communication with the 7th Division by the Peyrchorade road. Hope's Army Corps, including the Spaniards above mentioned, comprised 25,000 men and twenty guns. He had no siege train ; and his operations were consequently limited to a blockade.

An instance of the accurate marksmanship of our Riflemen is given in his reminiscences by Captain Gronow of the 3rd Guards, who was present at Bayonne.

' Several shots had been fired from the French piquets when the senior officer on duty came to me to inquire the cause of the firing, and desired me to make my way to the front and ascertain what had occurred. Having arrived near the ravine which separated us from the French, I stumbled upon our advanced sentry, a German, who was coolly smoking his pipe. I asked him whether the shots that had been heard came from his neighbourhood, upon which he replied in broken English, " Yes, zir, that feelow you see yonder has fired nine times at mine target " (meaning his body) " but has missed. I hope you, Capitaine, will let me have one shot at him."

The distance between the French piquet and ours could not have been less than 400 yards ; so without giving myself time to think, I said, " Yes, you can have one shot at him." He fired and killed his man ; whereupon a sergeant and two or three French soldiers who had seen him fall, ran down to the front and removed the body.'

On February 27 Captain J. W. Harrison commanding the Rifle Company attached to the Brigade of Guards was shot in the face and lost the sight of both eyes. He was retired on full pay for life.

4

Meanwhile Lord Wellington had not been idle. On February 25 he drove the French across the Gave de Pau at Orthez, and brought the 2nd, 6th, Light, and Portuguese Divisions up to the left bank while the 3rd reached Berenx, five miles lower down. Next day Sir W. Beresford crossed the river with the left wing consisting of the 4th and 7th Divisions, and advanced as far as the village of Baigts on the Orthez road. The 3rd Division, protected by Beresford's advance, then crossed the Gave ; and at daylight on the 27th the right wing of the army, with the exception of Hill's Corps, did the same. Wellington's available force was about 33,000 men and 48 guns.

Marshal Soult had taken up a strong position with his left at Orthez, his centre on the ridge separating that town from the village of St. Boës where his right was posted. His line extended over quite three miles, but with a force of about 37,000 men and forty guns, the Marshal was in a position to defeat in detail the British Divisions as they gradually emerged from the river on to the scene of action. The blow was not struck, and the light troops of the 3rd Division, skirmishing in front, helped to fill the gap which separated the 4th and 7th Divisions under Beresford from the 3rd, 6th, and Light Divisions which formed the centre of Wellington's line. Hill's Corps on the extreme right, opposite and above the town of Orthez, was still on the left bank of the river.

Beresford began the action by attacking St. Boës, where success would have rolled up the enemy's line. The 3rd and 6th Divisions simultaneously assailed Soult's centre, but each attack, although again and again renewed, proved fruitless. A new resource was evidently needed. The Light Division had not yet been engaged; and Lord Wellington directed Barnard's Brigade, which consisted of little more than the 52nd L.I., to advance against a point of the French line where Reille's Corps joined that of D'Erlon. The ground was marshy and the enemy's fire hot; but no rain had fallen for many days and the 52nd under its unrivalled commanding officer Colonel Colborne—in later life F.M. Lord Seaton—succeeded in crossing the bog. Its appearance at a point hitherto deemed unassailable, linked up the attacks of the 4th and 3rd Divisions, drove back Rouget's Division and went far to decide the battle—just as a year later the same magnificent regiment under the same Colonel decided the day at Waterloo.

In the battle of Orthez our regiment lost three O.R. killed, Captain Franchini, Lieutenant Currie, and 35 men wounded.

Marshal Soult retreated; first to St. Sever, then eastward up the Adour to Aire and Barcellonne; the right of his line—D'Erlon's Corps, consisting of the Divisions of D'Armagnac and Fririon (vice Foy wounded at Orthez)—being posted at Cazeres; the left—Reille's—at Aire. On March 2 Reille was suddenly attacked by Sir Rowland Hill, whose Corps formed the right of Wellington's army and whose rapid advance was unexpected; and after a very sharp contest was driven across the Adour; whereupon the

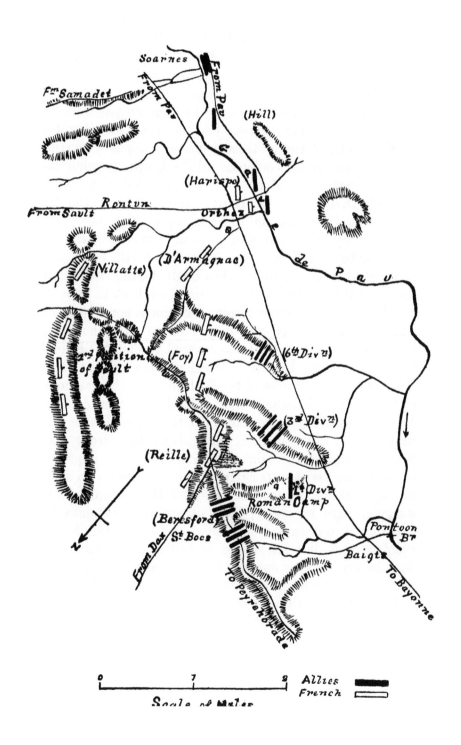

French Commander-in-Chief retired again eastward and took up an extended position on the Adour, from Tarbes on the south to Plaisance twenty miles northward. Stress of weather compelled Lord Wellington to halt about Aire.

In the middle of March Wellington was reinforced by the two Spanish Divisions which he had summoned from Sir John Hope's Army Corps. On the 18th he advanced southward from Aire by three columns in hopes of severing his opponent from Toulouse. Hill's Corps, forming the extreme right, moved on Conchez; the 3rd, 6th, and a Spanish Division, in the centre, upon Madiran, the 4th and Light Divisions, on the left, upon Plaisance. The French army had taken up a position between Simacourbe and Lembeye, and near the latter place Hill came into contact with the enemy. Then Soult, finding himself in danger of being outflanked on his right, fell back to Labatut and Lamayou. But the turning movement continued and next day the Marshal was compelled to retire hurriedly upon Tarbes. The retreat was covered by Count D'Erlon's Corps at Vic de Bigorre. Attacked here by the three Headquarter Companies of the 60th forming the vanguard of the 3rd Division about noon, a hard fight ensued. The French were strongly posted under the cover of walls and bridges, and the Riflemen, only about 130 strong, lost heavily in turning them out. But at 3 P.M., on the approach of the Light Division, D'Erlon withdrew Fririon's Division leaving that of D'Armagnac to hold the ground. 'The struggle,' says Fortescue, 'was exceedingly obstinate, for the country was exceedingly blind and the French sharp-shooters took full advantage of hedges and enclosures.' But the French were opposed to Riflemen who knew every turn of the game as well as they did themselves. The main body of the 3rd Division was also giving active support. Slowly but surely, bit by bit, the Riflemen gained ground and D'Armagnac was gradually forced back till night fell. In this combat Captain Kelly and Lieutenant Fourneret of our Regiment were wounded; the latter slightly, the former seriously. Four riflemen were killed and twenty-six wounded; but so well had they done their task in covering the advance of the main body that the remainder of the whole Division had little more than 200 casualties.

On the 20th the three battalions of the 95th greatly distinguished themselves in a successful attack on Clausel at Tarbes. Part of the 2nd and 3rd Divisions were also slightly engaged.

Wellington marched eastward in three columns and on the 25th found himself in presence of the French army drawn up behind the Touch, a tributary of the Garonne, and covering Toulouse, which was a great military arsenal and an important junction of roads by which Soult could march either northward down the Garonne to join General Decaen and the Army of the Gironde or southward to meet Marshal Suchet. His army was composed as follows :—

| Count D'Erlon | { | 1st Division | | General Darricau |
|---|---|---|---|---|
| | | 2nd   ,, | | ,,   D'Armagnac |
| Count Reille | { | 4th   ,, | | ,,   Taupin |
| | | 5th   ,, | | ,,   Maucune |
| Baron Clausel | { | 6th   ,, | | ,,   Villatte |
| | | 8th   ,, | | ,,   Harispe |
| | | Reserve of Conscripts | | ,,   Travot |
| | | Light Cavalry Division | | ,,   P. Soult |

The army comprised 42,000 men, with 56 field guns and about 30 guns of position. There were in addition about 7000 Conscripts.

On April 4 Wellington threw a pontoon bridge over the Garonne near Grenade fifteen miles below Toulouse. When three Cavalry Brigades and the 3rd, 4th, and 6th Divisions had crossed, a sudden rise of the river stopped the operation, which was not completed until the 8th. The troops (excepting Hill's Corps, which remained on the left bank of the Garonne) then marched southward towards Toulouse.

On the west side the city of Toulouse (excepting the suburb of St. Cyprien) was covered by the Garonne; on the north and east by the Languedoc Canal, beyond which the Mont Rave and the detached feature known as La Pujade formed strong defensive positions which were occupied thus:—

In Clausel's Corps, General Harispe occupied the Mont Rave; Villatte, La Pujade with one Brigade; the other being held in Corps Reserve. D'Erlon held the line of the canal from the bridge of Matabiau to the Garonne with Darricau's Division—now commanded by General Fririon—that of D'Armagnac being in reserve. The Conscripts manned the city walls, and Reille occupied St. Cyprien.

Wellington's army comprised the Cavalry Brigades of Bock, Ponsonby, Fane, Vivian, and Lord E. Somerset, about 7000 of all ranks. His infantry included the 2nd, 3rd, 4th, 6th, and Light Divisions in addition to the Portuguese Division of General Le Cor and the Spaniards of Freyre and Morille. Of this force of 45,000 officers and men, over 18,000 were British, about 13,000 Portuguese, and 14,000 Spaniards. His artillery included 64 guns, with 1500 gunners. Grand total of the Allied Army, 53,400 men and 64 guns.

On the left bank of the Garonne Sir R. Hill threatened Reille at St. Cyprien. The 3rd and Light Divisions were directed to demonstrate against Fririon; Freyre, to attack the Brigade posted on La Pujade; and by so doing to cover the march of Beresford, who was ordered to make a flank march southward along the Hers rivulet with the 4th and 6th Divisions past the Mont Rave, and then having

wheeled to the right to assail its southern spur. Beresford's
outer flank was protected by two brigades of cavalry.

At 10 A.M. on Sunday, April 10—Easter Day—the
allied columns opened the attack, and the enemy's advanced
posts were captured without difficulty. But Beresford's
progress in marshy ground was necessarily slow; and
Freyre, who had easily occupied La Pujade, either from
error or impatience attacked the northern end of the Mont
Rave. The attack, although made and renewed with spirit,
was met by a heavy fire and hurled back in confusion, and
the enemy's counter-attack checked only by a brigade of
the Light Division. The repulse of the Spaniards seems
to have decided Sir Thomas Picton to turn the feint of the
3rd Division into a real attack upon the strongly fortified
bridge of Jumeaux, for he feared that the French would
now fall in overwhelming numbers upon Wellington's
left wing.

The attack was made by Brisbane's Brigade (45th,
74th, and 88th) led by Galiffe, now Bt. Lt.-Colonel, with the
four Headquarter Companies of the 60th. But although
effected with magnificent élan, and assisted by the guns of
Hill's Corps on the further side of the Garonne, the bridge,
covered by a strong redoubt and flanked on either side by
artillery, proved impregnable. The assault was repulsed
with great loss—the casualties of the Riflemen being some
40 per cent. of their numbers; and since Hill was not
strong enough to carry the river line at St. Cyprien, and
the Light Division was immobilised by being compelled
to fill the gap left by Freyre's defeat, the action languished
at all points, the musketry fire almost ceased, and but for
that of the guns the action would have ceased.

Beresford was by this time—1 P.M.—approaching his
goal with the 4th and 6th Divisions, some 11,000 officers
and men; but Soult brought up not only his three Brigades

in Reserve, but three others from Reille's camps, and the cavalry, a force of 15,000 men—more than enough to strike a decisive blow on Beresford's lengthening column. The Marshal failed, however—not for the first time—to make use of the opportunity. Instead of striking with his whole available force, the cavalry and two infantry brigades only were despatched, under General Taupin, to fall upon Beresford. Even so a delay took place, for Taupin's artillery had not arrived, and when that General advanced to the attack, the British Divisions had completed their march and deployed opposite St. Sypière; the 6th on the right, the 4th on the left covered by the Hussar Brigade.

Taupin was in consequence received with a terrible musketry fire aided by the rocket battery. Lambert's Brigade counter-attacked with signal dash; Taupin was killed and his troops retired to the upper ground in confusion. The 4th Division captured the St. Sypière redoubt and a foothold was established on the Mont Rave.

Soult rallied the retiring troops, covering them with the reserve brigade of D'Armagnac; but the 6th Division, wheeling to the right, assailed Clausel. A desperate fight ensued; redoubts were taken and again lost. The British Division became exhausted. At 4 P.M. the attack came to a standstill, and Soult had another great opportunity to counter. But after a time the 6th Division advanced once more, while Picton still contained Fririon. The Spaniards also were now prepared to attack again. Under these circumstances Soult at 5 P.M. abandoned the contest and withdrew the whole of his army behind the canal. He had lost a gun and his casualties amounted to 3239; those of the Allies being nearly half as many again; 400 of which were sustained by the 3rd Division.

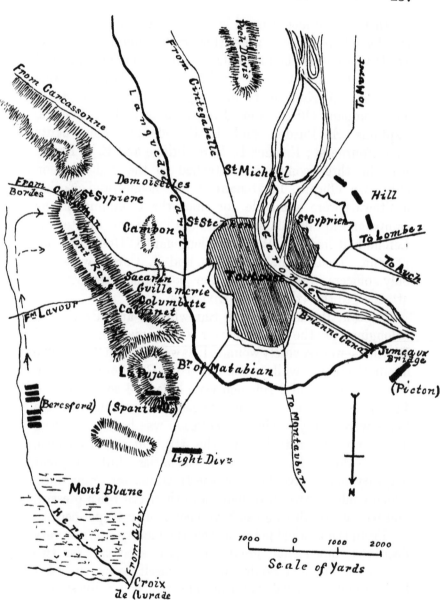

SKETCH OF TOULOUSE.

In this, the last general action of the Peninsular War, the battalion officers were thus distributed: With the 3rd Division were the four H.Q. Companies under command of Brevet Lieut.-Colonel Cycliffe, whose Adjutant was Lieutenant J. Kent, erstwhile Sergeant-Major in the 43rd. The Company Officers were Brevet Major J. H. Schöedde, Captains E. Purdon and I. Franchini; Lieutenants C. A. Fourneret; Ensigns H. Shrewbridge and E. J. Bruce. For the three companies attached to the 2nd Division only two officers, Lieutenant A. F. Evans and Ensign J. Stewart were available, for since the beginning of the campaign Captain Blassière had been killed and Lieutenant Lerche had lost a limb; and Lieutenant O'Hehir was not present. Lieutenants J. Moore and J. Currie were the only officers present with the companies respectively attached to the 4th and 6th Divisions.

Captain Purdon, Ensigns Shrewbridge and Bruce—all belonging to the Headquarter Companies in the 3rd Division—were severely wounded; as was Captain Scott Lillie of our Regiment, while in command of a Caçadore battalion. Two sergeants and 9 riflemen were killed; 4 sergeants and 44 riflemen wounded. One rifleman was missing. These losses were heavy. The 2nd Division was hardly engaged. The companies attached to the 4th and 6th Divisions, may, no doubt, have had some casualties; but the great majority must have been in the Headquarter Companies, which could hardly have had more than 120 rifles present, and whose casualties must have been nearly 40 per cent.

From the tactical point of view the victory was not very decisive; but it had great strategical success; for two days later Marshal Soult finding himself being encircled by Wellington's cavalry and in danger of being invested in Toulouse, evacuated the city with its great stores and arsenal, and marched southward, once more entertaining the hope of forming a junction with Marshal Suchet at Carcassonne.

On the same day, Lord Wellington entered Toulouse, and had hardly done so when news arrived that the Emperor Napoleon had abdicated, and that the war was at an end. This intelligence was at once forwarded to Soult who at first declined to credit it, for his latest news of the Emperor led him to believe that his master was still successfully holding the Russians, Austrians, and Prussians at bay, albeit on French soil. Wellington then pushed forward to Castelnaudary on the road to Carcassonne, but on the 18th the doubts of his opponent were set at rest and hostilities ceased.

The Campaign of 1814 had been expensive to the 5th Battalion. Since the advance had begun in February, 1 officer had been killed, 10 wounded, and the casualties of other ranks had been 211, sustained almost entirely by the companies attached to the 2nd and 3rd Divisions. In view of the fact that in February the number of rifles present in the nine companies engaged, had been only 420, the losses must be considered distinctly heavy.

### 5

Meanwhile at Bayonne Sir John Hope, having been deprived of his two Spanish Divisions, found it difficult to maintain the blockade of the fortress. In the absence of a siege train it was impossible to begin siege operations against the Citadel. The General's preparations to that end were, however, nearly complete when news reached the General of Napoleon's abdication and the cessation of hostilities. This news was communicated to General Thouvenot, commander of the garrison, as deserving weight although not official. Thouvenot replied that he would give an answer without delay; and so he did in a most unexpected manner. At 3 A.M. on the following morning—April 14—the French to the number of 3–4000 burst forth out of the Citadel along the Bordeaux road, and on both sides of it. Despite a warning received only an hour or two earlier the outposts of General Hay's Brigade were taken by surprise and the whole village of St. Etienne, excepting one house, was lost.

Still pouring along the road, driving piquets and supports before them, the French separated the two wings of the British force, and penetrating in rear of the right threw the whole line into confusion.

The K.G.L. Brigade in support had happily been kept under arms. At the critical moment, in conjunction with a battalion of Bradford's Portuguese Brigade, it came to the rescue and recovered the village of St. Etienne.

On the British right the Brigade of the 1st Guards taken in rear was no less severely handled by the assailants; but here also timely support was afforded by Stopford's Brigade, comprising the Coldstream and 3rd Guards. To this Brigade was attached the Rifle Company, commanded after Captain Harrison's wound by Lieutenant Richard Pasley. The conflict was, however, desperate, and the losses of the Guards were heavy.

'The battle,' says Sir W. Napier, 'was most confused and terrible; for on both sides the troops, broken into small bodies by the enclosure, and unable to recover their order, came dashing together in the darkness, fighting often with the bayonet; and sometimes friends encountered, sometimes foes; all was tumult and horror.'

At length the moon rose and day began to dawn. General Howard, in command of the 1st Division, brought up his reserves, and the French were driven back into the

Citadel with the loss of over 900 men.  Our own losses, including 236 prisoners, were nearly as heavy.  The casualties of the Rifle Company included both the officers, Lieutenants R. Pasley and John Hamilton, and 4 riflemen wounded ;  one sergeant and 4 riflemen were missing.

During the action Sir John Hope had disappeared.  It afterwards transpired that after ordering up his reserves from Boucan the General attempted to reach St. Etienne.  But the road was already in possession of the enemy.  A shot struck him in the arm ;  his charger was shot under him, and Sir John, unable to extricate himself, was made prisoner, with his Aide-de-Camp, William Moore, a nephew of Sir John Moore, and at a later date Colonel Commandant of our Regiment.

Lieutenant Hamilton died almost immediately of his wounds, and was buried in the lovely grove of oak and bracken enclosing the cemetery of the Coldstream Guards, who, with their wonted generosity, afforded their Rifle comrade a last resting-place amid their own officers.

A few days later General Thouvenot accepted the intelligence, by this time official, of the termination of hostilities.

## 6

Thus ended the campaigns, in the nick of time so far as the 5th Battalion was concerned.  The company at Bayonne had lost all its officers and was reduced to about 40 rifles.  The companies with Wellington had only 9 officers and about 250 rifles remaining.  The whole duration of the Peninsular War, which had been conducted on both sides with splendid gallantry and chivalrous feeling from start to finish, had been shared with our

Riflemen by two regiments only, viz. the 1st Battalion 40th and the 45th.

With the exception of the episode in 1809 when its reputation was tarnished by the conduct of a few individuals, who, being prisoners of war, ought never to have been allowed to join it, the career of the 5th Battalion bears comparison with that of any other British unit engaged, and there is not one whose individuality made itself felt in a greater degree.   The confidence reposed in the Riflemen by the Duke of Wellington has been abundantly shown ;   and the practice of detachment into companies, disadvantageous in some respects as a severe test upon their discipline and morale, served only to bring those features to greater prominence.   The glorious record of the 95th was almost exclusively confined to the Light Division ;   our 5th Battalion can claim the honour of furnishing riflemen for a large proportion of the whole army.

In this capacity they attracted the notice of foe no less than friend ; and it is remarkable that notwithstanding the exceptionally arduous nature of their duties their regimental system was so good that few regiments had a smaller proportion of absentees on account of sickness.

The battalion had sailed from Cork with 33 officers, 1007 N.C.Os. and riflemen.   During the last five and a half years it had received drafts from home amounting to 11 officers and 558 of O.R.   It returned home with 18 officers, 232 N.C.Os. and men.   The total number of casualties cannot be ascertained, for the official returns do not include men wounded on outpost duty or petty combats, but the following are known :—

Killed or died of wounds within a few days : officers, 9 ; O.R., 168.   Wounded : officers, 59 ; O.R., 599.   Missing : officers, 2 ; O.R., 225.   At the armistice, 6 officers and

178 O.R. were prisoners of war. Of the officers, 4 were taken (when the Spaniards abandoned the hospitals) at Talavera. The fifth was Colonel FitzGerald, captured in the Pyrenees when badly wounded. The sixth was Lieutenant Van Dyke, taken at St. Pierre.

The battalion was eventually awarded sixteen battle honours ; fifteen for definite actions and the sixteenth, the word " Peninsula," to cover the skirmishes and innumerable other minor affairs in which it had taken part between 1808 and 1814.

| | |
|---|---|
| Roliça. | Vimieiro. |
| Talavera. | Bussaco. |
| Fuentes de Oñoro. | Albuera. |
| Ciudad Rodrigo. | Badajoz. |
| Salamanca. | Vitoria. |
| Pyrenees. | Nivelle. |
| Nive. | Orthez. |
| Toulouse. | Peninsula. |

It has been said that the bestowal of so many honours was due to the fact of the companies being scattered throughout the army, and that they were given for actions in which a very small part of the battalion was engaged. It is true that at Albuera only three companies attached to the 2nd Division were engaged, and that the share of the battalion in the storming of Badajoz was confined to four companies, the remainder being attached to the covering divisions. At Fuentes de Oñoro, six companies were engaged, seven at the siege of Ciudad Rodrigo, and the same number at the battle of Salamanca. At Orthez and Toulouse, nine companies were on the field ; and at Roliça, Vimieiro, Talavera, Bussaco, Vitoria, the Pyrenees, Nivelle, and the Nive, viz. eight out of the whole fifteen, the entire battalion of ten companies was present.

The following distinctions were awarded : Colonels

Anderson, Williams, Keane, FitzGerald, and Galiffe received the Gold Medal and the Gold Cross; Colonels Woodgate, Gustavus Brown, Majors Davy, Schöedde, Andrews, Captains Scott Lillie and Power, the Gold Medal. Colonel Keane was also promoted Major-General and received the K.C.B. Colonels Anderson, Williams, Fitz-Gerald, Galiffe, and De Salaberry were given the C.B.

The Gold Medal was, however, bestowed only on Staff officers of high grade and regimental officers in command of units. The question will naturally be asked what reward was given to company officers, N.C.Os. and men for the seven campaigns which had occupied nearly six years. A few captains received brevet promotion, and a few sergeants were given commissioned rank. Apart from this no reward of any sort was granted. For Waterloo a medal was immediately issued to all ranks; but it was not till 1847 that a silver medal was awarded to the surviving veterans of the Peninsular War. Yet, in the words of Sir William Napier, the historian :—

'Those veterans had won nineteen pitched battles and innumerable combats; had made or sustained ten sieges and taken four great fortresses; had twice expelled the French from Portugal, once from Spain; had penetrated France, and killed, wounded, or captured two hundred thousand enemies—leaving of their own number, forty thousand dead, whose bones whiten the plains and mountains of the Peninsula.'

Colonel Keane quitted the Regiment on promotion, and was despatched to Jamaica to take command of a brigade in the force destined to operate against New Orleans. Landing near the mouth of the Mississippi on December 23, he repulsed the attack of a large American force which attempted to stop his advance. At the subsequent attack on New Orleans, Keane—described

as a dashing young officer—was badly wounded in command of a brigade.[1]  From 1823 to 1830, he commanded
the troops in Jamaica, and in 1833 became C. in C. of the
Bombay Presidency.  In 1839, consequent on a dispute
with Afghanistan, Sir J. Keane occupied Quetta, and
afterwards captured Ghuznee.  In the following month
—August—Keane occupied Kabul, and was raised to the
peerage as Baron Keane of Ghuznee.  He died in England
in 1844.

7

The good conduct, self-reliance, and resource of the
Riflemen have already been noticed ;  and the efficiency
of the regimental system of interior economy and discipline is illustrated by the small number of sick in proportion to that of other units of the army.

During or at the close of the campaigns the brevet
rank of Lieut.-Colonel had been bestowed on Majors
Woodgate, Galiffe, Gustavus Brown, and De Salaberry ;
that of Major on Captains Macmahon, De Wendt,
Schöedde, Purdon, and Friess.

As to their conduct in the field, the words of Wellington,
who was not in the habit of bestowing excessive praise,
are worth repetition :—

' The Commander of the Forces recommends the Companies of
the 5th Battalion 60th Regiment to the particular care and
attention of the General Officers commanding the Brigades of
Infantry to which they are attached ;  they will find them to be
most useful, active, and brave troops in the field, and that they

---

[1] In the course of this attack, which ended in a sanguinary repulse, Major-
General Samuel Gibbs, late of our Regiment, was killed.  In 1813, Gibbs had
earned distinction as commander of a small British force operating in the north
of Germany.  A statue of him is to be seen in the eastern transept of St.
Paul's Cathedral.

will add essentially to the strength of the Brigades.' . . . 'Certainly, if the services of any battalion could give to their C.O. a claim to promotion, the services of the 5th Battalion 60th in the late service in Portugal'—*i.e.* Vimieiro campaign—'entitled their C.O. to this advantage' . . . 'I am delighted with your account of the 5th Battalion 60th Regiment. Indeed everything that I have seen and known of that excellent corps has borne the same stamp.'

The battalion appears to have been well commanded throughout the war; but although Davy, FitzGerald, and Woodgate played their part well, the names of Williams and Galiffe will ever be remembered as those of its most distinguished Commanding Officers.

The principal Commanders on either side deserve a word of notice. First of all must, of course, come Wellington, whose successful conduct of the campaigns in the face of incredible difficulties and annoyances, marks him as a General and Administrator of the highest order. A stern, unlovable, solitary, almost unhuman character; yet he became the idol of his countrymen and with the exception of one brief period retained their unfailing admiration throughout his life. As an authority on his own exploits Wellington is unreliable; events succeeded one another so rapidly that his memory no doubt became blurred.

Of his lieutenants, Sir John Hope and Sir Rowland Hill best deserved the thanks of their Commander: but the name of Sir Thomas Picton, Commander of the 3rd Division to which our Battalion H.Q. belonged, should not be forgotten in the 60th.

On the side of the French, the best man was certainly Marshal Masséna: but Soult deserves great credit for the way in which he fought an uphill struggle. Of the others General Souham must be remembered as the only French General who contended successfully with Wellington.

## NOTES ON OFFICERS OF THE REGIMENT WHO SERVED IN THE PENINSULA, 1808 TO 1814

| Name. | Rank. | Battalion. | Period and Actions. | Remarks. |
|---|---|---|---|---|
| Acton, John Phillips | Lieutenant, 5.11.1813 | 8th | Jan.–Apr. 1814 | |
| Altenstein, Henry Baron v. | Ensign, 19.1.1809 Lieutenant, 10.11.1809 | 5th | May 1809–Apr. 1814 : Talavera | Wounded and taken prisoner |
| Anderson, Paul . | Lieut.-Col. 14.1.1808 Bt. Colonel, 4.6.1813 | Staff | Aug. 1808–Jan. 1809 : Coruña | A.A.G. Also served in Corsica, West Indies, St. Lucia, Irish Rebellion, Holland, Egypt, 1801, Sicily, 1806, Walcheren. Died as a Lieut.-Gen., Dec. 1851 |
| Andrews, Alexander | Captain, 2.6.1803 Major, 17.1.1811 | 5th | Aug. 1808–May 1811 : Roliça, Vimieiro, The Douro, Talavera, Bussaco | To 1st Battalion, May 1811. Died, 1823 |
| Antrobus, John C. . | Lieutenant, 7.12.1807 | 2nd | Oct. 1808–Jan. 1809 : Coruña | Died, 1810 |
| De Bree, Chevalier Charles | Ensign, 10.12.1807 Lieutenant, 11.10.1809 | 5th | Aug. 1808–Dec. 1812 : Roliça, Vimieiro, Talavera | Died, 1813 |
| Barbaz, John L. . | Ensign, 8.12.1807 Lieutenant, 10.10.1809 | 5th | Aug. 1808–Nov. 1813 : Roliça, Vimieiro, The Douro, Talavera, Bussaco, Albuera, Badajoz (1st and 2nd sieges), Arroyo Molinos, Almaraz, Vitoria, Pyrenees | Wounded and taken prisoner, 25.7.1813. Died of wounds |
| Bathurst, James . | Major, 1.10.1803 Bt. Lt.-Col. 10.10.1805 | Staff | Aug. 1808–Dec. 1810 : Roliça, Vimieiro, Coruña, The Douro, Talavera, Bussaco | A.Q.M.G. Also served in Egypt, 1801, Hanover, 1805, Copenhagen, 1807, Poland, 1807. Died, 1850, as Lieut.-Gen. |
| Berger, Christopher G. | Asst.-Surg., 18.10.1810 | 5th | Jan. 1811–Feb. 1814 : Salamanca, Vitoria, Pyrenees, Nivelle, Nive | Died, 23.2.1814 |

| Name. | Rank. | Battalion. | Period and Actions. | Remarks. |
|---|---|---|---|---|
| Biggs, Henry . | Paymaster, 22.10.1812 | 5th | Mar. 1813–Apr. 1814: Nivelle, Nive, Orthez, Vic-en-Bigorre, Toulouse | |
| Blassière, Frederick Peter | Captain, 4.12.1805 | 5th | Aug. 1808–Feb. 1814: Roliça, Vimieiro, The Douro, Talavera, Bussaco, Badajoz (1st and 2nd Sieges), Albuera, Arroyo Molinos, Almaraz, Villa Tranca (3.1.1812), Alba de Tormes, Vitoria, Pyrenees, Nivelle, Nive, Hellette | Severely wounded, 14.2.1814. Died of wounds, 22.2.1814 |
| Boardman, John . | Captain, 29.10.1807 | 2nd | Oct. 1808–Jan. 1809: Coruña | Died, 1810 |
| Boeek, Julius de . | Ensign, 7.12.1807 Lieutenant, 9.10.1809 | 5th | Aug. 1808–Apr. 1811: Roliça, Vimieiro | |
| Brœtz, Peter . | Ensign, 12.7.1810 Adjutant, 14.3.1811 Lieutenant, 9.10.1811 | 5th | Sept. 1810–Sept. 1812: Bussaco, Fuentes de Oñoro, Ciudad Rodrigo, Badajoz | Severely wounded, assault of Badajoz, 6.5.1812 |
| Breuney or Brunnig, Anthony Francis de | Ensign, 26.7.1810 Lieutenant, 25.7.1811 (York Light Infantry) | 5th | Oct. 1810–Apr. 1812 | Killed as Captain in 3rd Caçadores in storming Badajoz, 6.4.1812 |
| Brown, Gustavus . | Captain, 30.12.1797 Bt. Major, 25.7.1810 Major, 3.10.1811 Bt.Lt.-Col., 17.8.1813 | 5th | Apr. 1810–Dec. 1813: Bussaco, Salamanca, Pyrenees, Nivelle, Nive | Severely wounded a Nive, in command o 9th Caçadores |
| Bruce, Edward John | Ensign, 6.2.1813 | 5th | Aug. 1813–Apr. 1814: Nivelle, Nive, Orthez, Vic Bigorre, Tarbes, Toulouse | Severely wounded |

| Name. | Rank. | Bat-talion. | Period and Actions. | Remarks. |
|---|---|---|---|---|
| Brunton, Richard . | Captain, 10.11.1813 | 5th | May 1809–Dec. 1813 : Coa, Bussaco, Fuentes de Oñoro, Arroyo, Almaraz, Vitoria, Pyrenees, Nivelle, Nive (with 5th Caçadores), Waterloo, D.A.Q.M.G. | Slightly wounded, Pyrenees, 30.7.1813. Severely wounded, Nive, 13.12.181 . Died, Bath, 1846 |
| Burghaagen, Edward | Ensign, 24.1.1810 Lieutenant, 28.3.1811 | 5th | May–Sept. 1810 | |
| Cairnes, Allan Billinghurst or Bellingham | Ensign, 1.10.1811 Lieutenant, 21.12.1812 | 5th | Sept. 1812–Apr. 1813 | Died, 19.1.1814 |
| Campbell, John . | Lieut.-Col. | | | |
| Campbell, Colin . | Captain | 7th | | Afterwards F.M. Lord Clyde |
| Carroll, John . . | Surgeon, 22.8.1805 | 2nd | Oct. 1808–Jan. 1809 : Coruña | |
| Chatterway, Joseph | Qr.-Mr. 3.7.1800 | 2nd | Oct. 1808–Jan. 1809 : Coruña | |
| Cockburn, Francis . | Captain, 23.4.1807 | Staff | May 1809–June 1810 | A.A.A.G. Also served in S. America, 1807, Canada, 1811–1814 |
| Codd, Edward . | Lieut.-Col., 26.10.1804 Bt. Col., 4.6.1813 | 2nd | Oct. 1808–Jan. 1809 : Coruña | Also served in N. America and West Indies |
| Currie, John . . | Ensign, 2.10.1811 Lieutenant, 22.12.1812 | 5th | Apr. 1812–Apr. 1814 : Salamanca, Vitoria, Pyrenees, Nivelle, Nive, Orthez, Tarbes, Toulouse | |
| Dahlmann, John William van | Ensign, 28.9.1811 Lieutenant, 18.12.1812 | 5th | Feb. 1812–July 1813 : Badajoz, Salamanca, Burgos, Vitoria, Pyrenees | Killed, Pyrenees, 25.7.1813 |
| D'Arcy, Isaac Raboteau | Ensign, 15.4.1807 Lieutenant, 2.10.1809 | 5th | Aug. 1808–June 1810 : Roliça, Vimieiro | |
| Davy, William Gabriel | Major, 5.2.1807 | 5th | Aug. 1808–Dec. 1810 : Roliça, Vimieiro, Douro, Talavera | |

II.　　　　　　　　　　　　　　　　　　　　　　　　　　S

| Name. | Rank. | Bat-tallon. | Period and Actions. | Remarks. |
|---|---|---|---|---|
| Dibbley, Henry | Adjutant, 5.2.1800 Lieutenant, 23.5.1804 | 2nd | Oct. 1808–Jan. 1809 : Coruña | |
| Dickson, Hugh | Ensign, 1.4.1812 Lieutenant, 4.8.1813 | 5th | Aug. 1813–Jan. 1814 : Nivelle, Nive | Slightly wounded, Nive, 9.12.1813 |
| Drumgold, Nathaniel or Nicholas | Surgeon, 5.10.1809 | 5th | May–Oct. 1809 : Douro, Talavera ; June 1810–Dec. 1811 : Bussaco, Badajoz (2 sieges) | |
| Du Châtelet, Maximilian | Lieutenant, 3.12.1807 | 5th | Aug. 1808–May 1813 : Roliça, Vimieiro, Talavera, Bussaco, Fuentes de Oñoro | Slightly wounded, Fuentes |
| Dumoulin, James | Asst.-Surg., 13.4.1804 Surgeon, 9.9.1813 | 5th Staff | Aug. 1808–Apr. 1814 : Roliça, Vimierio, Douro, Talavera, Fu-entes de Oñoro, Bada-joz, Salamanca, Vi-toria, Pyrenees | |
| Eberstein, Francis Baron | Lieutenant, 10.12.1807 | 5th | Aug. 1808–Sept. 1812 : Roliça, Vimieiro, Douro, Talavera, Bussaco | Severely wounded Bus-saco, 27.9.1810. To 13th Royal Vet. Bn. Lt., 25.1.1813. |
| Eccles, Thomas | Lieutenant, 8.4.1813 | 5th | May 1813–Nov. 1813 : Vitoria, Pyrenees, Nivelle | Killed, Nivelle 10.11.1813 |
| Erskine, Hon. Esmé Stuart | Captain, 27.12.1810 | 2nd | Feb.–Sept. 1813 : Vi-toria, Pyrenees | Also served in Holland, 1813, as D.A.Q.M.G. |
| Evans, Augustine F. | Ensign, 10.11.1808 Lieutenant, 9.11.1809 | 5th | Nov. 1812–Apr. 1814 : Vitoria, Pyrenees, Ni-velle, Nive, Orthez, Toulouse | |
| Fitzgerald, Henry | Major, 11.11.1813 | 8th | Jan.–Apr. 1814 : Cadiz | |
| FitzGerald, John Forster | Major, 9.11.1809 Bt. Lt.-Col., 25.7.1810 | 2nd | Mar. 1812–Apr. 1814 : Badajoz, Salamanca, Vitoria, Pyrenees | Slightly wounded, Ba-dajoz. Died, 1877 |
| Forneret, Cuthbert Augustus | Ensign, 17.10.1811 Lieutenant, 6.2.1813 | 5th | Aug. 1813–Apr. 1814 : Nivelle, Nive, Orthez, Toulouse | |

| Name. | Rank. | Bat-talion. | Period and Actions. | Remarks. |
|---|---|---|---|---|
| Franchini, Ignace . | Lieutenant, 8.11.1804 Captain, 28.1.1813 | 5th | Nov. 1808–Apr. 1814: Douro, Talavera, Bussaco, Badajoz (1st and 2nd sieges), Albuera, Vitoria, Nivelle, Nive, Orthez, Toulouse | Slightly wounded at Bussaco, Vitoria, and Orthez |
| Franciosci, Charles Joseph de | Ensign, 8.3.1810 Lieutenant, 1.10.1811 | Staff | Oct. 1810–Apr. 1814 | |
| Friess, William . | Lieutenant, 8.8.1800 Captain, 24.5.1810 | 5th | May 1809–Apr. 1814: Douro, Talavera | Taken prisoner, Talavera, to end of war |
| Furst, Matthew . | Ensign, 23.6.1808 Lieutenant, 4.11.1809 | 5th | Aug. 1808–Jan. 1812: Roliça, Vimieiro, Talavera, Bussaco, Fuentes de Oñoro, Ciudad Rodrigo | |
| Galiffe, John . . | Captain, 30.12.1797 Bt.-Major, 25.4.1808 Major, 15.3.1810 Bt. Lt.-Col., 3.3.1814 | 5th | Aug. 1808–Apr. 1814: Roliça, Vimieiro, Talavera, Bussaco, Fuentes de Oñoro, Albuera, Ciudad Rodrigo, Badajoz, Salamanca, Pyrenees, Nive, Vitoria, Nivelle, Orthez, Toulouse | Slightly wounded, Talavera, 28.7.1809. Severely wounded, Salamanca, 22.7.1812 |
| Gilbert, George . | Paymaster, 6.2.1806 | 5th | Aug. 1808–May 1812: Roliça, Vimieiro | Died, 22.5.1812 |
| Gilse, Frederick de | Ensign, 6.2.1806 Adjutant, 15.5.1806 Lieutenant, 26.9.1809 | 5th | Aug. 1808–Sept. 1812: Roliça, Vimieiro, Douro, Talavera, Bussaco, Fuentes de Oñoro, Ciudad Rodrigo, Badajoz, Salamanca | Slightly wounded, Roliça and Badajoz |
| Green, Charles Dixon | Lieutenant, 14.8.1800 | 2nd | Oct. 1808–Jan. 1809: Coruña | To 5th Royal Veteran Bn., Lt., 14.9.1809 |
| Hames, Thomas | Captain, 12.11.1803 | 5th | Aug. 1808–Apr. 1810: Roliça, Vimieiro, Talavera | Retired, 1813 |
| Hamilton, John . | Ensign, 7.10.1811 Lieutenant, 31.12.1812 | 5th | Apr. 1812–Apr. 1814, Salamanca, Vitoria, San Sebastian, Nivelle, Nive | Killed, Bayonne, 14.4.1814 |

| Name. | Rank. | Bat-talion. | Period and Actions. | Remarks. |
|---|---|---|---|---|
| Harrison, John William | Captain, 12.8.1812 | 5th | Oct. 1810–Mar. 1814 : Cadiz, Barossa, Vitoria, San Sebastian, Nivelle, Nive, Adour, Bayonne | Wounded and deprived of sight, Bayonne, 27.2.1814 |
| Hazen, Robert . | Captain, 31.5.1803 Major, 14.6.1810 | 2nd | Oct. 1808–Jan. 1809 : Coruña | Died, 1813 |
| Holmes, Francis . | Lieutenant, 13.8.1802 Captain, 19.9.1811 | 5th | Aug. 1808–Feb. 1813 : Roliça, Vimieiro, Douro, Talavera, Badajoz (1st siege), Badajoz, Salamanca (siege), Salamanca | |
| Hope, John . . | Maj.-Gen., | 2nd | Salamanca | |
| Howard Thomas . | Captain, 9.10.1806 | 2nd and 5th | Oct. 1808–Jan. 1809 : Coruña | |
| Im Thurm, Lewis . | Captain, 25.4.1805 | 2nd | Oct. 1808–Jan. 1809 : Coruña | Taken prisoner, 14.1.1809 |
| Ingersleben, John Leopold de | Ensign, 14.12.1809 Lieutenant, 31.10.1810 | 5th | June 1810–Oct. 1813 : Bussaco, Badajoz (1st and 2nd sieges), Albuera, Vitoria, San Sebastian, Waterloo | Died, 21.11.1834 |
| Johnston, James . | Captain, 10.6.1813 | 5th | Oct. 1811–Apr. 1814 : Ciudad Rodrigo, Badajoz, Nivelle, Orthez | |
| Joyce, John . | Ensign, 2.7.1807 Lieutenant, 5.10.1809 | 5th | Aug. 1808–Oct. 1813 : Roliça, Vimieiro, Talavera, Bussaco, Fuentes de Oñoro (uncertain but probable), Salamanca, Vitoria, Pyrenees | Severely wounded Bussaco, slightly Vitoria, severely and taken prisoner Pyrenees, 25.7.1813 |
| Keane, John . . | Lieut.-Col., 25.6.1812 Bt.-Col., 1.1.1812 Maj.-Gen., 4.6.1814 | 5th | Apr. 1813–Apr. 1814 : Vitoria, Pyrenees, Nivelle, Orthez, Vic-en-Bigorre, Tarbes, Toulouse | Also served in Egypt, 1801. A.D.C. to Earl of Cavan. N. America, 1814. Attack on New Orleans, seriously wounded |
| Kelly, Robert . | Lieutenant, 17.10.1800 Captain, 16.8.1810 | 5th | Feb. 1813–Apr. 1814 : Vitoria, Pyrenees, Nivelle, Nive, Orthez, Vic-en-Bigorre. | Severely wounded, Bigorre, 19.3.1814 |

| Name. | Rank. | Battalion. | Period and Actions. | Remarks. |
|---|---|---|---|---|
| Kemmeter, Adolphus J. | Qr.-Mr., 30.12.1797 | 5th | Aug. 1808–Dec. 1812: Roliça, Vimieiro, Douro, Talavera | To 3rd Garrison Bn., 31.12.1812 |
| Kent, James . | Ensign, 11.4.1811 Lieutenant, 2.9.1812 Adjutant, 31.12.1812 | 5th | May 1811–Apr. 1814: Ciudad Rodrigo, Badajoz, Salamanca, Vitoria, Pyrenees, Nivelle, Nive, Orthez, Toulouse | Slightly wounded, Pyrenees, 30.7.1813 |
| Koch, Charles W. H. | Lieutenant, 2.8.1800 Captain, 12.10.1809 | 5th | Aug. 1808–July 1811: Roliça, Vimieiro, Douro, Talavera, Bussaco | Wounded, Vimieiro |
| Kruger, Frederick Peter Nicholas de | Ensign, 30.11.1809 Lieutenant, 30.10.1810 | 5th | May 1810–Mar. 1811 | |
| Lahrbusch, Frederick de | Ensign, 16.11.1809 Lieutenant, 29.10.1810 | 5th | May 1810–Dec. 1810: Lisbon | |
| Lawless, James . | Ensign, 10.10.1811 Lieutenant, 21.11.1813 | 5th | Apr. 1812–May 1813: Salamanca | |
| Lerche, Gottlieb . | Ensign, 14.5.1812 Lieutenant, 19.8.1813 | 5th | July 1812–Apr. 1814: Salamanca, Burgos, Pyrenees, Nivelle, Nive, Garris | Garris, severely wounded, 15.2.1814 |
| Leslie, Charles . | Captain, 5.11.1813 | 8th | Jan.–Apr. 1814: Cadiz | |
| Lillie, J. Scott . | Captain, 11.11.1813 | 7th | Nive, Orthez, Toulouse | Toulouse, severely wounded |
| Linstow, William . | Lieutenant, 9.12.1807 | 5th | Aug. 1808–Apr. 1814: Roliça, Vimieiro | |
| Livingstone, Alexander | Captain, 8.8.1805 | 5th | May 1811–May 1812 | |
| MacArthur, Edward | Ensign | 2nd | Egypt, Martinique, Ciudad Rodrigo, Vitoria, Pyrenees, Nivelle, Orthez, Toulouse (with 39th Regiment), Coruña | |
| McGilivray, William | Surgeon, 12.8.1809 | 5th | May–Oct. 1809: Doura, Talavera | Died, 11.10.1809 |

| Name. | Rank. | Battalion. | Period and Actions. | Remarks. |
|---|---|---|---|---|
| Mackenzie Alexander | Lieutenant, 25.6.1802 | 5th | Aug. 1808–May 1809: Roliça, Vimieiro | |
| Mackenzie, Alexander Wedderburn | Captain, 23.1.1812 | 5th | July 1812–Oct. 1812: Salamanca, Burgos | |
| McKenzie, Charles | | | | |
| Macmahon, John . | Captain, 20.8.1803 Bt.-Major, 4.6.1814 | 5th | Nov. 1808–Feb. 1813: Oporto, Bussaco, Badajoz (1st, 2nd, and 3rd sieges), Albuera, Arroyo Molinos, Almaraz | |
| Martin, Christopher Bernard | Ensign, 31.3.1812 Lieutenant, 3.8.1813 | 5th | Nov. 1812–Sept. 1813: Vitoria, Pyrenees | Severely wounded |
| Mitchell, Robert . | Lieutenant, 22.9.1808 | 5th | Apr. 1809–Apr. 1814: Talavera | Severely wounded and prisoner to April 1814 |
| Moore, John . . | Lieutenant, 22.7.1813 | 5th | Sept. 1813–Apr. 1814: Nivelle, Nive, Orthez, Vic-en-Bigorre, Toulouse | |
| Morgenthal, William | Ensign, 21.4.1808 Lieutenant, 1.11.1809 | 5th | Aug. 1808–Oct. 1812 | D.A.Q.M.G. in Portugal. To 2nd Royal Veteran Batt. Lieut., 30.12.1812 |
| Muller, Henry . | Lieutenant, 8.12.1807 | 5th | Aug. 1808–Feb. 1814: Roliça, Vimieiro, Douro, Talavera, Bussaco, Fuentes de Oñoro, Arroyo Molinos, Pyrenees, Nivelle, Nive | |
| Nugent, Pierse . | Lieutenant, 3.3.1814 | | | |
| O'Hehir, Sylvester . | Lieutenant, 14.6.1810 | 5th | Sept. 1810–Apr. 1814: Bussaco, El Bodon, Ciudad Rodrigo, Badajoz, Salamanca, Burgos, Nive, Orthez | Severely wounded, 19.9.1812 |
| Parker, Michael Edward | Surgeon, 12.6.1806 | 5th | Aug. 1808–June 1809: Roliça, Vimieiro, Douro | Died, 4.6.1809 |
| Passley, John Panton | Lieutenant, 5.12.1807 | 5th | Mar. 1813–Feb. 1814: Vitoria, Pyrenees, Nivelle | Severely wounded, 10.11.1813 |

| Name. | Rank. | Bat-talion. | Period and Actions. | Remarks. |
|---|---|---|---|---|
| Pasley, Richard . | Ensign, 8.9.1808 Lieutenant, 6.11.1809 | 5th | Bayonne | Wounded |
| Perry, John . . | Lieutenant, 12.2.1806 | 2nd | Oct. 1808–Jan. 1809 : Coruña | Died, 1812 |
| Perston, David . | Asst.-Surg., 1.2.1810 | 5th | May 1810–Mar. 1813 : Salamanca, Burgos | |
| Phelan, George . | Captain, 13.11.1813 | | Jan. 1812–Apr. 1814 : Bussaco, Albuera, Nive | With Portuguese Army, 6th Regt. Infantry. Severely wounded, Nive |
| Plenderleath, George | Ensign, 27.4.1806 | 2nd | Oct. 1808–Jan. 1809 : Coruña | |
| Prevost, James . | Captain, 26.2.1807 | 5th | Sept. 1810–Sept. 1811 : Bussaco, Badajoz (2nd siege), Aldea de Pontea | Severely wounded, 10.5.1811. Severely wounded and died, 27.9.1811 |
| Purdon, Edward . | Captain, 27.8.1807 Bt. Major, 12.4.1814 | 5th | May–Dec. 1811 ; Aug. 1813 ; April 1814 : Nivelle, Nive, Orthez, Toulouse | Severely wounded, Tou-louse |
| Pym, J. Barton . | Captain, 4.11.1813 | 8th | Jan.–Apr. 1814 : Cadiz | Died, 1815 |
| Rafler, William . | Ensign, 22.9.1808 Lieutenant, 2.10.1811 Captain, 10.2.1814 | 5th | Nov. 1812–June 1813 | |
| Reckney, Charles . | Qr.-Mr. 31.12.1812 | 5th | Feb. 1813–Apr. 1814 : Nivelle, Nive, Orthez, Vic-en-Bigorre, Tou-louse | |
| Ritter, Lewis . . | Lieutenant, 24.3.1804 | 5th | Aug. 1808–Jan. 1812 : Roliça, Vimieiro, Talavera, Bussaco, Fuentes de Oñoro | |
| Roepel, George (or Govert, R.) | Cornet, 21.1.1809 Lieutenant, 28.2.1812 | 5th | Apr. 1812–Dec. 1812 | Also served in Holland, 1814, as Interpreter to the Forces |
| Romance, Godfrey de | Lieutenant, 12.12.1805 | 2nd | Oct. 1808–Jan. 1809 : Coruña | Died, 1810 |
| Rutledge, William . | Ensign, 1.8.1813 | 5th | Mar. 1813–Dec. 1813 : Vitoria, Pyrenees, Nivelle, Nive | Severely wounded, Nive, St. Pierre, 13.12.1813. Died, 22.12.1813 |

| Name. | Rank. | Battalion. | Period and Actions. | Remarks. |
|---|---|---|---|---|
| Salaberry, C. de | Bd.-Major | | Walcheren | Bd.-Major to Bn. Gen. de Rothenberg |
| Sawatzky, Charles | Lieutenant, 25.11.1805 | 5th | Aug. 1808–Mar. 1811: Roliça, Vimieiro, Talavera, Bussaco | Killed, 15.3.1811 |
| Schaw, Charles | Captain, 10.1.1811 | 5th | July 1811–Apr. 1814: Ciudad Rodrigo, Badajoz, Salamanca | Also served in N. America, New Orleans, 1814. Died, 5.3.1874 |
| Schmidt, James Godley | Lieutenant, 5.5.1808 | 2nd | Oct. 1808–Jan. 1809: Coruña | Died, 1811 |
| Schöedde, James Holmes | Captain, 19.9.1805 Bt.-Major, 21.6.1813 | 5th | Aug. 1808–Apr. 1814: Egypt, Roliça, Vimieiro, Douro, Talavera, Bussaco, Fuentes de Oñoro, El Bodon, Ciudad Rodrigo, Badajoz, Salamanca, Vitoria, Pyrenees, Nivelle, Nive, Orthez, Toulouse | |
| Sewell, William Henry | Captain, 29.4.1813 Bt.-Major, 3.3.1814 | 5th | Mar. 1809–Apr. 1814: Coruña, Talavera, Bussaco, Ciudad Rodrigo, Badajoz, San Sebastian, Nivelle, Nive, Orthez, Toulouse | Served as A.D.C. to Marshal Beresford. Died, 13.3.1862 |
| Shewbridge, Henry | Ensign, 23.12.1812 Lieutenant, 12.12.1814 | 5th | Aug. 1813–Apr. 1814: Nivelle, Nive, Orthez, Vic-en-Bigorre, Toulouse | Severely wounded Toulouse, slightly Nivelle |
| Sprecher de Bernegg, John | Ensign, 22.1.1807 Lieutenant, 1.10.1809 | 5th | Aug. 1808–Nov. 1811: Roliça, Vimieiro | Died, 1812 |
| Steitz, Frederick | Lieutenant, 9.6.1804 | 5th | Aug. 1808–June 1810: Roliça, Vimieiro, Douro, Talavera | Slightly wounded, Roliça. Died, 1813 |
| Sterne, John H. | Lieutenant, 17.2.1810 | 5th | Feb. 1812–Apr. 1812: Badajoz | Killed |
| Stevenson, William | Asst.-Surg., 25.11.1813 | 5th | Dec. 1813–Apr. 1814: Orthez, Vic-en-Bigorre, Toulouse | |
| Stewart, Joseph | Ensign, 3.9.1812 Lieutenant, 24.1.1814 | 5th | Feb. 1813–Apr. 1814: Pyrenees, Nivelle, Nive, Orthez, Toulouse | |

| Name. | Rank. | Bat-talion. | Period and Actions. | Remarks. |
|---|---|---|---|---|
| Stopford, James | Captain, 9.4.1812 Major, 10.2.1814 | 5th | Oct. 1812–Feb. 1814: Vitoria, Pyrenees, Nivelle | Severely wounded, 10.11.1813 |
| Tressider, Samuel | Lieutenant 6.11.1813 | 8th | Jan.–Apr. 1814 | |
| Tripp, Ernest Otto | Major, 10.11.1813 | Staff | Mar. 1810–Jan. 1814 | D.A.A.G. to the Cavalry Division. Also served at Waterlooo as A.D.C. to Prince of Orange. Died, 1816 |
| Van Dyck, Peter R. A. | Lieutenant, 2.9.1813 | 5th | Oct. 1813–Apr. 1814: Nivelle, Nive | |
| Vieusseux, Andrieu | Ensign, 22.3.1810 | 5th | May 1810–June 1811: Bussaco, Fuentes de Oñoro | |
| Watson, John | Lieutenant, 24.5.1804 Captain, 29.12.1812 | 2nd | Oct. 1808–Jan. 1809: Coruña | Died, Jan. 1813 |
| Weatherley, James Dent | Captain, 7.11.1813 | 8th | Nov. 1813–Apr. 1814: Cadiz | |
| Wehsarg, Charles | Asst.-Surg. 6.2.1805 | 5th | Aug. 1808–Apr. 1814: Roliça, Vimieiro, Douro, Bussaco, Fuentes de Oñoro, Ciudad Rodrigo, Badajoz, Vitoria, Orthez, Bidassoa, Bordeaux | |
| Wendt (or Wend) Michael de | Captain, 7.12.1804 Bt.-Major, 4.6.1814 | 5th | Aug. 1808–Mar. 1811: Roliça, Vimieiro, Douro, Talavera, Bussaco | |
| Wilkinson, William | Captain, 16.12.1813 | 5th | April 1814: Ciudad Rodrigo, Badajoz, Vitoria, Pyrenees, Nivelle, Nive | Died, 14.3.1848 |
| Williams, William | Lt.-Col., 15.11.1809 | 5th | Dec. 1808–Jan. 1809; June 10–July 1812: Bussaco, Fuentes de Oñoro, Ciudad Rodrigo, Badajoz, Salamanca | |
| Wincary, Charles | Asst.-Surg. | 5th | Roliça, Vimiero, Bussaco, Fuentes de Oñoro, Ciudad Rodrigo, Badajoz, Vitoria | |

| Name. | Rank. | Bat-talion. | Period and Actions. | Remarks. |
|---|---|---|---|---|
| Wolff, John Anthony | Captain, 5.6.1806 | 5th | Aug. 1808–Apr. 1814: Roliça, Vimieiro, Douro, Talavera | Severely wounded and prisoner to 1814 |
| Wolfe, Joseph Alexander | | 5th | Roliça, Vimieiro, Talavera, Bussaco, Fuentes de Oñoro, Ciudad Rodrigo, Badajoz, Salamanca, Vitoria, Pyrenees, Orthez, Toulouse | |
| Woodgate, William . | Major, 13.8.1807 Bt. Lt.-Col., 30.5.1811 Lt.-Col., 16.6.1814 | 5th | Aug. 1808–Mar. 1812: Roliça, Vimieiro, Douro, Talavera, Bussaco, Fuentes de Oñoro, Ciudad Rodrigo | Slightly wounded, Fuentes de Oñoro, 5.5.11 |
| Wynne, Abraham William | Ensign, 23.4.1807 | 5th | Aug. 1808–Nov. 1813: Roliça, Vimieiro, Talavera, Bussaco, Fuentes de Oñoro, Ciudad Rodrigo, Badajoz, Salamanca, Vitoria | |
| Zuhleke, George Henry | Lieutenant, 6.8.1800 Captain, 11.1.1810 Major and Lt.-Col. in Portuguese Service, 2.6.1814 | 5th | Aug. 1808–Apr. 1814: Roliça, Vimieiro, Oporto, Talavera,* Fuentes de Oñoro, Ciudad Rodrigo, Badajoz, Salamanca, Vitoria, Pyrenees, Nivelle, Nive, Orthez | * Severely wounded and prisoner, but escaped, 26.4.1810 |

## Alphabetical List of Officers present with the 5th Battalion at the Principal Engagements

### Roliça

| | |
|---|---|
| Andrews. | Koch. |
| Barbaz. | Linstow. |
| Blassière. | Mackenzie, A. |
| D'Arcy. | Muller. |
| Davy. | Ritter. |
| De Bree. | Sawatzky. |
| De Gilse. | Schöedde. |
| Du Châtelet. | Sprecher. |
| Eberstein. | Steitz. |
| Furst. | Wendt. |
| Galiffe. | Wolff. |
| Hames. | Woodgate. |
| Holmes. | Wynne. |
| Joyce. | Zuhleke. |

### Vimieiro

| | |
|---|---|
| Andrews. | Koch. |
| Barbaz. | Linstow. |
| Blassière. | Mackenzie, A. |
| D'Arcy. | Muller. |
| Davy. | Ritter. |
| De Bree. | Sawatzky. |
| De Gilse. | Schöedde. |
| Du Châtelet. | Sprecher. |
| Eberstein. | Steitz. |
| Furst. | Wendt. |
| Galiffe. | Wolff. |
| Hames. | Woodgate. |
| Holmes. | Wynne. |
| Joyce. | Zuhleke. |

### Talavera

| | |
|---|---|
| Altenstein. | Holmes. |
| Andrews. | Joyce. |
| Barbaz. | Koch. |
| Blassière. | Mitchell. |
| Davy. | Muller. |
| De Bree. | Ritter. |
| De Gilse. | Sawatzky. |
| Du Châtelet. | Schöedde. |
| Eberstein. | Steitz. |
| Franchini. | Wendt. |
| Friess. | Wolff. |
| Furst. | Woodgate. |
| Galiffe. | Wynne. |
| Hames. | Zuhleke. |

### Bussaco

| | |
|---|---|
| Andrews. | McMahon. |
| Barbaz. | Muller. |
| Blassière. | O'Hehir. |
| De Gilse. | Prevost. |
| Du Châtelet. | Ritter. |
| Eberstein. | Sawatzky. |
| Franchini. | Schöedde. |
| Furst. | Vieusseux. |
| Galiffe. | Wendt. |
| Ingersleben. | Williams. |
| Joyce. | Woodgate. |
| Koch. | Wynne. |

### Fuentes de Oñoro

| | |
|---|---|
| De Gilse. | Ritter. |
| Du Châtelet. | Schöedde. |
| Furst. | Vieusseux. |
| Galiffe. | Williams. |
| (? Joyce). | Woodgate. |
| Muller. | Wynne. |

### Aroyo

| | |
|---|---|
| Barbaz. | McMahon. |
| Blassière. | Muller. |
| Franchini. | |

### Ciudad Rodrigo

| | |
|---|---|
| De Gilse. | Schaw. |
| Furst. | Schöedde. |
| Galiffe. | Williams. |
| Kent. | Woodgate. |
| Livingstone. | Wynne. |
| O'Hehir. | |

### Albuera

| | |
|---|---|
| Barbaz. | Galiffe. |
| Blassière. | Ingersleben. |
| Franchini. | McMahon. |

### Badajoz

Dahlman.
De Gilse.
FitzGerald.
Galiffe.
Holmes.
Kent.
McMahon.

O'Hehir.
Schaw.
Schöedde.
Sterne.
Williams.
Wynne.

### Salamanca

Currie.
Dahlman.
De Gilse.
FitzGerald.
Galiffe.
Hamilton.
Holmes.
Joyce.
Kent.

Lawless.
Lerche.
A. W. Mackenzie.
O'Hehir.
Schaw.
Schöedde.
Williams.
Wynne.

### Vitoria

Barbaz.
Blassière.
Currie.
Dahlman.
Eccles.
Erskine.
Evans.
FitzGerald.
Franchini.
Galiffe.
Hamilton.

Harrison.
Ingersleben.
Joyce.
Kelly.
Kent.
Martin.
Passley, J. P.
Rutledge.
Schöedde.
Stopford.
Wynne.

### Pyrenees

Barbaz.
Blassière.
Currie.
Dahlman.
Eccles.
Erskine.
Evans.
FitzGerald.
Galiffe.
Joyce.

Kelly.
Kent.
Lerche.
Martin.
Muller.
Passley, J. P.
Rutledge.
Schöedde.
Stewart.
Stopford.

### Nivelle

Blassière.
Bruce.
Currie.
Dickson.
Eccles.
Evans.
Fourneret.
Franchini.
Galiffe.
Hamilton.
Harrison.
Johnston.
Kelly.

Kent.
Lerche.
Moore.
Muller.
Passley, J. P.
Purdon.
Rutledge.
Schöedde.
Shewbridge.
Stewart.
Stopford.
Van Dyck.

### Nive

Blassière.
Bruce.
Currie.
Dickson.
Evans.
Fourneret.
Franchini.
Galiffe.
Hamilton.
Harrison.
Kelly.

Kent.
Lerche.
Moore.
Muller.
O'Hehir.
Purdon.
Rutledge.
Schöedde.
Shewbridge.
Stewart.
Van Dyck.

### Orthez

Bruce.
Currie.
Evans.
Fourneret.
Franchini.
Galiffe.
Kelly.

Kent.
Moore.
O'Hehir.
Purdon.
Schöedde.
Shewbridge.
Stewart.

### Toulouse

Bruce.
Currie.
Evans.
Fourneret.
Franchini.
Galiffe.

Kent.
Moore.
Purdon.
Schöedde.
Shewbridge.
Stewart.

### Bayonne

Hamilton.

R. Passley.

APPROXIMATE ALLOCATION OF THE OFFICERS OF THE 5TH
BATTALION TO DIVISIONS, DURING THE PENINSULAR
WAR, WITH DATES

### 1st Division

| | |
|---|---|
| Capt. T. Hames . . . | Aug. 1808—April 1810. |
| Capt. J. A. Wolff . . . | Aug. 1808—July 1809. |
| Ensign A. Vieusseux . . | May 1810—June 1811. |
| Lieut. J. Joyce . . . | Aug. 1808—March 1812. |
| Lieut. C. Sawatzky . . | June 1810—March 1811. |
| Lieut. P. Broetz . . . | Sept. 1810—Feb. 1811. |
| Ensign E. Burghaagen . . | May—Sept. 1810. |
| Capt. T. Howard . . . | { Oct. 1806—Jan. 1809 ; and Nov. 1810—April 1811. |
| Lieut. S. O'Hehir . . . | Sept. 1810—Jan. 1811. |
| Lieut. M. Du Châtelet . . | March 1811—May 1813. |
| Ensign A. W. Wynne . . | May 1811—March 1812. |
| Capt. C. Schaw . . . | July 1811—Dec. 1812. |
| Lieut. S. O'Hehir . . . | June 1812—Feb. 1813. |
| Lieut. T. V. Van Dahlmann . | Oct. 1812—March 1813. |
| Capt. J. W. Harrison . . | Feb. 1813—Feb. 1814. |
| Lieut. J. L. Ingersleben . . | Feb. 1813—Oct. 1813. |
| Ensign A. W. Wynne . . | March—Nov. 1813. |
| Lieut. J. Hamilton. . . | April 1812—April 1814. |
| Capt. E. Purdon . . . | Aug. 1813—Oct. 1813. |
| Lieut. H. Dickson . . . | Aug. 1813—Oct. 1813. |
| Lieut. R. Pasley . . . | Aug. 1813—Oct. 1813. |

### 2nd Division

| | |
|---|---|
| Lieut.-Col. J. F. FitzGerald . | March 1812—July 1813. |
| Capt. J. McMahon . . . | Nov. 1808—Feb. 1813. |
| Lieut. Barbaz . . . | June 1810—Nov. 1813. |
| Capt. F. O. Blassière . . | June 1810—Feb. 1814. |
| Lieut. R. Mitchell . . . | April 1809—July 1809. |
| Lieut. J. Sprecher de Bernegg. | June—July 1810. |
| Capt. M. De Wendt . . | May 1809—March 1811. |
| Lieut. J. L. Ingersleben . . | Nov. 1810—July 1811. |
| Lieut. H. Muller . . . | Nov. 1810—Nov. 1811. |
| Capt. I. Franchini . . . | Jan. 1811—March 1813. |
| Lieut. G. Roepel . . . | April—Dec. 1812. |

Lieut. J. Currie . . . April 1812—April 1813.
Lieut. A. F. Evans. . . Nov. 1812—April 1814.
Lieut. J. Joyce . . . April—Aug. 1813.
Lieut. A. B. Cairnes . . Sept. 1812—April 1813.
Lieut. J. V. Van Dahlmann . April 1813—July 1813.
Lieut. T. Eccles . . May 1813—Nov. 10, 1813.
Lieut. G. Lerche . . Sept. 1813—April 1814.
Capt. Hon. E. S. Erskine . Feb.—Sept. 1813.
Ensign W. Routledge . Oct.—Dec. 1813.
Lieut. P. R. A. Van Dyck . Oct. 1813—April 1814.
Lieut. H. Dickson . . . Nov. 1813—Jan. 1814.
Lieut. S. O'Hehir . . . Feb. 1814—April 1814.
Lieut. J. Stewart . . . Feb.—April 1814.

### 3RD DIVISION

Capt. A. Andrews . . . Aug. 1808—Dec. 1810.
Lieut. C. De Bree . . . Aug. 1808—Dec. 1810.
Ensign J. R. D'Arcy . .
Major J. H. Schöedde . . Aug. 1808—April 1814.
Ensign A. W. Wynne . . June 1810—April 1811.
Lieut. J. L. Von Ingersleben . June 1810—Oct. 1810.
Capt. J. Franchini . . . June 1810—Dec. 1810.
Capt. C. W. H. Koch . . June 1810—July 1811.
Ensign M. Furst . . . June 1810—Jan. 1812.
Lieut. Adj. F. De Gilse . . June 1810—Sept.1812.
Lieut. F. Baron Eberstein . Sept. 1810.
Lieut. S. O'Hehir . . . Sept. 1810—Oct. 1810.
Lieut. S. O'Hehir . . . Feb. 1811—May 1812.
Lieut. P. Broetz . . . March 1811—April 1812.
Capt. E. Purdon . . . May—Dec. 1811.
Capt. A. Livingstone . . May 1811—May 1812.
Lieut. Adj. J. Kent . . May 1811—April 1814.
Lieut. J. H. Sterne . . Feb. 1812—April 1812.
Ensign A. W. Wynne . . April 1812—Feb. 1813.
Lieut. J. Lawless . . . April 1812—May 1813.
Lieut. J. W. Van Dahlmann . April 1812—July 1812.
Major J. Galiffe . . . August 1812—April 1814.
Lieut. A. B. Cairnes . . Sept. 1812—Feb. 1813.
Capt. F. Holmes . . . Oct. 1812—Feb. 1813.
Lieut. J. L. Ingersleben . . Nov. 1812—Jan. 1813.

| Major J. Galiffe | . | . | . | Nov. 1812—April 1814. |
| G. C. Reckney | . | . | . | Feb. 1813—April 1814. |
| Capt. R. Kelly | . | . | . | Feb. 1813—April 1814. |
| Ensign W. Routledge | . | . | March 1813—Sept. 1813. |
| Lieut. J. P. Passley | . | . | March 1813—Feb. 1814. |
| Capt. I. Franchini | . | . | . | April 1813—April 1814. |
| Lieut. J. Currie | . | . | . | May 1813—Jan. 1814. |
| Lieut. J. Stewart | . | . | . | July 1813—Jan. 1814. |
| Lieut. C. A. Fourneret | . | . | Aug. 1813—April 1814. |
| Capt. E. Purdon | . | . | . | Nov. 1813—April 1814. |
| Lieut. S. O'Hehir | . | . | . | Dec. 1813—Jan. 1814. |
| Lieut. H. Shewbridge | . | . | Dec. 1813—Jan. 1814. |
| Ensign E. J. Bruce | . | . | March 1814. |

## 4TH DIVISION

| Major J. Galiffe | . | . | . | June—Sept. 1810. |
| Lieut. L. Ritter | . | . | . | June—Sept. 1810. |
| Lieut. H. Muller | . | . | . | June—Oct. 1810. |
| Capt. F. Holmes | . | . | . | June 1810—Sept. 1812. |
| Lieut. M. Du Châtelet | . | . | Sept. 1810. |
| Capt. J. Prevost | . | . | . | Sept. 1810—Sept. 1811. |
| Lieut. J. Joyce | . | . | . | April 1812—July 1812. |
| Major J. Stopford | . | . | . | Oct. 1812—Feb. 1814. |
| Lieut. C. B. Martin | . | . | Nov. 1812—Sept. 1813. |
| Capt. W. Rafler | . | . | . | Feb.—May 1813. |
| Lieut. J. Moore | . | . | . | Sept. 1813—April 1814. |

## 6TH DIVISION

| Lieut. L. Ritter | . | . | . | Oct. 1810—Jan. 1812. |
| Lieut. M. Du Châtelet | . | . | Oct. 1810—Feb. 1811. |
| Major J. Galiffe | . | . | . | Oct. 1810—July 1812. |
| Lieut. C. J. De Franciosé | . | Oct. 1810—April 1814. |
| Lieut. H. Muller | . | . | . | Dec. 1811—Feb. 1814. |
| Capt. A. W. Mackenzie | . | . | July 1812—Oct. 1812. |
| Lieut. G. Lerche | . | . | . | Aug. 1812—Aug. 1813. |
| Capt. J. W. Harrison | . | . | Nov. 1812—Jan. 1813. |
| Lieut. H. Shewbridge | . | . | Sept. 1813—Nov. 1813. |
| Ensign E. J. Bruce | . | . | Nov. 1813—Feb. 1814. |
| Lieut. J. Currie | . | . | . | Feb. 1814—April 1814. |

CASUALTIES DURING THE PENINSULAR CAMPAIGNS

| | Officers. | | | Other ranks. | | |
|---|---|---|---|---|---|---|
| | Killed. | Wounded. | Missing. | Killed. | Wounded. | Missing. |
| Obidos, Aug. 15, 1808 . | — | — | — | 1 | 5 | 8 |
| „ Aug. 16, 1808 . | — | — | — | 1 | — | 1 |
| Roliça, Aug. 17, 1808 . | — | 3 | — | 8 | 39 | 6 |
| Vimieiro, Aug. 21, 1808 . | — | 2 | — | 14 | 23 | 1 |
| The Douro, May 12, 1809 | — | — | — | — | 4 | — |
| Talavera, July 27, 1809 . | — | 1 | 1 | 3 | 4 | 19 |
| „ July 28, 1809 . | — | 6 | — | 7 | 25 | 12 |
| Bussaco, Sept. 27, 1810 . | — | 5 | — | 3 | 16 | 5 |
| Sobral, Oct. 12, 1810 . | — | — | — | — | 6 | — |
| „ Oct. 14, 1810 . | — | — | — | 1 | 6 | 4 |
| Redinha, March 12, 1811 | — | — | — | — | 10 | 5 |
| Casal Nova, March 14, 1811 . . | — | 1 | — | — | 3 | 1 |
| Foz de Arouce, March 14, 1811 . . | 1 | — | — | 3 | 8 | — |
| Sabugal, April 3, 1811 . | — | — | — | 2 | 2 | 1 |
| Fuentes de Oñoro, May 3, 1811 . . | — | 2 | — | 3 | 9 | 8 |
| Fuentes de Oñoro, May 5, 1811 . . | — | 2 | — | — | 12 | — |
| Badajoz (2nd siege of), May 8–15, 1811 . . | — | 1 | — | 1 | 8 | — |
| Albuera, May 16, 1811 . | — | 1 | — | 2 | 18 | — |
| Aldea de Ponte, Sept. 27, 1811 . . . . | 1 | — | — | — | — | — |
| Ciudad Rodrigo, Jan. 18, 1812 . . . . | — | — | — | — | 2 | — |
| Ciudad Rodrigo, Jan. 19, 1812 . . . . | — | — | — | — | 1 | — |
| Ciudad Rodrigo (Assault of), Jan. 19, 1812 . | — | 1 | — | 1 | 4 | — |
| Badajoz, March 18–22, 1812 . . . . | — | — | — | 2 | 3 | — |
| Badajoz, March 23–5, 1812 . . . . | — | — | — | — | 1 | — |
| Badajoz, March 31— April 2, 1812 . | — | — | — | 1 | 4 | - |
| Badajoz (Assault of), April 6, 1812 . . | 1 | 4 | — | 6 | 24 | — |
| Salamanca Forts, June 18–24, 1812 . . | — | — | — | 1 | 2 | — |
| Castrejon, July 18, 1812. | — | — | — | — | 1 | 2 |
| Battle of Salamanca, July 22, 1812 . . | — | 3 | — | 6 | 21 | 3 |
| Burgos, Sept. 19, 1812 . | — | 1 | — | 3 | 7 | 1 |
| „ Sept. 27—Oct. 3, 1812 . . | — | — | — | 1 | 1 | — |

| | Officers. | | | Other ranks. | | |
|---|---|---|---|---|---|---|
| | Killed. | Wounded. | Missing. | Killed. | Wounded. | Missing. |
| Burgos, Oct. 4–5, 1812 . | — | — | — | 1 | 2 | — |
| „ Oct. 6–17, 1812 . | — | — | — | 4 | 1 | — |
| „ Oct. 18–21, 1812. | — | — | — | — | 2 | — |
| Alba de Tormes, Nov. 10–11, 1812 . . | — | — | — | — | 8 | — |
| Retreat from Burgos, Oct.—Nov. 1812 . . | — | — | — | — | — | 115 |
| Vitoria, June 21, 1813 . | — | 3 | — | 2 | 47 | — |
| „ July 4–8, 1813 . | — | — | — | — | 6 | — |
| Pyrenees, July 25, 1813 . | 2 | 1 | — | 5 | 11 | 25 |
| „ July 26, 1813 . | — | 1 | — | — | — | — |
| „ July 27, 1813 . | — | — | — | 1 | 4 | — |
| Sorauren, July 28, 1813 . | — | 1 | — | — | — | — |
| „ July 30, 1813 . | — | 1 | — | 2 | 28 | — |
| Pyrenees, July 30—Aug. 1, 1813 . . . | — | 1 | — | — | — | — |
| Bidassoa, Oct. 7–8, 1813 | — | — | — | — | 1 | 2 |
| Various actions in Oct. 1813 . . . . | — | — | — | 8 | — | — |
| Nivelle, Nov. 10, 1813 . | 1 | 3 | — | 9 | 58 | 2 |
| Nive, Dec. 9, 1813 . . | — | 1 | — | — | 13 | — |
| „ Dec. 12, 1813 . | — | — | — | 1 | 5 | — |
| „ Dec. 13, 1813 . | 1 | 3 | 1 | — | 18 | 1 |
| Various actions, Dec. 1813 . . . . | — | — | — | 5 | — | — |
| Helette, Feb. 14, 1814 . | 1 | — | — | — | 3 | — |
| Garris, Feb. 15, 1814 . | — | 1 | — | 3 | 8 | — |
| St. Palais, Feb. 16, 1814 . | — | — | — | 1 | 1 | — |
| Rirareyte, Feb. 17, 1814 . | — | — | — | 1 | 2 | — |
| Various actions, Feb. 1814 . . . . | — | — | — | 2 | — | — |
| Orthez, Feb. 27, 1814 . | — | 2 | — | 4 | 35 | 1 |
| Bayonne (blockade of), Feb. 27, 1814 . . | — | 1 | — | — | — | — |
| Aire, March 2, 1814 . | — | — | — | — | 6 | — |
| Lembeye, March 18, 1814 | — | — | — | — | 6 | — |
| Vic de Bigorre, March 19, 1814 . . . . | — | 2 | — | 4 | 26 | — |
| Tarbes, March 20, 1814 . | — | — | — | — | 3 | — |
| Various actions, March 1814 . . . . | — | — | — | 8 | — | — |
| Toulouse, April 10, 1814 . | — | 4 | — | 11 | 48 | 1 |
| Bayonne (sortie from), April 14, 1814 . . | 1 | 1 | — | — | 4 | 5 |
| Various actions, April 1814 . . . | — | — | — | 27 | — | — |
| Total . . | 9 | 59 | 2 | 169 | 615 | 229 |

CASUALTY LIST OF OFFICERS DURING THE PENINSULAR
CAMPAIGNS

| Name. | Killed. | Wounded. | Missing. | Remarks. |
|---|---|---|---|---|
| Von Altenstein, Lieut. Baron H. | | Talavera | Talavera | |
| Andrews, Major A. . | | Talavera Bussaco | | |
| Barbaz, Lieut. J. L. | Pyrenees | | | |
| Blassière, Capt. P. P. | Hellette | | | Wounded Feb. 14, died Feb. 22, 1814. |
| Broetz, Lieut. and Adjt. P. | | Badajoz (capture of) | | |
| Brown, Lieut.-Col. Gustavus | | Nive | | Commanding 9th Caçadores. |
| Bruce, Ensign E. J.. | | Toulouse | | |
| De Brunnig (or Breuney), Ensign A. F. | Badajoz (capture of) | | | Attached 3rd Caçadores and transferred to York Light Infantry, July 25, 1811. |
| Currie, Lieut. J. . | | Orthez | | |
| Van Dahlmann, Lieut. J. W. | Pyrenees | | | |
| D'Arcy, Ensign I. R. | | Roliça | | |
| Dickson, Lieut. H. . | | Nive | | |
| Du Châtelet, Lieut. M. | | Fuentes de Onoro | | |
| Eberstein, Lieut. Baron F. | | Bussaco | | |
| Eccles, Lieut. T. . | Vivelle | | | |
| FitzGerald, Lieut.-Col. J. F. | | Badajoz (capture of) Pyrenees | | Taken prisoner when wounded. |
| Fourneret, Lieut. C. A. | | Vic-en-Bigorre | | |
| Franchini, Capt. I. . | | Bussaco, Vitoria, Orthez | | |
| Friess, Lieut. W. . | | | Talavera | |
| Galiffe, Lieut.-Col. J. | | Talavera, Salamanca | | |
| De Gilse, Lieut. and Adjt. F. | | Roliça, Badajoz (capture of) | | |

| Name. | Killed. | Wounded. | Missing. | Remarks. |
|---|---|---|---|---|
| Hamilton, Lieut. J. . | Bayonne | | | |
| Harrison, Capt. J. W. | | Bayonne | | |
| Im Thurn, Capt. L. (2nd Battn.) | | | Coruña | |
| Ingersleben, Lieut. J. L. | | Albuera | | |
| Joyce, Lieut. J. . | | Bussaco, Vitoria, Pyrenees | | Severely wounded in the Pyrenees and taken prisoner. Reported killed in dispatches (erroneously). |
| Kelly, Capt. R. . | | Vic-de-Bigorre | | |
| Kent, Lieut. and Adjt. J. | | Pyrenees | | |
| Koch, Capt. C. W. H. | | Vimieiro | | |
| Lerche, Lieut. G. . | | Salamanca, Nive, Garris | | |
| Lillie, Capt. J. Scott | | Toulouse | | |
| Livingstone, Capt. A. | | Ciudad Rodrigo | | |
| Mitchell, Lieut. R. . | | Talavera | | |
| Martin, Lieut. C. B.. | | Pyrenees | | |
| Nugent, Lieut. P. . | | Orthez | | |
| O'Hehir, Lieut. S. . | | Burgos | | |
| Passley, Lieut. J. P. | | Nivelle | | |
| Pasley, Lieut. R. . | | Bayonne | | |
| Prevost, Capt. J. . | Aldea de Ponte | Badajoz 1811 | | |
| Purdon, Capt. E. . | | Toulouse | | |
| Phelan, Capt. G. . | | Nive | | Attached to 6th Portuguese Infy. |
| Ritter, Lieut. L. . | | Vimieiro Talavera | | |
| Routledge, Ensign W. | St. Pierre | | | Died of wounds, Dec. 22, 1813. |
| Stepney, Lieut. S. (2nd Batt.) | | Garris | | |
| Sawatzky, Lieut. C.. | Casal Novo | | | |
| Shewbridge, Ens. H.. | | Nivelle, Toulouse | | |
| Steitz, Lieut. F. . | | Roliça | | |
| Sterne, Lieut. J. H. | Badajoz (capture of) | | | |
| Stopford, Capt. J. . | | Nivelle | | |

| Name. | Killed. | Wounded. | Missing. | Remarks. |
|---|---|---|---|---|
| Van Dyck, Lieut. P. R. A. . . | | | St. Pierre | |
| Williams, Lieut.-Col. W. | | Bussaco, Fuentes de Oñoro, Badajoz (capture of), Salamanca | | |
| Wolff, Capt. J. A. . | | Talavera | | |
| Woodgate, Major W. | | Fuentes de Oñoro | | |
| Wynne, Lieut. W. . | | Condeixa, Fuentes de Oñoro | | |
| Zuhleke, Capt. G. H. | | Talavera | | |

## Gold Medals and Crosses awarded to Officers of the Regiment

| Name. | Qualification. | Decorations. | Actions. |
|---|---|---|---|
| Brevet Colonel Paul Anderson. . | | Medal | Coruña |
| Major Alexander Andrews . . | Command of Detachment | Medal | Talavera |
| Brevet Colonel James Bathurst . | | Medal and 2 Clasps, Cross | Roliça and Vimieiro Coruña Talavera Bussaco |
| Brevet Lieut.-Col. Gustavus Brown . | Command of 9th Caçadores | Medal and 2 Clasps, Cross | Salamanca Pyrenees Nivelle Nive |
| Colonel William G. Davy . . | Battalion Commander | Medal and 1 Clasp | Roliça and Vimieiro Talavera |
| Lieut.-Col. James Forster FitzGerald | Brigade Commander | Medal and 2 Clasps, Cross | Badajoz Salamanca Vitoria Pyrenees |

| Name. | Qualification. | Decorations. | Actions. |
|---|---|---|---|
| Colonel John Galiffe . . . | Battalion Commander | Medal and 2 Clasps, Cross | Vitoria Nivelle Orthez Toulouse |
| Major-General J. Hope . . . | Command of 7th Division | Medal | Salamanca |
| Brevet Colonel Sir John Keane . | Command of Brigade | Medal and 2 clasps, Cross and 2 Clasps | Martinique Vitoria Pyrenees Nivelle Orthez Toulouse |
| Capt. J. Scott Lillie . . . | Command of Caçadore Batt. | Medal and 2 Clasps, Cross | Nivelle Nive Orthez Toulouse |
| Colonel G. Mackie . . . . | Command of the 3rd Batt. 60th | Medal | Martinique |
| Capt. T. Power . . . . | | Medal | San Sebastian |
| Brevet Major J. H. Schöedde . . | Command of H.Q. Coys. 5th Batt. | Medal | Nivelle |
| Lieut.-Col. W. Williams . . . | Command of Batt. | Medal and 2 Clasps, Cross and 1 Clasp | Coruña (with 81st Regt.) Fuentes de Oñoro Ciudad Rodrigo Badajoz Salamanca |
| Lieut.-Col. W. Woodgate . . | Command of H.Q.Coys.of Batt. | Medal | Fuentes de Oñoro |
| Capt. G. Zuhleke . . . . | Command of 2nd Caçadores | Medal and 2 Clasps | Vitoria Pyrenees Orthez |

*N.B.*—The Gold War Medal and Cross were given only to commanders of units or to Staff officers of high grade. The Medal alone was awarded for one action, and clasps added for the two succeeding actions. The fourth action was rewarded with a Gold Cross, and subsequent actions by clasps. In the above list, officers whose sole connection with the Regiment was that of being a Colonel Commandant (*e.g.* Sir H. Clinton) are not included ; nor is Colonel Gabriel Gordon, who received the Medal for Martinique and Guadeloupe, the year after he had given up command of the 4th Battalion.

## RECIPIENTS OF THE SILVER MEDAL ISSUED IN 1847 TO THE SURVIVORS OF THE PENINSULAR CAMPAIGN

### 5TH BATTALION

### OFFICERS

| Names. | Clasps. |
|---|---|
| Von Altenstein, Ensign, Baron H. | Talavera. |
| D'Arcy, Ensign I. H.. | Roliça, Vimieiro. |
| Evans, Lieut. A. F. | Vitoria, Pyrenees, Nivelle, Nive, Orthez, Toulouse. |
| Fo(u)neret, Ensign C. A. W. | Nivelle, Nive, Orthez, Toulouse. |
| Furst, Lieut. M. | (Egypt), Roliça, Vimieiro, Talavera, Bussaco, Fuentes de Oñoro, Ciudad Rodrigo. |
| Galiffe, Lieut.-Col. J. P. | Roliça, Vimieiro, Talavera, Bussaco, Fuentes de Oñoro, Albuera, Ciudad Rodrigo, Badajoz, Salamanca, Pyrenees, Nive. |
| De Gilse, Lieut. Adjt. F. | Roliça, Vimieiro, Talavera, Bussaco, Fuentes de Oñoro, Ciudad Rodrigo, Badajoz, Salamanca. |
| Johnstone, Capt. J. | Ciudad Rodrigo, Badajoz, Nivelle, Orthez. |
| Kelly, Capt. R.. | Vitoria, Pyrenees, Nivelle, Nive, Orthez. |
| Von Linstow, Lieut. W. | Roliça, Vimieiro. |
| Livingstone, Capt. A. | (Egypt), Martinique, Ciudad Rodrigo. |
| McKenzie, Lieut. A. | Roliça, Vimieiro. |
| Mackenzie, Capt. A. W. | (Bussaco, Fuentes de Oñoro), Salamanca. |
| McMahon, Capt. J. | Bussaco, Albuera, Badajoz. |
| Schaw, Capt. C. | Ciudad Rodrigo, Badajoz, Salamanca, (San Sebastian, Nivelle, Nive). |
| Schöedde, Major J. H. | (Egypt), Roliça, Vimieiro, Talavera, Bussaco, Fuentes de Oñoro, Ciudad Rodrigo, Badajoz, Salamanca, Vitoria, Pyrenees, Nive, Orthez, Toulouse. |
| Sewell, Capt. W. H. | (Coruña, Talavera, Bussaco, Ciudad Rodrigo, Badajoz), San Sebastian, Nivelle, Nive, Orthez, Toulouse. |
| Stevenson, Assist. Surgeon W. | Orthez, Toulouse. |
| Viesseux, Ensign A. | Bussaco, Fuentes de Oñoro. |
| Wehsarg, Assist. Surgeon C. | Roliça, Vimieiro, Bussaco, Fuentes de Oñoro, Ciudad Rodrigo, Badajoz, Vitoria, (Orthez). |
| Wilkinson, Capt. W. | (Ciudad Rodrigo, Badajoz, Vitoria, Pyrenees, Nivelle, Nive.) |
| Wolff, Sergt.-Major, afterwards Ensign J. A. | (Egypt), Roliça, Vimieiro, Talavera, Bussaco, Fuentes de Oñoro, Ciudad Rodrigo, Badajoz, Salamanca, Vitoria, Pyrenees, Orthez, Toulouse. |
| Woodgate, Lieut.-Col. W. | Roliça, Vimieiro, Talavera, Bussaco, Ciudad Rodrigo. |

| Names. | | | Clasps. |
|---|---|---|---|
| Wynne, Lieut. A. W. | . | . | Roliça, Vimieiro, Talavera, Bussaco, Fuentes de Oñoro, Ciudad Rodrigo, Badajoz, Salamanca, Vitoria. |

*N.B.*—Recipients of the Gold Medal did not receive a silver one, unless they had taken part in actions other than those for which they had received the former. When actions are shown above in brackets, it means that the officer was not in the 60th when he took part therein.

## OTHER RANKS

| | | | |
|---|---|---|---|
| Androwitz, Pte. A. | . | . | Roliça, Vimieiro. |
| Androwitz, Pte. A. | . | . | Talavera, Bussaco, Albuera, Badajoz, Salamanca. |
| Belmont, Pte. J. | . | . | Talavera, Vitoria. |
| Bougart, Pte. P. | . | . | Ciudad Rodrigo, Salamanca. |
| Brandt, Pte. F.. | . | . | Toulouse. |
| Danel, Pte. C. | . | . | Ciudad Rodrigo, Salamanca. |
| Esch, Pte. M. | . | . | Vitoria, Pyrenees. |
| Esch, Pte. P. | . | . | Ciudad Rodrigo, Salamanca, Pyrenees. |
| Gabriel, Pte. F. | . | . | Albuera, Salamanca, Vitoria. |
| Graff, Pte. J. | . | . | Vitoria. |
| Hesear, Pte. J. | . | . | Vitoria, Pyrenees. |
| Hollenhorst, Pte. J. | . | . | Toulouse. |
| Joop, Pte. J. | . | . | Salamanca, Vitoria, Pyrenees, Toulouse. |
| Knabline, Pte. F. | . | . | Vitoria, Orthez, Toulouse. |
| Kruger, Pte. G.. | . | . | Ciudad Rodrigo, Salamanca. |
| [1] Loochstadt, Pte. Daniel | . | . | Roliça, Vimieiro, Talavera, Bussaco, Fuentes de Oñoro, Albuera, Ciudad Rodrigo, Badajoz, Salamanca, Vitoria, Pyrenees, Nivelle, Nive, Orthez, Toulouse. |
| Luschen, Pte. D. | . | . | Vitoria, Pyrenees, Orthez, Toulouse. |
| Mammert, Sergt. L. | . | . | Toulouse. |
| Miller, Armourer Sergt. T. | . | . | Vimieiro, Talavera. |
| Moller, Pte. H. | . | . | Vitoria, Toulouse. |
| Rissa, Pte. F. | . | . | Salamanca, Vitoria, Pyrenees, Nivelle, Nive, Orthez. |
| Schabe, Pte. C.. | . | . | Salamanca, Vitoria. |
| Schonten, Pte. P. | . | . | Roliça, Vimieiro, Talavera, Bussaco, Fuentes de Oñoro, Ciudad Rodrigo, Badajoz, Salamanca, Vitoria. |
| Schultz, Pte. F.. | . | . | Badajoz. |
| Schultz, Pte. Joseph | . | . | Ciudad Rodrigo, Badajoz, Vitoria, Toulouse. |
| Schultze, Pte. F. | . | . | Talavera, Salamanca, Vitoria, Toulouse. |
| Volmer, Pte. F.. | . | . | Vitoria. |
| Warth, Pte. A. | . | . | Ciudad Rodrigo, Badajoz. |
| Wasserman, Sergt. G. | . | . | Vitoria, Pyrenees, Toulouse. |
| Zeitlen, Sergt. C. | . | . | Vimieiro, Talavera, Bussaco, Albuera, Badajoz, Vitoria, Pyrenees. |

[1] This Rifleman joined the 5th Battalion from the Foreign Depôt, Lymington, on Sept. 2, 1807, and was posted to Blassière's Company. Later on he was transferred to No. 6 Company—Erskine's.

2ND, 3RD, 4TH, AND 7TH BATTALION

### OFFICERS

| Names. | Clasps. |
|---|---|
| Adair, Ensign J. H., 4th . . | Martinique, Guadeloupe. |
| Booth, Pte. (later Q.M.) J., 2nd . | Coruña. |
| Hearn, Lieut. J. P., 2nd . . | Martinique, Guadeloupe. |
| MacArthur, Ensign E., 2nd . | Coruña (Vitoria, Pyrenees, Nivelle, Nive, Orthez, Toulouse). |
| Ramsay, Ensign G. A., 3rd . | Martinique (Guadeloupe). |

### OTHER RANKS

| | |
|---|---|
| Baker, Pte. J., 2nd . . . | Coruña. |
| Brockman, Pte. O., 3rd . . | Guadeloupe, Martinique. |
| Cobley, Pte. T., 3rd . . . | Martinique. |
| Dore, Pte. M., 2nd . . . | Guadeloupe. |
| Fogarty, Pte. D., 3rd. . . | Martinique. |
| Heincke, Pte. H., 3rd . . | Martinique, Guadeloupe. |
| Lindeboom, Sergt. P., 2nd . | Coruña. |
| Massler, Pte. G., 2nd . . | Guadeloupe. |
| Mulligan, Pte. J., 3rd. . . | Martinique, Guadeloupe. |
| Peters, Sergt. J., 2nd. . . | Guadeloupe. |
| Quadlen, Drummer R., 2nd . | Coruña, Guadeloupe. |
| Radusch, Pte. J., 2nd . . | Guadeloupe. |
| Schop, Pte. A., 3rd . . . | Martinique. |
| Thièle, Sergt. A., 3rd. . . | Martinique, Guadeloupe. |
| Thierman, Pte. M., 3rd . . | Martinique. |
| Vanwort, Pte. M., 3rd . . | Martinique, Guadeloupe. |
| Watske, Pte. G., 3rd . . . | Martinique, Guadeloupe. |
| Westwood, Pte. S., 3rd . . | Martinique. |

The following claims for a Medal were referred back for further information; with what ultimate result is unknown :—

| | |
|---|---|
| Bucher, Pte. N., 5th . . . | Vitoria, San Sebastian, Toulouse. |
| Doering, Pte. M., (?) . . . | Martinique. |
| Erben, Pte. G., 7th . . . | North America. |
| Foris, Pte. J., 2nd . . . | Guadeloupe. |
| Kirschgens, Pte. H., 3rd . . | Martinique, Guadeloupe. |
| Knuffer, Pte. F., 5th. . . | Salamanca, Vitoria. |
| Niemayer, Pte. H., 2nd . . | All actions in which battalion took part. |
| Osstead, Pte. J., (?) . . . | Portugal and West Indies. |
| Pontren, Pte. W., 5th . . | Vitoria, Toulouse. |
| Rohr, Pte. T., 5th . . . | Vitoria, San Sebastian. |
| Scholzen, Sergt. J., 2nd . . | Guadeloupe. |
| Schurren, Corpl. C., 7th . . | North America. |

## CHAPTER XI

THE war with France being at an end, the British Government found it possible to prosecute that against the United States with greater vigour. Orders were consequently sent directing Wellington—whose services had been rewarded by £400,000 and a dukedom—to embark a force of his veteran infantry for America. For this force was detailed our 5th Battalion; but although its strength was estimated at only 350 rank and file its losses during the recent campaign had been so heavy, and since the termination of hostilities the number of men discharged had been so great, that the battalion was unable to find even the modest quota desired.[1] The idea of sending it to America was consequently abandoned, and on July 5 the eighteen officers and 232 N.C.Os. and men representing the Rifle battalion, under command of Colonel Galiffe, embarked at Pauliac in H.M.S. *Clarence*. On the 25th they landed at Cork—whence six years previously the battalion had sailed over a thousand strong.

The officers on board the *Clarence* were the following :—

Brevet Lieut.-Colonel J. Galiffe.
Brevet Major J. H. Schöedde.
Captain F. Franchini.
    ,,    W. Wilkinson.
Lieutenant F. Muller.
    ,,    A. F. Evans.

---

[1] At the Armistice 82 men had been discharged ;  166 took ' French leave.'

Lieutenant S. O'Hehir.
,,         P. N. de Kruger.
,,         J. Kent (Adjutant).
,,         J. Currie.
,,         C. Fourneret.
,,         J. Moore.
,,         J. Stewart.
Ensign H. Shewbridge.
  ,,    Bruce.
Paymaster H. Biggs.
Quartermaster Reckney.
Assistant Surgeon W. Stevens.

Of these officers, Lieut.-Colonel Galiffe, Major Schöedde, and Lieutenant Muller had started from Cork with the battalion in 1808 and now returned therewith.

The State, dated Cork, July 25, 1814, gives the following figures : 1 major, 3 captains, 10 subalterns, 2 staff, 43 sergeants, 4 buglers, 249 riflemen.  Wanting to complete establishment, 573.

2

The scene of action now changes to the further side of the Atlantic where (as shown in Vol. I. p. 277) since June 1812 we had been at war with the United States.  At sea the Americans appeared to be under enormous disadvantage, both in regard to experience and numerical inferiority.  But to the astonishment of the British nation the American frigates won victory after victory over our own ; and caused so great an impression in this country that, when in the summer of 1813, H.M.S. *Shannon* captured the U.S.S. *Chesapeake* outside Boston harbour, the British public fell into transports of joy ; and the Admiralty granted to the Fourth Lieutenant, who brought the *Shannon* out of action, the right to remain upon full

pay during the rest of his life. As Admiral Sir Provo
Wallis he lived to be almost a centenarian and died only
in 1892.[1]

On land, and particularly on the Canadian frontier, all
the advantage seemed to be on the side of the United
States ; for the regular force in Canada consisted of hardly
3000 men ; and the militia comprised only a few battalions,
one of which was commanded by Patrick Murray, son no
doubt of his namesake, who had served with our 4th
Battalion in the American War of Independence and whose
historical memoir forms an Appendix in Vol. I. of these
Annals ; but the attempts of the Americans upon British
territory met at first with very partial success.[2] After a
few raiding expeditions during the early months of the war,
in one of which Colonel de Salaberry of our 5th Battalion,
whom we last saw as Brigade Major to General de
Rothenburg, gained distinction at the head of a body of
Voltigeurs by repulsing, in December 1812, the attack of
the advance guard of an army under General Wilkinson
at La Colle, matters remained quiet for the moment.

In the defence of Canada the 60th was well represented,
for the Governor was Lieut.-General Sir George Prevost—
son of Augustin—formerly Lieut.-Colonel of our 4th Battalion, and at the present time Colonel Commandant of the
5th ; while Major-General Baron F. de Rothenburg, the
father of British riflemen, commanded the Montreal
district.

The Voltigeurs, as stated above, were commanded by

[1] Relics from the *Chesapeake*, and pictures of the action, are to be seen at the
Royal United Service Institute, Whitehall.

[2] The first attack upon British territory was made near Detroit by General Hull,
who on July 12 issued a proclamation to the English settlers in which among other
flowers of speech he observed that his success was certain since he was at the head
of an army which ' would look down all opposition.' Hull's force consisted of 2500
men ; that of the British was 750 with 600 Indians. The proclamation was dated
July 12. On that day five weeks, Hull and the whole of his army surrendered !

Colonel de Salaberry, whose previous career deserves a passing notice.

Charles Michel d'Irumberry de Salaberry, descendant of an old Basque family, was born in November 1778 at Beaufort in Canada; and having at the age of fourteen served as a Volunteer in the 44th Regiment, was in April 1793 given a commission in the 60th at the instance of H.R.H. the Duke of Kent, who had been sent out to inaugurate parliamentary government in Canada. He joined a battalion—apparently the 4th—at Dominica.

In 1794 he saw active service under General Prescott at the gallant defence of Fort Matilda, Guadeloupe, where his cool courage was conspicuous. Almost the whole of his detachment were killed or wounded; and on the evacuation of the fort on December 6 only three men of it marched out under arms.

'This great handsome boy of 16,' says the historian of the Canadian Militia, 'as strong as a Hercules, lithe in body and of happy countenance, who spoke English with the grace of a Briton of old family, showed in the highest degree the fascinating refinement characteristic of the Canadians at that period.'

In August 1794 De Salaberry, still well under 16, was promoted to the rank of lieutenant; and in July 1799 to that of captain. Among the officers of the regiment were some Germans who were incorrigible duellists. One day at luncheon one of these swashbucklers boasted that he had just despatched a Canadian—one of the Des Rivières whose brother was in the 60th. 'Another is at your service at this moment,' retorted De Salaberry. Immediately after luncheon they crossed swords in the garden and the German quickly found he had met his match. Infuriated by receiving a slight wound on the forehead De Salaberry hurled himself on his antagonist and 'clove him in two just

LIEUT.-COLONEL C. M. DE SALABERRY, C.B.

like an apple ' !   The astonishing strength of this Samson
made him a superman in the eyes of his soldiers.

In the early years of the nineteenth century our hero
was employed for the most part on recruiting duty in
England until selected by General de Rothenburg as his
Staff officer.   He accompanied the General to Ireland and
on the Walcheren Expedition, and returned with him to
Canada.

Yet another former officer of our Regiment deserves
mention.   Louis Joseph Fleury Deschambault (termed in
the Army List Fred. de Chambault) had in boyhood been a
page of honour to the King of France, Louis XV.   Shortly
after the death of that monarch Deschambault came to
Canada, and in 1788—being then thirty-one years of age—
was given a captain's commission in the newly formed
4th Battalion of the 60th in which he afterwards became
major.   At the present time he was the efficient Q.M.G. to
the militia in Lower Canada.

During the summer of 1813 the Americans again took
the offensive, and after defeating the British flotilla on
Lake Erie gained possession of the whole of Upper Canada,
excepting Kingston.   The moment was therefore highly
critical when two armies under the respective command of
Generals Hampton and Wilkinson threatened an immediate
attack on Lower Canada.

At the end of September General Hampton with a force
estimated by British historians at 7–8000 men ;  but stated,
and no doubt correctly, by the Americans at 180 cavalry,
5720 infantry, and 10 guns, threatened the frontier with a
view to attacking Montreal, the capital.   By a beautiful
illustration of light infantry tactics De Salaberry with
three hundred Voltigeurs and half as many Red Indians
held them in check for a month ;  and it was not till
October 21 that Hampton crossed the frontier and marching

down the left bank of the Chateauguay River made for the Isle Perrot.   De Salaberry thereupon took post on the left bank of the river, commanding the road to Perrot, with a force of Canadian militia and light troops to the number of about 400 men and 150 Indians ; but without artillery.

In compliance with his orders Colonel de Salaberry carried out the principles of field fortification with considerable skill, throwing up breastworks and blocking the enemy's line of approach with felled trees.   The left of his main body was protected not only by the river but by wooded and marshy ground beyond.   The other flank was covered by similar ground.   Seventy men posted on the right bank of the river guarded a ford.   Sixty men in addition, under command of Michel Duchénay (brother-in-law to De Salaberry, and formerly in the 60th), were posted *en potence* still further beyond the river, with a view to countering a movement which a detachment of the enemy was known to be attempting on that side.   A hundred and thirty Indians were held in reserve.   To a point a mile and a half in rear Colonel McDonnell with a body of militia had just arrived after a forced march ; forming a most useful support.

At 10 A.M. on October 26 the advance guard of the enemy drove in De Salaberry's outposts and afterwards his main body from their first position.   De Salaberry had no mind to allow a further approach and seizing a bugler by the collar ordered him to sound the advance for an immediate counter-attack.   The bugle was heard by Major Macdonell, and he, divining that De Salaberry was hard pressed, also sounded the advance ; while in order to deceive the enemy as to his numbers, he gave orders for a dozen buglers to extend widely and repeat the call.   The ruse had the desired effect.   The Americans, observing the advance of the rallied Voltigeurs and believing themselves to be in the presence of a force of 7000 men, were by no means anxious to try conclusions with them. De Salaberry's counter-attack was successful; and although General Purdy of the U.S.A. Army, after losing his way all the previous night in the woods, had succeeded in bringing up his men upon our left flank as intended, De Salaberry reinforced the flank guard under Duchénay and after hard fighting the assault was repelled and Purdy's men fell into confusion and fired incontinently on friend not less than foe.

The attack was over at 2.30 P.M.; but during the whole day a considerable amount of desultory firing took place, and, owing chiefly to the fact of our men fighting under cover, very much to our advantage. The enemy fell back in some confusion which was so far increased after dark that two brigades of his infantry fired upon one another. Before daylight the Americans abandoned the attack and retraced their steps. Early on the 27th De Salaberry followed them. He found large graves, and the bodies of 90 American soldiers unburied. His own losses amounted to only 2 killed, 10 wounded, and 4 missing.

The Americans returned to their own country and gave up the enterprise. Colonel de Salaberry, fearing the vengeance of his Indians upon prisoners, offered a reward for all brought in unharmed.

Meanwhile on October 26 General Wilkinson had concentrated a force of 8800 men at Grenadier Island, Lake Ontario, opposite Kingston, where Major-General de Rothenburg had assembled 3000 men.

With a view to co-operating with his colleague, General Hampton, Wilkinson cleverly embarked his men in boats and despatched them down the St. Lawrence. As soon as de Rothenburg knew what had happened, he sent a detachment of 800 men in pursuit. But this small force, made up of details from the British regular army, encountered, and after a stern combat defeated, 3000 of the enemy near Chrysler Farm; after which General Wilkinson, who had sustained a loss of 450 men and three guns, retired into American territory; and hearing of Hampton's repulses made no further attempt at invasion of British soil.

Few actions on a small scale have led to more important results than those of Chateauguay and Chrysler Farm. Lower Canada was not only saved but was never attacked again. De Salaberry became a national hero, and gained the soubriquet of ' the Leonidas of Canada.' His fame is still fresh in that country; and monuments have been erected to his memory at Chambly and Chateauguay. After the war he took part in local politics, but died in about 1823.

### 3

In April 1814 the newly formed 7th Battalion of our Regiment, under command of Lieut.-Colonel Henry John, embarked at Guernsey for Halifax, Nova Scotia, where it landed 811 rank and file strong on May 28, and was quartered in the South Barracks.

A new field of action, albeit on a minor scale, soon presented itself. The American State of Maine, separated only by the Bay of Fundy from the western boundary of Nova Scotia, was in sore distress; for the war with England had ruined her trade, and her inhabitants were greatly tempted to secede from the United States. Under these circumstances policy pointed to a British occupation of at all events a portion of Maine.

### EXPEDITION TO THE PENOBSCOT

On August 24, 1814, the two Rifle companies of the 7th Battalion with one company R.A., the 29th, 1/62nd, and

98th Regiments under Lieut.-General Sir J. Sherbroke, sailed from Halifax for Penobscot Bay on the coast of Maine, escorted by a small squadron under Rear-Admiral Griffiths. It had been intended to take possession of Machias near the mouth of the Machias River *en route ;* but very early on the 30th the expedition fell in with a British cruiser which gave information that the U.S. frigate *John Adams* had got into the Penobscot, and being apprehensive of an attack had gone up the river due northward as far as Hampden to land her guns and arm the batteries on shore for her protection. It was therefore decided to proceed at once to the Penobscot ; and very early on September 1 the expedition arrived at the mouth of the river. A little after sunrise, the Fort of Custine, situated on a peninsula near the left bank of the mouth of the river, was surrounded ; but the commandant refusing to surrender opened fire with 24-pounders upon a small schooner in which Lieut.-Colonel Nicholl, C.R.E., had gone to reconnoitre. Before the troops could, however, be landed the garrison blew up their magazine and escaped up the Majetaquaclony River in boats, carrying off with them two field-pieces. A detachment R.A., the companies of the 60th and the 98th, under Colonel Douglas, were then disembarked in rear of the peninsula on which the fort stood, with orders to secure the isthmus and occupy the heights commanding the town in case any American force should remain on the peninsula ; but the militia there assembled dispersed as soon as the troops began to land and no opposition was experienced.

The next object of the British commanders was to effect the destruction or capture of the *John Adams*, for which purpose a detachment R.A., the flank companies of the 29th, 62nd, 98th, and a Rifle company of the 60th were detailed under Lieut.-Colonel John of our 7th Battalion to accompany and co-operate with a naval force under

Captain Barrie, R.N.  As the distance between Custine and Hampden was, however, twenty-seven miles, it was deemed a necessary precaution to occupy in the first place a post on the right or western bank with a view to giving support if needed to the force proceeding up the river, as well as to hold in check the militia of the districts south and westward and prevent it harassing the troops on their march.  To this end the 29th Regiment under Major-General Gosselin was ordered to occupy Belfast, commanding a bridge on the high-road from Hampden to Boston, and to hold it till further orders.  Whilst therefore the remainder of the troops landed in the evening at Custine, the troops destined for Hampden sailed thence ; and on the morning of September 2 arrived above the town of Frankfort, where a detachment of Americans was discovered marching up the left bank in the direction of Hampden.  Brevet Major Crosdaile with a detachment 98th and some riflemen of the 60th under Lieutenant Wallace was forthwith ordered to land and intercept it. A skirmish ensued, in which the Americans were driven back with loss of one man killed and several wounded, and prevented joining their friends assembled at Hampden ; the British then re-embarked without loss.

In the evening on arriving about 5 P.M. off Balar Head Cove three miles below Hampden, Colonel John had begun to land his troops when the enemy's piquets were discovered disadvantageously posted on the north side of the Cove. Major Crosdaile with the Grenadier Company 62nd, and the Rifle Company 60th under Capt. Ward, was directed to dislodge them and get possession of the ground.  By 7 o'clock this was completely effected, and before 10 P.M. the whole of the troops, reinforced by eighty marines under Captain Carter, and a detachment of seamen with one 6-pounder, a $6\frac{1}{2}$-inch howitzer, and a rocket apparatus, were on shore. They bivouacked for the night, during which it rained incessantly.

At 5 A.M. next morning, September 3, the troops were under arms and in motion : the Rifle Company forming the advance guard and the Light Company 62nd bringing up the rear, while the marines covered the left flank of the column and on the right the ships and gunboats under Captain Barrie moved up the river towards Hampden. Owing to a thick fog the ground could not be properly reconnoitered, but between 7 and 8 A.M. the skirmishers of the 60th in advance became sharply engaged with the enemy. Supported by half of the Light Company of the 29th under Captain Carter the riflemen pushed on, and after advancing a short distance discovered an American force under Brig.-General Blake, some 1400 strong, drawn up in line on a very strong and advantageous position covering Hampden. The American left rested on a high hill on which several heavy guns were mounted to command the road and river, and their right extended considerably beyond the British left ; while in advance of the centre an 18-pounder and some light field-pieces were so placed as completely to enfilade the road and a narrow bridge at the foot of a hill over which the British column was obliged to approach their position. Ward and his riflemen, cleverly passing the bridge by twos and threes, advanced and captured some of the guns ; but upon the appearance of Colonel John's main body, a heavy and well-sustained fire of grape and musketry was opened upon it, despite which the troops passed the bridge, deployed and charged up the hill to get possession of the remaining guns.

The fire of the Americans now began to slacken and they were rapidly driven at all points from their position, while Captain Carter, with the Light Company 29th, carried the hill on his left, from which the *Adams* was seen to have been set on fire, and the battery erected to protect her deserted by the enemy. While the troops thus gained complete possession of the enemy's position above the hill, Captain Barrie with the gunboats had secured that below it. Twenty guns were taken in the action, and the combined force continued the pursuit up to Bangor, which was entered without opposition. Two brass 3-pounders and 3 stands of colours were here captured, while Brig.-General Blake and other prisoners to the number of 121 surrendered and were released on parole. Their losses at Hampden were said to have been about 30 or 40 killed, wounded, or missing, and 80 prisoners. On the side of the British 2 men were killed ; 1 officer and 7 men wounded ; 1 man missing ; there were no casualties among the Riflemen. The name of Captain Ward was mentioned among others in Colonel John's dispatch, and so was that of Lieutenant du Châtelet acting as Brigade Major.

4

While Lieut.-Colonel John was thus successfully engaged up the river, Lieut.-General Sherbroke, on hearing that a large force of militia from the neighbouring townships had assembled on the road to Blue Hill about four miles from Custine, sent out a strong patrol before daybreak on September 3, which ascertained that the militia had indeed assembled there when alarm guns were fired from the fort at Custine on the first appearance of the British expedition, but that the main body had since dispersed and gone home. With a view to gaining information of Colonel John's movements, General Sherbroke accompanied by Rear-Admiral Griffiths marched at 2 A.M. on the 5th with about 700 men and two light field-pieces to Buckston on the left bank of the Penobscot, about eighteen miles above Custine.

There was reason to believe that the light guns carried

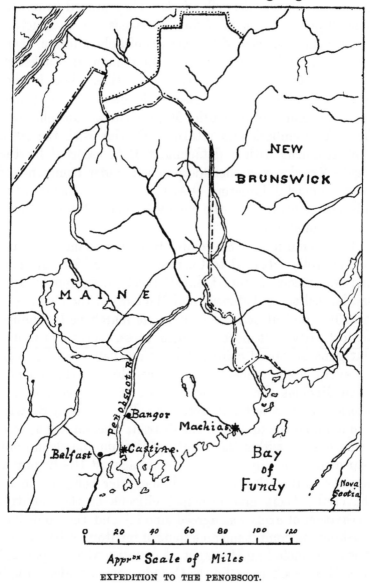

EXPEDITION TO THE PENOBSCOT.

away by the garrison of the fort at Custine were secreted

hard by, and a threat to destroy the town unless they were given up had the desired effect. They proved to be two brass 3-pounders on travelling carriages. A satisfactory report being received of the success of John's forces, there was no necessity for Sherbroke to remain at Buckston, and on the 6th he returned to Custine. The further occupation of Belfast was also no longer necessary ; and on the evening of the 6th, Major-General Gosselin was directed to embark and return to Custine with the 29th. That regiment with a detachment R.A. escorted by the frigate *Tenedos* was then despatched under command of Lieut.-Colonel Pilkington, D.A.G., to occupy Machias, the only post remaining to the enemy between Penobscot and Passamaquoddy Bay.

On the morning of September 9 the troops under Lieut.-Colonel John returned to Custine. On the 12th information was received from Lieut.-Colonel Pilkington that he had landed some distance from Machias on the evening of the 10th, and after a difficult night march had occupied the fort early next morning without loss, capturing twenty-four guns of varied calibre, more than half of which had, however, been rendered unserviceable by the enemy. From Machias the force was about to advance into the interior when a letter arrived from Brig.-General Brewer, commanding the militia of the district, wherein he engaged that the force within the county of Washington should not bear arms or in any way serve again against His Britannic Majesty during the war. A similar offer being made by the civil officers and principal citizens of the country, a cessation of arms was agreed upon ; and the county of Washington, comprising an extent of territory of 100 square miles and including the tract of country which separated New Brunswick from Lower Canada passed under British dominion, and so remained, apparently not to the dissatisfaction of its inhabitants, until the conclusion of peace, a few months later.

5

In 1814 Colonel William Williams, late of our 5th Battalion, who had exchanged with Colonel Keane into the 13th Regiment, was at St. John's, New Brunswick, in command of the advanced posts on the Richelieu River when he gained the public approbation of the Commander-in-Chief in Canada ' for the judgment, zeal, and unwearied assiduity in his arrangement ' of their defence. Williams took part in the subsequent advance by Lake Champlain. In 1813 he had been awarded the Portuguese Order of the Tower and Sword, and in 1815 he received the K.C.B. In 1830 he was promoted to the rank of Major-General, but died two years later at the comparatively early age of 56. As a commander of light troops in the field, Sir William Williams has had few equals.

General de Rothenburg also was doing good work. Before the end of 1813 he had been entrusted with the command in the upper provinces, and had been sworn in as President of Upper Canada. In 1814 he became Commander of the left division of the army in Canada, was present as second in command to Sir G. Prevost in the advance to Plattsburg, and at the peace received knighthood of the Hanoverian Guelphic Order (K.C.H.).

Almost simultaneously with our success on the Penobscot, Sir George Prevost, whose available force had been made up to 16,000 men by the arrival of a portion of Wellington's victorious army, began an invasion of the province of New York by way of the Little Lakes. He was assisted by a naval flotilla on Lake Champlain ; but when opposite Plattsburg, the flotilla was defeated by the

American vessels ; and Prevost, deeming its aid an indispensable factor in the campaign, decided to abandon the whole enterprise.

This decision, whether right or wrong, gave rise to much hostile criticism ; and Prevost, resigning his appointment, sailed for England where he demanded a Court Martial. But before it could be assembled Sir George died, and the rights of the case were never ascertained. It was, however, felt that he deserved much credit for his generally successful defence of the Canadian frontier against superior numbers, during upwards of two years ; and after his decease a warm expression of approbation was issued by the Prince Regent, who also conferred honours on the General's family. A mural tablet to his memory may be seen on the south wall of the nave in Winchester Cathedral.

At the sanguinary repulse of the British force under Sir Edward Pakenham which took place at New Orleans in January 1815, two of three brigadiers in Samuel Gibbs and John Keane had belonged to the 60th. The former was killed, the latter badly wounded. A life-size statue of Gibbs is to be found in the south transept of St. Paul's Cathedral. He is shewn as sharing a cloak with Pakenham who was also killed. It seems probable that the likeness of both is good.

At least two other monuments to old 60th officers are to be found in the Cathedral : one for Moore, the other for Craufurd. They serve chiefly to demonstrate the depth of degradation to which so-called art can be prostituted.

The termination of the war in Europe not unnaturally led to negotiations between Great Britain and the United States, and on December 24, 1814, a treaty of peace was signed at Ghent. Its terms showed that although the contest had not been fought to a finish and the Americans were flushed with their success at Plattsburg and New Orleans, the substantial advantage rested with Great Britain, whose frontier not only remained intact but was rectified to her advantage, while her claims for right of search to apprehend deserters was tacitly admitted.

The United States had suffered severely in the war. Their foreign trade had fallen from £50,000,000 to less than £4,500,000 ; their marine casualties amounted to 1400 ships of war and merchantmen ; two-thirds of their mercantile classes became bankrupt ; and so great was the discontent in the New England states that a considerable section of the people contemplated secession from the Union and reincorporation in the British Empire.

It is much to be regretted that the war took place at all ; a little mutual forbearance should have prevented its outbreak ; and that it may be the last ever to sever the relations which should exist between the two countries must be the fervent hope of the Anglo-Saxon race on both sides of the Atlantic.

## 6

The Monthly State of the 5th Battalion, dated Cork, July 25, 1814, showed it as already mentioned reduced to only 16 officers and 296 of other ranks. No further States are to be found until 1816 ; but discharges were no doubt granted liberally ; and when the spring of 1815 brought Napoleon back to his throne the 5th Battalion, consisting of only 189 rifles, was obviously too weak to be included in the army under the Duke of Wellington's command which gained the crowning victory of Waterloo. A few officers of the Regiment were, however, present on the Staff, at the battle, viz.—

Lieut.-Colonel Baron E. O. Tripp, A.D.C. to Prince of Orange, commanding 1st Army Corps.

Captain Lord John Somerset, A.D.C. to the Prince of Orange.

Captain Horace Seymour, A.D.C. to Lieut.-General the Earl of Uxbridge, G.C.B., commanding the Cavalry Corps.

Captain R. Brunton, D.A.A.G.

J. L. de Ingersleben, with 2nd Light Battalion K.G.L. to which he had been recently transferred.

John Brunnan, who twenty years later became Quartermaster.

Two of the Colonels Commandant, Lieut.-General Sir Henry Clinton and Major-General Sir James Kempt, were also present in command of Divisions ; the latter succeeding Sir Thomas Picton, who was killed in the battle.

Major the Honourable E. S. Erskine, D.A.A.G., was present at Quatre Bras, but was wounded and taken prisoner next day during the retirement of the army to the position at Waterloo. He was taken before Napoleon, and found to his surprise that the Emperor was intimately acquainted with all the details of Wellington's army. Erskine's arm was amputated by a French surgeon.

Of Captain (afterwards Sir Horace) Seymour, Captain Gronow in his Reminiscences says :

' Sir John Elley, colonel of the Blues, and Horace Seymour, who was on the Staff, two of the most powerful men in the army, performed deeds worthy of the Paladins of the olden time. Horse and man went down before them as they swept onward in their headlong career, and neither helmet nor cuirass could stand against swords wielded by such strong arms.'

During the ' Hundred Days '—as Napoleon's short period of restoration was termed—Colonel Galiffe was in Switzerland, his native country, and commanded the 2nd Federal Contingent. Its services were, however, not required in the field.

Waterloo practically ended the great war with France,

which except for two short intervals had lasted since 1793.
Some of us may think that in view of more recent events
Britain might have done better to co-operate with Napoleon
rather than with Prussia and Austria.


## 7

During the war, or at its conclusion, the following
distinctions were conferred on officers of the Regiment other
than in the 5th Battalion :—

*Brevet of Colonel :* Edward Codd, G. Mackie, P. Anderson.
*Brevet of Lieut.-Colonel :* W. Marlton, J. Grant, C. Bouverie,
W. Batteley, J. Jordan, E. O. Tripp.
*Brevet of Major :* J. W. Aldred, L. de Mangon, P. Mauriage,
C. Vigny, H. Rennells, A. Liebertwood, C. Gibbons, C. Mackenzie,
W. H. Sewell, W. H. Savill, W. Friess, Hon. E. S. E. Erskine, Lord
J. Somerset, E. Purdon, J. Macmahon.
*C.B.* and *War Medal :* Colonels P. Anderson, G. Mackie,
G. Gordon.

## CHAPTER XII

WE left the 1st Battalion (Vol. I. p. 283) at Grahamstown in the Cape Colony, whither it had gone in 1811. It was still in South Africa when the peace came, and remained there for some years longer, but of its doings no record remains.

The 2nd Battalion went to Barbados in 1810 and remained there till July 1817, when it moved to Halifax, Nova Scotia ; and went on thence in September to Quebec.

The 3rd Battalion was in 1815 distributed between Guadeloupe, Martinique, Dominica, and St. Lucia. In the following year we find it at Halifax, N.S., Annapolis, and Prince Edward's Island ; in 1817 the battalion was concentrated at the first-named place.

The peace found the 4th Battalion at Dominica whence in 1816 it proceeded to Demerara.

In 1817, the battle honour ' Martinique ' was granted to the Regiment by W.O. letter dated September 17, in commemoration of the services of the 3rd Battalion under Lieut.-Colonel Mackie at the capture of that island in 1809.

### 2

The overthrow of the Emperor Napoleon, and the prospect of a long cessation from war not unnaturally led to an impolitic demand in England for the reduction of the land forces. It was decided that regiments of the line should be reduced to one battalion and the two Rifle Regiments to two apiece. The 95th, whose title was now changed to that of the Rifle Brigade, having had only three battalions, suffered the loss of one. Our Regiment was doomed to be deprived of six, and the two selected for survival were the 2nd and 3rd.

In November 1815 a party of sixty men of the 8th Battalion, under Captain C. Leslie, was embarked at Cadiz and brought to England, with a view to being sent home to Germany. Their green rifle uniform coupled with the fact that none of the detachment could speak a word of English, caused great astonishment as they marched through London *en route* to their port of embarkation.

On May 20, 1816, the Parade State of the 8th Battalion showed a strength of 1 major, 7 captains, 17 subalterns, 5 staff, 36 sergeants, 16 buglers, and 482 riflemen, present. On the 31st about 334 N.C.Os. and men were transferred to the 5th Battalion at Gibraltar, and the remainder were taken to Portsmouth where the battalion was disbanded.

3

On the signature of peace between the United States and Great Britain, the 7th Battalion quitted the State of Maine and returned to Halifax, Nova Scotia. Here it remained until its disbandment in June 1817.

Its last Parade State, dated the 25th of that month, gives the following figures :—

One major, 4 captains, 25 subalterns, 6 staff, 45 sergeants, 20 buglers, 620 men.

A Memorandum dated June 24 gives the strength and distribution of the battalion as follows :—

|  | Sergeants. | Buglers. | Other ranks. |
|---|---|---|---|
| In Nova Scotia . . . . | 46 | 19 | 617 |
| In Europe . . . . . | – | – | 6 |
| Total . . | 46 | 19 | 623 |
| Enlisted for the 7th Battalion exclusively. . . . . | 26 | 2 | 150 |
| Invalids . . . . . | 2 | 2 | 58 |
| Casualties . . . . . | 2 | – | 10 |
| In Europe . . . . . | – | – | 6 |
| Total . . | 30 | 4 | 224 |
| To be drafted into the 2nd Battalion . . . . . | 8 | 8 | 200 |
| To be drafted into the 3rd Battalion . . . . . | 8 | 7 | 199 |

4

At the end of October 1817 the 6th Battalion, which had spent practically the whole of its existence at Jamaica, embarked for England under command of Major Aldred, and landed at Portsmouth, where in February 1818 it was disbanded and the men sent to the 3rd Battalion at Halifax, N.S. Its State, dated the 25th of that month, shows, 2 majors, 2 captains, 7 subalterns, 1 staff, 26 sergeants, 22 drummers, 405 privates.

At the time of its disbandment nine-tenths of the men were foreigners, but almost all the officers were British.

5

Leaving for the moment the fortunes of the 5th Battalion the one next to claim attention is the 1st, which after being quartered for some years at Simon's Town, quitted the Cape of Good Hope early in January 1819, and was brought to Portsmouth by the somewhat circuitous route of Demerara—commonly spelt in those days, ' Demerary.'

Up to the end of the war the battalion had been maintained on a high establishment, but since the peace its numbers had gradually dwindled and its Distribution State, dated April 27, 1819, and signed by Lieut.-Colonel Alexander Andrews, gives the following figures :—

|  |  | Sergeants. | Corporals. | Buglers. | Privates. |
|---|---|---|---|---|---|
| To be discharged | At home | 3 | 5 | – | 64 |
|  | On the Continent | 12 | 7 | 4 | 56 |
|  | Foreign in England | 1 | – | – |  |
|  |  | 16 | 12 | 4 | 120 |
| To be retained | To the 2nd Battalion | 6 | 7 | 2 | 64 |
|  | To the 3rd Battalion | 15 | 12 | 10 | 212 |
|  | Total | 21 | 19 | 12 | 276 |

A few days afterwards the 1st—Bouquet's—Battalion was disbanded, its numeral being assumed by the 2nd Battalion, while the 3rd became the 2nd.

During the sixty years of its existence Bouquet's battalion had been engaged in only two really desperate contests—the one at Ticonderoga, the other at Bushey Run. But in Forbes' toilsome march against Fort Duquesne, the pestiferous swamps of Niacaragua and elsewhere, the battalion had shown its mettle.[1]

---

[1] The Colours would seem to have been retained by Lieut.-Colonel Andrews, the C.O., and in about 1882 were given by one of his descendants to Lieut.-Colonel Walter Holbech of our Regiment. Some years after Colonel Holbech's death, his widow most kindly handed them to the writer, who with her approval retained the King's Colour on behalf of the Regiment, and at the request of the Vicar and Church-wardens of Trinity Church, New York, sent the Regimental Colour to be kept in their charge ; that church having been the place of worship for the Garrison, in the days preceding the War of Independence when the Royal American Regiment was quartered in New York.

6

During the winter of 1813–14 the headquarters of the 4th Battalion, under command of Lieut.-Colonel J. Lomax, were sent from Dominica to Surinam and quartered at Fort Zelandia. At the general peace of 1815 Surinam was restored to the Dutch, and now forms part of Dutch Guiana. In June 1816 the State of the battalion, consisting of 23 officers and 888 men, was signed at Barbados, but if it actually went there it very quickly returned to the South American continent, for in January 1817 we find the battalion at Demerara, in British Guiana, and commanded by Lieut.-Colonel W. Woodgate late of the 5th Battalion.

Here with slowly diminishing numbers the battalion remained until June 1819 when—once more under command of Colonel Lomax it sailed for England, reaching Portsmouth on July 16. On the 24th the battalion, the third edition of which, as will be remembered, had been raised in 1787, was disbanded.

7

In 1814 the 2nd Battalion, under Colonel E. Codd, was at ' Demerary ' with a strength of 22 officers and 1000 other ranks. At the peace its numbers rapidly diminished, and at Barbados in 1817 had fallen to 19 officers and 381 other ranks.

But the long and dreary sojourn of the regiment in the West Indies or Guiana was now about to terminate. The regiment had been raised for service in North America. In 1765 Parliament had decided that the West Indian islands were included in North America. Later on it decided that the South American continent was also in North America ! Whether the British M.Ps. were now sufficiently educated to realise their error of elementary geography is doubtful, but at all events the service of the 60th in those pestilential regions now came to an end.

In June the battalion embarked for Quebec, where it arrived in July with only 269 men. Reinforced by others from disbanded battalions, it had in January 1818 regained a strength of 440 men, but soon fell away again.

Just at this time, however, came official recognition of the status which the *raison d'être* of its existence had claimed since the day in which the regiment was originally raised : viz. that of light infantry in the widest sense of the term ; for on July 16, 1818, the Prince Regent gave his approval to and signed the following Memorandum : ' That in consequence of the 5th or Rifle Battalion being about to be disbanded, the 2nd Battalion of that Regiment be clothed, equipped, and trained as a Rifle Corps.'

8

The 3rd Battalion under Colonel Mackie was in January 1814 quartered at St. Pierre, Martinique—happily without experiencing a volcanic eruption from Mont Pelée such as that which destroyed the town nearly a century later. The battalion showed a strength of 26 officers and 1053 other ranks present. Next year we find it at Morne Fortuné, St. Lucia ; whence it moved to Halifax, N.S., where in June 1816, still under command of Colonel Mackie, it mustered 32 officers and 774 other ranks. At Halifax the battalion remained for some years ; but by this time the misery of thirty years in the West Indian islands had produced its inevitable result. To an officer who joined it from the 8th Battalion during the autumn of this year the majority of the subalterns seemed the concentration of all the worst elements from the other battalions, while

II.                                                                    X the men were demoralised by habits of drink inseparable from service in the West Indies. 'There were several officers in arrest for every crime in the Calendar of Newgate, yet they dared their C.O., Colonel B., to bring them to trial, alleging that they would produce charges of embezzlement against himself!' From force of circumstances a new era was happily at hand. Mackie resumed command and B. departed.

### 9

By the authority of W.O. letter dated April 6, 1815, the 5th Battalion was granted the war honour 'Peninsula,' the precursor of many to follow at no distant date.

From Cork the Riflemen went on to Buttevant. On February 16, 1816, they quitted Ireland for Sandown in the Isle of Wight. The battalion was still numerically weak, counting only 334 rifles ; but on reinforcement by a large draft from the foreign depôt at Harwich, sailed on May 4 for Gibraltar. The battalion embarked on board three transports as follows :—

In the *Minerva* : Headquarters, consisting of 13 officers, 316 N.C.Os. and men, 21 women and 13 children. Names of the officers :—

| | |
|---|---|
| Lieut.-Colonel J. Galiffe. | Ensign K. Supple. |
| Captain J. Franchini. | „  T. Keale. |
| Lieutenant F. Muller. | „  J. Robinson. |
| „  J. Currie. | Lieutenant and Adjutant J. Kent. |
| „  H. Shewbridge. | Paymaster H. Biggs. |
| | Assistant Surgeon W. Stevenson. |

The *Isabella* sailing on May 14 conveyed Brevet Major Schöedde, Captain Stepney, Lieutenants Pictet and Hakewell, with 175 N.C.Os. and men, 16 women, and 13 children.

The *Duncombe* carried Captains McKenzie, Kelly, Wilkinson; Lieutenants O'Hehir, Evans, Serjent, Van Dyck, Cochrane, Bruce; Ensigns Bernard, Colclough, Dickson; Quartermaster Reckney.

On arrival at Gibraltar the N.C.Os. and men of the disbanded 8th Battalion were incorporated in the 5th, and the State of June 25 showed 31 officers, 1094 N.C.Os. and riflemen present. Shortly after arrival command was taken over from Brevet Lieut.-Colonel Galiffe by the senior Major, Lieut.-Colonel J. F. FitzGerald, who had been made prisoner of war in the Pyrenees while in command of a brigade of the 2nd Division.

## 10

In September the battalion received another draft, this time from England, but there was no further reason for the enlistment of foreigners and the strength gradually dwindled.

Life at Gibraltar was uneventful; but Captain Colin Campbell, afterwards Field Marshal Lord Clyde, was at the time of which we are speaking, an officer in the battalion, having been transferred thereto on the disbandment of the 7th. Forty years later he described his service in the regiment as the happiest of his life.[1] Like most periods of happiness it was short, for early in 1818 the battalion was

---

[1] Colin Campbell—'war worn Sir Colin,' as Sir Charles Napier at a later date termed him—was the first of the three officers who, after doing regimental duty in the 60th, have eventually received the Field Marshal's baton. FitzGerald, Campbell's C.O. in the 5th Battalion, was the second; and Lord Grenfell, our senior Colonel Commandant, the ideal occupant of such a position, the third. Sir John Michel, who died a Field Marshal in 1886, was also for a very short time in the 60th, but could hardly have joined the regiment for duty. Sir John in early days was captain of the Eton Eleven; and among others of our regiment who have attained that distinction, the names of Colonel F. V. Northey and Lieutenant E. O. H. Wilkinson will be recalled. Both met their death on active service in South Africa.

bidden to prepare for disbandment, and on May 10 it
embarked, still under command of Colonel FitzGerald,
for England.  On June 18 it landed at Cowes, Isle of Wight.
A week later its State showed 23 officers, 47 sergeants,
21 buglers, and 543 riflemen present.  On July 25, 1818,
the battalion was disbanded ; the list of its officers—
whether present or absent—being apparently the
following : —

*Lieut.-Colonel*
J. Stopford.

*Majors*
Brevet Lieut.-Colonel J. F. FitzGerald, C.B.
Brevet Lieut.-Colonel J. Galiffe, C.B.

*Captains*
Brevet Lieut.-Colonel W. H. Sewell ; Brevet Majors J. Schöedde,
E. Purdon, C. MacKenzie ; Captains R. Kelly, I. Franchini,
A. C. Bowers, W. Wilkinson, Colin Campbell, J. Hewitt, S. R. Stepney.

*Lieutenants*
H. Muller, A. F. Evans, H. Baron Altenstein, S. O'Hehir,
P. Eason, G. Cochrane, J. Kent, J. Currie, C. Fourneret, J. Moore,
C. Ross, S. Tresider, F. Pictet, J. Cochrane, J. Stewart, J. H. Craw-
ford, H. Goldicut, R. Hall, S. Ridd, H. Shewbridge, B. Clare.

*Ensigns*
T. Keal, J. Robinson, R. Newman, D. McKay, T. Adams (Adjutant), H. Colclough, D. Dickson (or Dixon), C. H. Couper, F. Moore.

*Paymaster*
H. Biggs.

*Quartermaster*
A. Reckney.

The N.C.Os. and riflemen were still of foreign extraction,

but of the officers all but six were British. Among those not present with the battalion was Captain John Hewett, a student at the Royal Military College, Farnham, corresponding with the Staff College of the present day ; and it is noticeable that of thirty-three students in 1817 no less than four belonged to the 60th ; the other three being Major E. FitzGerald, Captains C. M. Smith and A. Leitch. The first of these was a relation of the well-known Irish leader Lord Edward FitzGerald, whose attainder was reversed in 1819. Speaking on this occasion in the House of Lords the Duke of Wellington, referring to Major Fitz-Gerald, kindly remarked that :

' as one of the individuals on whom the act of grace would confer so incalculable a benefit had served for some time under his command, he would not let the present opportunity pass by without bearing testimony to the brave, honourable and excellent conduct of the young man in question during the time he had been acquainted with him.'

Four hundred of the N.C.Os. and men of the 5th Battalion, under command of Brevet Lieut.-Colonel J. F. Fitz-Gerald, sailed in the autumn for Quebec and were drafted into the 2nd Battalion, which was not much more than a cadre ; and even after this reinforcement could show only 634 N.C.Os. and riflemen on the half-yearly State dated January 25, 1819. The remainder of the 5th Battalion was discharged.[1] It had hitherto been the practice, on the

---

[1] Captain Colin Campbell now disappears from the regiment. As a Lieut.-Colonel he was present in the China War of 1842. The second Sikh War found him in India ; and in command of a Division he gained distinction at Chillianwallah and Gujerat. In 1851-2 he commanded an expedition—which included our 1st Battalion—against the hill tribes beyond the Indus. In 1854 Campbell commanded the Highland Brigade in the Crimea, and reached the rank of Major-General. In 1857, on the arrival of the news of the outbreak of the Indian Mutiny, he was sent to the scene of action as Commander-in-Chief. Having successfully quelled the revolt, he was raised to the peerage as Lord Clyde and promoted to the rank of Field Marshal. He died in 1863.

disbandment of a battalion to place the officers on half-pay ; but on the present occasion the following officers were also transferred to the 2nd Battalion, Brevet Lieut.-Colonel J. F. FitzGerald, Brevet Majors Schöedde and Mackenzie ; Captains Kelly and Bowers ; Lieutenants Muller, Eason, Currie, Ross, Pictet, and Stewart ; Ensigns Robinson, Newman, McKay, and Dickson ; Ensign and Adjutant Adams, Quartermaster Reckney, and Assistant Surgeon Stevenson.

In accordance with G.O. dated Quebec, September 23, 1819, the appointments of the 5th Battalion were handed over to the 2nd ; and those of the 2nd Battalion to the 3rd. Next year the variations between the uniforms of the officers in the two battalions were abolished.

On the disbandment of the 1st (Bouquet's) Battalion in the following year, its C.O. Lieut.-Colonel Andrews, also as will be remembered an old officer of the 5th Battalion, was posted vice Colonel Codd to command the 2nd Battalion, which now became the 1st. This (Haldimand's) is the 1st Battalion of the present day, representing the amalga-mation of the old 2nd and the 5th Battalions.

But to those who belonged thereto in 1819 it seemed as clear as daylight that the 5th Battalion had absorbed the 2nd and had merely been re-numbered. The composition of the 2nd—now the 1st—lent colour to this view, for which indeed official sanction may be quoted from a letter of the Adjutant-General dated Horse Guards, February 25, 1826, which speaks of 'the 1st—late 5th—Battalion.' It was actually known for several years as ' the 1st Rifle Corps, 60th Regiment,' a designation which, perhaps, conveyed the hint of a claim to seniority among the rifle battalions of the army. The customs of the 5th Battalion were adopted by the 1st ; among others that of carrying rifles with ' butts up ' when marching at ease ; a practice which although at the present common to the whole regiment was for many years confined to the 1st Battalion. For many years the members of the 1st Battalion were in the habit of designating it as the 5th.

## 11

On the disbandment of the 1st (Bouquet's) Battalion, the 3rd of course became the 2nd. It was still commanded by Colonel Mackie, C.B., who had been its Lieut.-Colonel since 1808. Its two Majors were the redoubtable John Galiffe and Henry FitzGerald who, like his two namesakes, had served in Spain during the war. The battalion remained at Halifax, N.S., until 1820, when it was ordered to Annapolis, with a half-battalion at Bermuda. In 1821 the right half-battalion returned to Halifax, whence in the year following it proceeded to Newfoundland. In 1824 the battalion was concentrated at Demerara.

On a voyage to rejoin his battalion Galiffe was shipwrecked. Writing in 1819 to congratulate him on his escape H.R.H. the Duke of Kent incidentally remarks that the Duchess had just presented him with ' a fine girl.' The baby became in due course H.M. Queen Victoria. The intimacy of the Duke with three officers of the 5th Battalion, *i.e.* De Rothenburg, Galiffe, and De Salaberry is somewhat remarkable. In the case of Galiffe it originated in the fact that the Duke had studied under his father in Switzerland.

## CHAPTER XIII

On the sweeping reduction of the Regiment to two battalions, many officers found themselves, of course, on half-pay without occupation or prospects. Among these was Baron de Rothenburg, by this time a Major-General and K.C.H. In 1813 he had been made Colonel Commandant of De Roll's, a foreign regiment in the British pay. In 1816 De Roll's regiment was disbanded, and the General lost in consequence an income of about £1200 a year. Although long naturalised, De Rothenburg's foreign birth seems to have debarred him from being Colonel of a regiment in the British service, or holding command of a fortress. Despite his fine services he was placed on half-pay with the wretched pittance of nine shillings and sixpence per diem. This was afterwards augmented to sixteen shillings, upon which—so far as military emolument is concerned—he subsisted in his declining years until his death at Portsmouth in 1832, aged 75. Nevertheless we find him in 1819 writing a cheery letter to De Salaberry, and expressing his perfect readiness to lend his sword to Canada in the event of further hostilities on the part of the United States.

A tablet in honour of De Rothenburg, who had been promoted to the rank of Lieut.-General in 1819, may still be seen in the Garrison Chapel at Southsea; and although allowed by the country of his adoption almost to starve, yet the Army even to the present day reaps the benefits of his system and reforms; and the bugle-calls of the

barrack square, culled from his treatise on the training of
light troops, should perpetuate his memory.[1]

## 2

Other Rifle officers still in the full vigour of youth
sought adventure where it might be found.   At the time of
which we are speaking, the Spanish colonies of vast extent
in Central and South America had repudiated the rule of
the mother country and were seeking political independence.
Their cause excited interest in England, and among others
who volunteered their services was the celebrated Admiral
Lord Dundonald, who went out in 1818 and gained brilliant
success in command of the Chilian Navy.

The unfortunate adventures of some Rifle officers which
occurred in 1819 during this struggle are narrated as
follows by Colonel Charles Leslie of our Regiment in his
' Military Journal ' :—

' Not content with the deeds of valour performed in the Peninsula,
many officers who had distinguished themselves on various occasions
by gallant conduct in the field during that arduous struggle, disregarding the ease and indolence of retired life, availed themselves
of the advantages held out in the New World by the South American
States, then in arms to assert their rights and free themselves from
the oppressions and exactions to which they had so long been
subjected while under the dominion of Spain.

' Amongst the many who volunteered their services to aid these
oppressed people in the recovery of their freedom were three who
had served in the corps to which I belonged : Captain Rafler,
Lieutenant Acton, and Lieutenant Ross of the 50th Rifles.   The
latter, who possessed a very determined courage with cool judgment,
had by his bravery particularly distinguished himself at the sieges
of Ciudad Rodrigo and Badajoz ; and it was principally owing to
his talents on the occasion I am now alluding to that the expedition

---

[1] It has been stated that a portrait of De Rothenburg exists in Canada.   The
writer has been unable to verify the assertion.

to which these three officers were attached succeeded in the capture (afterwards reversed) of Porto Bello in the States of Colombia, on the Isthmus of Panama, although this operation was carried on under the nominal command of Sir Gregor M'Gregor, under the high-sounding title of Cacique of Poyais. The said Cacique, deeming this a place of some security, established here his headquarters, dubbed himself Governor, and in that capacity took possession of an excellent house, lived in luxurious indolence, and had a sloop of war in the offing at his command. His force, although principally consisting of British, being all new levies, required much drilling and exercise to make them effective, their numbers not being sufficient to admit of a proper distribution for defence and instruction separately. The advanced piquets were posted every night in the principal roads, at some little distance from the town ; and as no movement of the enemy to attack the place was apprehended, they were withdrawn every morning after sunrise for the purpose of being drilled. They went quietly on in this manner until lulled into a fatal security.

'One beautiful, balmy spring morning the luxurious shrubs ornamenting the neighbouring heights were yielding their fragrance to the gentle breeze, which playfully waved their tops to and fro, while the clear waters of the adjacent river, shaded by the verdant foliage of the acacia and the weeping willow, seemed by their delightful coolness to invite those who had been on the midnight watch to refresh their wearied limbs. The piquets had been withdrawn as usual ; the men had piled arms, and, along with those off duty, were at drill in squads in front of the Government House in the Grand Placa, when, before a sound was heard or any alarm given, many of the best and bravest were by a murderous volley laid low. This most unexpected vicissitude staggered the whole for a moment, but they soon recovered themselves ; the former piquet siezed their arms, and, being joined by the Governor's guard, a stand was made to enable the others to retreat to a fort at no great distance. This was effected after considerable loss. During this disastrous attack the vainglorious Cacique of Poyais, instead of gallantly heading his troops and sharing their fortunes, on the first alarm jumped out of bed, leaped out of a back window, plunged into the sea, swam on board his ship of war, got under way, and thus saved himself by an ignominious flight, to the horror and mortification of his brave people in the fort, who beheld him making his escape to sea in a vessel which was their only effectual means of defence. They were thus abandoned to the mercy of a relentless enemy ; whereas, if he had brought the broadside of the ship to bear on the town, so as to give a cross-fire to that of the fort, the enemy would probably have been driven out again without much difficulty.

'The fort, which had been constructed more to protect the harbour and lower part of the town than to form any defence against an attack from the interior of the country, was surrounded by some houses, which the Spanish army got possession of, and picked the men off who were working the guns. Ross cleverly got some guns turned round and dislodged them. But the place not having been provisioned, any further defence was only maintained with a view of getting terms. They, therefore, offered to capitulate, on being provided with shipping to quit the country and return to England. The Spaniards readily acceded to the proposition, and further promised to treat them with respect, but the instant the men laid down their arms, upon marching out of the fort, they were surrounded by an armed force, the officers separated from the men, and all divided into small parties, and marched in different directions far up the country.

'The officers were treated in the most degrading manner, being obliged to perform menial offices, and literally to become hewers of wood and carriers of water. Under a pretence that it had been discovered that they were forming a plan to make their escape, their legs were put into stocks every night, so that they could get but little rest after their daily labours. Ross complained of this to the Spanish commander, stating that such usage was ungenerous after the service he had performed for them in Spain. But all the redress he got was a grin, with the repetition of "tan peor," " so much the worse," to every particular action he mentioned. It was thus implied that if he had fought for the King of Spain then, he had no business to fight against him now.

'One morning, before they were relieved from the stocks, a corporal and a file of men marched up and began loading their firelocks in front of the officers. On being asked what they meant by this, they pointed to Captain Rafler, and said to him "We are going to shoot you ! " He earnestly insisted that there must be some mistake. They replied that there was no mistake, adding that he was the person who was plotting to escape, and that, in order to frighten the others, he must be made an example of. They accordingly released him, and made him proceed in front of them to some short distance, when they deliberately fired and shot him dead.

'The fate of poor Acton was particularly melancholy, and so, in its proportion, was that of his favourite poodle dog, Leo. This fine animal had been for years his most attached and faithful companion. Leo had been sheared and trimmed up into the shape and appearance of a most ferocious lion ; but this was only in outward resemblance, for Leo had a kind and playful spirit. He possessed wonderful instinct and sagacity, and performed many amusing pranks and tricks, such as fetching anything from his master's room which he desired him to bring. His forte, however, was in aquatic displays, particularly in diving or leaping overboard, and then scrambling up again by a rope thrown over the side. And Leo was the constant attendant on his master on all occasions of bathing. Acton had been an out-piquet on the night preceding the surprise, and, on returning, had remained to bathe. While enjoying this luxury after the fatigue of the night, being helpless and defenceless the enemy came rapidly on—escape was impossible—and these cruel agents of oppression bayoneted him in cold blood. His faithful Leo made a vigorous attack on the barbarous miscreants, and likewise fell covered with wounds, in the vain endeavour to defend his master.

'The successful surprise of this place by the Spaniards was owing to information of the manner in which the liberating forces carried on the duty, given by spies and those within the town friendly to the old *régime*. The Spaniards had with the utmost secrecy moved a large force on the fortress, and had concealed them behind the heights, covered with brushwood, as above described ; and it is supposed that on a signal from some traitor to his country within, and at the fitting moment, they dashed on and obtained their success.

'Lieutenant Ross continued for fifteen months under the control of his hard taskmasters ; when, reduced by ill-health and despair, he was fortunately relieved in exchange for some Spanish officers, on whom retaliation was about to be made.'

3

But to return to the Regiment. In 1821 the 1st Battalion went from Quebec to Montreal—a repetition of its march in 1760—and Colonel FitzGerald was appointed

Commandant of the District.   In Montreal the battalion
was joined by 200 German recruits.

In September the following battle honours were conferred on the Regiment :  ' Roleia,' ' Vimiera,' ' Talavera,'
' Fuentes D'Onor,' ' Ciudad Rodrigo,' ' Badajos,' ' Salamanca,' ' Vittoria,' ' Nivelle,' ' Orthes,' ' Toulouse.'   It
will be noticed that many of these names were mis-spelt
by the scribes at the Horse Guards.   Thus ' Roliça,' is
meant by ' Roleia ' ;  ' Vimieiro,' by ' Vimiera' ;  ' Fuentes
de Oñoro,' by ' Fuentes D'Onor ' ;  etc.


# 4

In November 1822 a letter from the Adjutant-General
directed Commanding Officers to revise and complete the
records of their regiments.   So far as the 60th was concerned the result was unsatisfactory, all records having
evidently been lost in both battalions.   In the 1st Battalion their compilation was entrusted to Colonel Charles
Leslie, already mentioned as the author of several works,
and some attempt was made to reconstruct past history ;
and the upshot is shown in its ' History of Services,' which
contain, however, little beyond the memoirs of Patrick
Murray noticed in Vol. I. of these annals.   Documents are,
however, extant showing that orders were given to seek
information on the subject from Baron de Rothenburg,
Sir William Williams, and other heroes of the 5th Battalion.
It may therefore be taken for granted that the information
given in the ' History of Services of the 1st Battalion '
regarding the formation of the 5th was derived from officers
still alive who had been posted to the battalion on its
earliest existence, and is correct.   That this information
is not more compendious may very probably be due
to the rather sudden death of Colonel Andrews; and

to the fact that during the following three months the battalion was either on the march or had embarked for England.

5

In 1823 the 1st Battalion proceeded to Kingston, Ontario, where it received a draft of 150 Irish recruits enlisted for general service.

At Kingston Colonel Andrews died after a short illness and was succeeded by Colonel J. F. FitzGerald, who wishing, however, to serve in India, exchanged early in 1824 with Lieut.-Colonel Thomas Bunbury of the 20th. Thus ended FitzGerald's memorable connections with the Rifles; but his life was prolonged for upwards of half a century. In 1877 he died at Pau in the south of France, and was escorted to his grave by French soldiers with all the honours due to a British Field Marshal.

6

The anomaly of the 60th being still, from the official point of view, a colonial regiment had become by this time so absurd that early in 1824 H.R.H. the Duke of York, Commander-in-Chief of the Army, as well as Colonel-in-Chief of the 60th, realising the fact that since foreigners could not legally serve in Great Britain in time of peace, the British officers belonging to his own regiment were obliged to spend the greater part of their lives abroad, ordered the Regiment home that it might be recruited by British subjects and placed on a footing similar to that of other units of the army. In June 1824 the 1st Battalion accordingly left Kingston for Quebec. Here it parted with all foreigners—about 150 in number. They had the option of transfer to the 2nd Battalion or, if officers, either to sell their commissions or accept half-pay; if of other rank, to take their discharge in Canada when every facility was afforded them of becoming settlers. The 'History of Services of the 1st Battalion' observes that:

' it ought to be recorded in justice to the German soldiers that they were a most quiet and orderly set of men, attentive to their duty and always particular in having their necessaries complete and everything in good order so that Captains of Companies had very little or no trouble with them.'

In name no less than in fact the Regiment now severed its connection with America; for a letter from the Adjutant-General dated Horse Guards, June 19, 1824, informed the Commanding Officer that:

' His Majesty has been pleased to direct that the 60th Regiment shall cease to bear the appellation of " The Royal American Regiment " and be termed the " 60th Regiment, The Duke of York's Own Rifle Corps and Light Infantry." '

Within little more than a month both battalions were placed on the same footing as riflemen. A War Office letter dated July 24, 1824, stated that:

' His Majesty has been pleased to approve of the 2nd Battalion of the 60th Regiment of Foot being equipped and trained as a Rifle Corps, and has also been pleased to direct, as both the battalions of that Corps are to be Rifle Battalions, the Ensigns Rank in future be made Second Lieutenants.'

## 7

On August 19 the 1st Battalion—under the command of Brevet Lieut.-Colonel Rumpler, the senior major—which had embarked at Quebec in the month previous, landed at Chatham, whither its depôt preceded it from the Isle of Wight. Colonel Bunbury joined the battalion and assumed command.

Colonel Bunbury was a man of what we should nowadays call 'the old school,' but in many respects a good commanding officer and one who knew his own mind. He was a friend of the Duke of York and considered like him in appearance, so much so indeed that he was once assaulted in the street by a man who mistook him for H.R.H. The late Captain G. H. Courtenay, who died in 1914 at the age of 98, served under Colonel Bunbury at Gibraltar in 1832, and told several anecdotes of him to the writer. An inspecting general having asked whether there was any gambling among the officers in the battalion, Colonel Bunbury's reply was ' No, for I have won all their money.' On another occasion the battalion being on parade close to the lines of San Roque close to the Spanish border, a man quitted the ranks and made his way over the frontier. The Colonel without the slightest hesitation galloped after him, regardless of violation of Spanish territory, and seized the man by the collar and brought him back to the British lines.

H.R.H. the Duke of York, Commander-in-Chief of the Army and Colonel-in-Chief of our Regiment, lost no time in coming down to inspect the battalion. His visit is thus described in the ' History of Services of the 1st Battalion ' :

' His Royal Highness came to Chatham to review the troops in garrison there, and to inspect the 1st Rifle Corps 60th Regiment, on which occasion he did them the honour to appear in the uniform of his own Rifle Corps ; and during the time he remained the 1st Rifle Corps 60th Regiment furnished the Guard of Honour on all occasions and was minutely inspected by H.R.H. on the 1st October.'

COLONEL J. W. DES BARRES.
60th Royal American Regiment, 1756–1803.

On the 5th the battalion marched to Canterbury, but returned to Chatham a few days afterwards.


### 8

In the autumn of 1824 official permission was given to the Regiment to resume the motto ' Celer et Audax ' which had evidently fallen into disuse.   The terms of the letter were as follows : —

'Horse Guards, 11th October, 1824.

' I have the honour to acquaint you by direction of the Commander-in-Chief that His Majesty has been pleased to permit the 60th Regiment, The Duke of York's Own Rifle Corps, to resume the motto " Celer et Audax " which was formerly borne by the Regiment in commemoration of its distinguished bravery while employed with the British Army in North America under Major-General Wolfe in the year 1759.   I have, etc. . . .

'JOHN MACDONALD, D.A.G.

' To the officers commanding 1st and 2nd Battalions 60th Regt.' [1]

In February 1825 the following additional battle honours were granted to the Regiment :—

    Albuhera (Albuera).

    Pyrenees.

    Nive.

' In consequence '—to quote the wording of the A.G.'s letter—' of the distinguished conduct of the 1st (late 5th) Battalion of that regiment ' at the three battles mentioned.


### 9

On April 15 the 1st Battalion left Chatham for Weedon by march route.   On marching through London the

---

[1] This year died Colonel J. F. W. Des Barres, the last original member of the 60th.  In Canada his fame as an Engineer is still cherished.

following day it was again inspected by the Commander-in-Chief in Hyde Park, who took the opportunity of complimenting Colonel Bunbury by desiring him to wear his button.[1] Weedon was reached on the 23rd, but in consequence of disturbances in the Isle of Man three companies were ordered to Stockport.

In December the whole battalion proceeded to Manchester, which in the following spring was the scene of great distress. The 'History of Services' remarks that :

'The people driven to desperation from want of food and employment began to hold riotous meetings which ended in great disturbance. They attacked and destroyed several of the principal power loom manufactures in different parts of the country, so that it was necessary to call in the aid of the troops, in consequence of which the following detachments were immediately marched to the undermentioned places where they were employed in protecting and defending the mills many of which had been fortified by the owners :—

'Blackburn, Bury, Middleton, Hyde, Ashton-under-Lyne.'

During these disturbances the several detachments into which the battalion was broken up were credited with having shown great steadiness under circumstances of unusual difficulty.

Unhappily the unrest did not end without bloodshed. In April Lieutenant D. Fitzgerald, a sergeant and 13 riflemen were summoned from Ramsbottom to Chadderton [2] where they were called upon for the protection of the power-loom mill belonging to a Mr. Aitken, which on the 26th was attacked by an infuriated mob. The troops present—which included the Queen's Bays—were pelted with stones and (according to the *Manchester Courier* of the 29th)

---

[1] This was a gilt button with the Duke's coronet above the letters D.Y.O.R. in old English characters. It was worn in plain clothes.

[2] Spelt in the 'History of Services of the 1st Battalion' 'Chattertown.'

a number were seriously hurt.  The magistrates present
called on the officer commanding the troops to disperse
the rioters by force.  The Riflemen were reluctantly compelled to open fire.  Four of the mob were killed, and so
was an unfortunate woman who was waiting for the
Manchester coach.  The crowd now began to disperse,
but another shot fired killed a man named Watterton
or Whatacre, and a coroner's jury subsequently returned
a verdict of wilful murder against the unknown rifleman who had fired the shot.  Colonel Bunbury wrote
a letter of indignant protest, which was supported by
the evidence of the officer commanding the Queen's
Bays.

During the course of this year a handsome silver snuff-
box, surmounted by a Rifle bugle in miniature, was presented
to Colonel Bunbury and the officers of the battalion by
Lieutenant H. T. Evans.  In 1877 when the battalion was
quartered at the Cambridge Barracks, Southsea, a civilian
entered the barrack square and watched the battalion on
parade.  He afterwards entered into conversation with
some of the officers, apologising for his presence but saying
that he had known the battalion well in former days.
He was accordingly invited to dinner and while dining
recalled the fact that on the last occasion he had dined at
the Officers' Mess a snuff-box was presented to the officers.
The snuff-box was on the table and proved to be that
presented more than half a century previously.

.

## 10

In June the battalion proceeded to Plymouth.  At
this period affairs in Spain and Portugal were greatly

disturbed, and as Portugal appeared to be in imminent danger of invasion from her more powerful neighbours the British Government made up its mind to send an armed force to the assistance of the old ally of England.

A brigade consisting of the 4th, 10th, 23rd, and our 1st Battalion under Major-General Sir Edward Blakeney, K.C.B., formed the advance guard of the British contingent. The 1st Rifle Corps accordingly embarked at Plymouth on December 18, comprising 1 lieut.-colonel, 2 majors, 6 captains, 12 subalterns, 5 staff, 28 sergeants, 10 buglers, 520 rank and file. After receiving a highly complimentary message of farewell from the General commanding the Western District, Major-General Sir John Cameron, the battalion sailed and reached the mouth of the Tagus on the 27th. On New Year's Day 1827 it disembarked at Lisbon. The brigade moved by forced marches to Coimbra, where it formed the advance guard of the army, which was commanded by Lieut.-General Sir William Clinton, who, during the latter part of the Peninsular War, had commanded the British contingent in the eastern provinces of Spain.

During this occupation of Portugal news arrived of the death of our Colonel-in-Chief H.R.H. the Duke of York. His demise was a loss to our Regiment in which as Colonel-in-Chief during the last thirty years he had always taken the keenest interest. The Duke of York was succeeded by his brother the Duke of Cambridge, a gallant old man who had fought in the campaign of 1793–4 and a most estimable prince but not entirely free from a suspicion of eccentricity. He had the engaging habit of speaking his thoughts aloud, particularly during Divine Service. It is stated of him, that when the officiating clergyman said, ' Let us pray,' the Duke would reply, ' By all means.' On one occasion at a funeral service on hearing the words, ' We brought nothing into this world, and it is certain we can carry nothing out,' he responded with the observation, ' Very true, very true ; too many demands on our purse for that.'

In April 1828 the executors of the Duke of York kindly presented to the 1st Battalion the sword, sabre-tasch and pouch belt worn by H.R.H. when he inspected the battalion at Chatham in 1824. They still hold an honoured place in the ante-room of the Officers' Mess of the battalion.

## 11

In February 1828 the British troops were withdrawn from Portugal.

The presence of the British Army in Portugal had been sufficient to stop any intention of a Spanish invasion. No fighting had taken place, but fever and dysentery had carried off about forty of our riflemen. The battalion embarked at Lisbon on March 30 and having disembarked at Cork on April 12, marched to Fermoy where it was to be quartered.

Serving with the Royal Welsh Fusiliers in Portugal at this time was a Lieutenant Arthur Lawrence, at a later date the distinguished commanding officer of the 2nd Battalion of the Rifle Brigade in the Crimea. Lawrence was never tired of recalling the smartness, good order, and discipline of our 1st Battalion, which to his mind appeared to be the ideal of a Rifle Corps. He was again associated therewith a few years later in the Ionian Islands, and bore it in mind as a model when he himself became a Rifleman.

During the stay of the battalion in Fermoy, a curious incident occurred for the details of which the reader must be referred to the 'History of Services of the 1st Battalion.' Suffice it to say that the higher military authorities took a very strong step which might easily have caused an undying feud between the Riflemen and another regiment. By the tact and good feeling of the officers of both regiments, not only was a breach averted but most friendly feeling has ever since existed between them.

In September 1828 headquarters and four companies marched to Limerick; the remaining six companies forming detachments at Glenduff, Newcastle, Ballingarry, Bruff, and Rathkeale.

This was the period immediately preceding Catholic Emancipation, an act of justice long overdue, and the country was in a condition of political excitement unusual even in Ireland. At the end of June the Headquarter Companies marched to Clare Castle for the purpose of keeping order during a parliamentary election; on which occasion we learn that the good humour of the officers and men overcame the initial hostility of the people. In August Headquarters returned to Limerick and remained there for just two years, at the end of which the outlying detachments were called in and the battalion was concentrated at Cork.

During the year 1829 Major J. H. Schöedde was promoted to an unattached Lieut.-Colonelcy and consequently quitted the Regiment in which indeed he had been born, for his father had also served therein. In due course he was appointed to the command, first of the 48th, afterwards of the 55th Regiments. In 1841–2 he served as a Major-General in the Chinese war; was mentioned in terms of high praise by Sir Hugh Gough the commander-in-chief, and received the thanks of both Houses of Parliament.

He also received the K.C.B. During the Russian war
1854-6, his name was submitted to the Queen for a
command. Sir James Schöedde died at Lyndhurst in
1871.

Orders for foreign service were received, and on October
6, 1830, the battalion having been conveyed to the Cove of
Cork in 'steam packets,' embarked for Gibraltar, where it
arrived on November 2.

### 12

We left the 2nd Battalion in Demerara. The right
wing was quartered at Berbice and the left probably at
Georgetown, where for several years they appear to have led
an uneventful life.

On June 18, 1825, Lieut.-Colonel Galiffe succeeded to
the command of the battalion ; Mackie, who had been its
commanding officer for nearly seventeen years, having been
presented to the rank of Major-General. Galiffe, however,
retired in the following December, handing over command
to Major Henry FitzGerald.

Colonel Galiffe returned to his native land, where he
spent his declining years, and died at Geneva in 1847.
He left a diary which has been unfortunately lost. If
it could be found it might be of the greatest interest to us.
Other papers belonging to him are, however, deposited in
the museum at Geneva.

On July 5, 1829, the following District Order was
published :—

'Berbice, 5th July, 1829.
'The Major-General commanding the District, in his annual
inspection yesterday of the Headquarter Division of the 2nd Battalion of the 60th Rifles, was justly gratified with every part of its
state and condition.

'The order of the barracks and the health and appearance of the men were very good.

'The field exercise deserved especial praise, and well showed how formidable a small body of troops may be rendered if they are well disciplined and judiciously managed. The general attention, the marching and firing, the squareness and precision of the different movements, their fitness for their respective objects, and their execution, were alike excellent, and the Major-General desires to express to Major Shee his unqualified approbation.'

Towards the end of the year the battalion received orders to proceed to England and the General bade farewell to them as follows :—

'Demerara, 15th December, 1829.

'Major-General Sir B. D'Urban, Commanding the District, takes leave of the 2nd Battalion 60th Regiment with every sentiment of sincere regard and attachment, which that old and excellent corps so well deserves from him, and he requests to offer his best thanks to Major Shee for the incessant care which he has devoted to it, and for the ability with which he has commanded it. He thanks Captain Slyfield for his great attention to the left wing. He thanks the battalion for their general good conduct, and he heartily wishes them all health and good fortune.'

The battalion accordingly embarked; the left wing, under Captain Slyfield, reached the Isle of Wight on February 20, 1830; and the right wing, commanded by Major Shee, reached the same destination on March 2. Here it was joined by its commanding officer Lieut.-Colonel the Honourable A. F. Ellis, M.P., who had succeeded Colonel Henry FitzGerald on December 28, 1828.

In the autumn they were ordered to Weedon, but before leaving the Isle of Wight received compliments from the civil not less than from the military authorities. That of the General was couched in the following terms :—

‘ Portsmouth, 25th September, 1830.

‘ The Major-General commanding the District (Sir Colin Campbell [1]) was much pleased at the steady and soldierlike appearance of
the 2nd Battalion 60th Rifles at their half-yearly inspection yesterday. It is evident that great attention has been paid by Lt.-
Colonel Ellis to the drill, discipline, and interior economy of the
regiment. Their arms, clothing, and appointments were in the best
order. They performed their evolutions with accuracy, precision,
and due celerity.

‘ The Major-General will have much pleasure in reporting the
efficiency of the regiment to the General Commanding in
Chief.

‘ Upon the departure of the 2nd Battalion 60th Regiment from
the South-West District, the Major-General cannot withhold the
expression of his satisfaction and approbation at the regularity and
good conduct of the regiment since it has been under his command,
and requests that Lt.-Colonel Ellis, the Officers, Non-Commissioned
Officers, and soldiers will accept his best acknowledgment, and
trusts that, wherever the regiment may be stationed, they will
continue to maintain that discipline, regularity, and good order which
is so highly creditable to themselves, and so essential to the good of
His Majesty’s service.

‘ (Signed) J. GURWOOD, Lieut.-Colonel,
‘ Major of Brigade.’

On September 27 the march commenced in three
columns, but the destination was soon changed from Weedon
to Manchester, where the battalion arrived in the first days
of November.

During the summer of this year—1830—the King had
died and was succeeded by his brother King William IV.,
who lost little time in bestowing on our Regiment a special
mark of royal favour and thereby placing it thenceforward
under the immediate protection of the reigning sovereign.
The following letter was received by the officers commanding
the 1st and 2nd Battalions :—

---

[1] *Not* the future Lord Clyde of the Indian Mutiny.

'Horse Guards, 12th November, 1830.

' SIR,

'I have the honour to acquaint you by the direction of the General Commanding in Chief, that His Majesty has been graciously pleased to approve of the 60th Regiment being in future styled " The 60th ; The King's Royal Rifle Corps " instead of " The Duke of York's Own Rifle Corps."

'I have, etc.,
'JOHN MACDONALD,
'A.G.'

## NOTES ON OFFICERS WHO SERVED WITH THE 5TH BATTALION IN THE PENINSULA

### FIELD MARSHAL LORD CLYDE, G.C.B., C.G.S.I., 1792–1863,

better known as Sir Colin Campbell, was the son of John M'Liver, a cabinet maker of Glasgow. The name of his mother, who had lost a brother in the American War of Independence, was Campbell. When the boy was fifteen years old a commission in the army was requested of the Duke of York by a maternal uncle, and in May 1808, the Commander-in-Chief believing young Colin to be a Campbell, gazetted him to the 9th Regiment under that name, which he accordingly appropriated during the rest of his life.

He went to Portugal in the autumn of 1808, and in January 1809 was present with his regiment at the Battle of Coruña. In the same year Campbell served in the expedition to Walcheren, and in 1810 returned to Spain where he was present at the battle of Barrosa and the defence of Tarifa. In the summer of 1813 he was present at the battle of Vitoria and distinguished himself at the siege of San Sebastian by leading the forlorn hope at the assault of the Convent of San Bartolomeo. He was rewarded by being given a company in the 7th Battalion of the 60th. Unfortunately the wounds which he had received not only prevented the continuance of his service in Spain but stopped him from joining his new battalion until October 1814. It was then quartered on the Penobscot River in the State of Maine; but although hostilities with the United States were not yet at an end there was no further fighting in this quarter. In July 1815 considerations of health compelled him to go home.

At the end of 1816 Captain Campbell was transferred to the 5th Battalion at Gibraltar; and this, many years after, he declared was the happiest time of his life. At Gibraltar he served until the disbandment of the battalion in 1818, when he was transferred to the 21st Fusiliers.

As a Lieut.-Colonel he was present in the China War 1842. The second Sikh War found him in India where in command of a division he served with great distinction in the battles of Chillianwallah and Goojerai. In 1851–2 Colonel Campbell commanded an expedition against the hill tribes on the right bank of the Indus, his force including our 1st Battalion under Colonel Bradshaw. In 1854 Campbell, by this time ' Sir Colin,' was appointed to the command of the Highland Brigade in the Crimea. During the same year he gained the rank of Major-General, and in course of time was appointed to the command of a division.

When the news of the Indian Mutiny in 1857 reached England Sir Colin was at once sent as Commander-in-Chief to the East, where he successfully quelled the revolt, and was raised to the peerage under the title of Baron Clyde of Clydesdale. In 1862 he received the rank of Field Marshal but died in the following year.

Of Lord Clyde it may be said that if not in the first rank of our nineteenth century generals, he was at all events high up in the second.

### MAJOR-GENERAL SIR WILLIAM WILLIAMS, K.C.B., K.T.S.
### 1776–1832

In 1792 this officer purchased an ensigncy in the 54th Regiment at that time quartered in Guernsey, an attack on which island was daily expected. During the two following years he served under the Duke of York in the Low Countries.

Promoted to the rank of Lieutenant in September 1793, Williams proceeded in 1795 to the West Indies and during the following January was wounded at St. Vincent. In 1796 he attained the rank of captain, and having returned to the British Isles took part in the suppression of the Irish Rebellion. In 1800 Captain Williams joined Sir James Pulveney's expedition to Ferrol, and afterwards went on to Egypt in the army under Sir Ralph Abercromby. He is said to have been the first soldier to land, and was engaged in every action. In 1802 he was given the brevet rank of major and two years later the substantive rank in the 81st Foot, with which he served at Coruña where he was wounded a second time. Later in the year he was in the expedition to Walcheren, and on November 15, 1809, was posted to the 5th Battalion of the 60th as Lieut.-Colonel in place of Baron de Rothenburg.

The headquarters of the battalion being, as we have already seen, attached to the 3rd Division, Williams commanded a battalion of its light troops, to which were sometimes added cavalry, artillery, and caçadores. At Bussaco he was twice wounded. Wounded again at Fuentes de Oñoro, this time in the neck. At Pastores he commanded the outposts. After being wounded for the sixth time at Salamanca, Colonel Williams exchanged with Colonel Keane into the 13th Regiment stationed in North America, and with that regiment fought at Lake Champlain and Plattsburg.

In March 1813 he had received the Portuguese Order of the Tower and Sword. At St. John's, New Brunswick, he was placed in command of the advanced posts on the Richelieu River ; and after the attack of the American General Wilkinson had been repulsed, Sir G. Prevost, Commander-in-Chief, mentioned in G.O. his most entire approbation of ' the judgment, zeal and unwearied assiduity displayed by Lieut.-Colonel Williams in his arrangement of the defence of the important posts placed under his immediate command.' In 1815 Williams became a K.C.B. In 1830 he was promoted to the rank of Major-General, but died at Bath two years later. As a commander of light troops in the field Sir William Williams has never been excelled.

### COLONEL JOHN GALIFFE, C.B.

John Galiffe served in the Royal Army of France from 1785 to 1792. In consequence, no doubt, of Royalist opinions, he then quitted the French and entered the Dutch service, joining the regiment of Timorman hussars with which he served in the Campaigns of 1793 and 1794. When the French overran Holland, Galiffe came over to England ; and having served for a few months in the 6th West India Regiment as Lieutenant, was made Captain in the York Rangers in October 1796. Upon the disbandment of that regiment just at the time when the 5th Battalion of the 60th was being raised, he was transferred thereto and no doubt served with it at Surinam and Halifax, N.S. In June 1808 he embarked for the Peninsula, and served continuously in the following campaigns. After the appointment of Colonel Keane to a brigade, Galiffe commanded the battalion during the remainder of the war, including the battles of Vitoria, the Pyrenees, Nivelle,

Nive, Orthez, and Toulouse, and shortly before its termination received the brevet rank of Lieut.-Colonel.

On the reduction of the regiment to two battalions Colonel Galiffe was posted to the 2nd, and in 1825 became its Lieut.-Colonel. His subsequent career is unknown, but he died either at the close of 1847 or at the beginning of 1848.

Galiffe received the Gold Medal and Cross for the Nive, Orthez, and Toulouse, and the Silver Medal with eleven clasps, for Roliça, Vimieiro, Talavera, Bussaco, Fuentes de Oñoro, Albuera, Ciudad Rodrigo, Badajoz, Salamanca, Pyrenees, and the Nive.

### Colonel William Woodgate, C.B.,
### 1780–1861,

received a commission as Lieutenant in the 6th Regiment in 1800 and in April 1803 was promoted to the rank of Captain in the 60th. He got his majority in August 1807 ; and after fighting at Roliça, Vimieiro, Talavera, and Bussaco, succeeded Colonel Williams in command of the battalion at Fuentes de Oñoro, being himself wounded shortly afterwards. Woodgate was also present at Badajoz and Salamanca ; received the brevet rank of Major in December 1811, and at the end of the war the Gold Medal and the C.B. On June 16, 1814, he became Colonel of the 4th Battalion. He gained the rank of Colonel in 1821 and then apparently retired from the army. He died at Paris in January 1861.

### Field Marshal Sir John Forster FitzGerald, G.C.B., etc.
### 1784–1877

This officer, son of Edward FitzGerald, M.P., of Carrigoran, was born either in 1784 or 1786 and was fortunate enough to receive his first commission in the army at the somewhat early age of seven or nine years. In 1794 he became a lieutenant and a month or two later a captain. In September 1802, when either seventeen or nineteen years old, his seniority as a captain was so pronounced that he received the rank of brevet major, and in November 1809 gained the substantive rank in the 60th. By the time he had joined the

5th Battalion in the Peninsula as major, his seniority entitled him to the brevet rank of lieut.-colonel, his commission being dated July 25, 1810.

At the capture of Badajoz, where he was wounded, and also at the battle of Salamanca, Colonel FitzGerald was mentioned in Lord Wellington's dispatches. After the departure of Colonel Williams, FitzGerald commanded the 5th Battalion until the arrival of Colonel Keane in April 1813, shortly after which he took command of the three Rifle Companies attached to the 2nd Division. His command included the Light Companies of the other battalions of that division, and was considered the equivalent of a brigade. He therefore remained with the 2nd Division when Colonel Keane was appointed to command a brigade, leaving Major Galiffe, his junior, to command the 5th Battalion. FitzGerald was mentioned in Lord Wellington's dispatch on Vitoria, and also in that of Sir R. Hill on the battles of the Pyrenees, where he was taken prisoner and not liberated till the end of the war in 1814 ; when he returned to his regiment and commanded the 5th and the (new) 1st Battalion.

FitzGerald became Major-General in 1830 and commanded a Division in the Bombay Presidency. His promotion to the rank of Lieut.-General took place in 1841 and to that of General in 1854. In 1862 he was created a G.C.B. and on May 29, 1875, received the rank of Field Marshal. He died at Tours in France in 1877 at the age of 91 or 93, when by the chivalrous order of the Minister of War his funeral received the military honours paid to Marshals of France and his body was escorted to the grave by troops of the nation against which he had so often been engaged in action.

GENERAL LORD KEANE OF GHUZNEE, G.C.B.
1781–1844

John Keane began his military career on the half-pay list of the 124th Regiment in November 1794. During the Egyptian Campaign of 1801 he was present as A.D.C. to Major-General Lord Cavan at the battles of Alexandria. In May 1802 he was appointed to the 60th as a Major, but in the following year was appointed as Lieut.-Colonel to the 13th Regiment. On January 1, 1812, Keane received the brevet rank of Colonel and in the following June exchanged into the 60th with Colonel Williams. He commanded the 5th

Battalion for only a short time and quitted it on being given command of a brigade in May 1813. On June 4, 1814, he was gazetted Major-General.

On the termination of hostilities in France, Keane, now a K.C.B., was ordered to Jamaica to take command of a body of troops assembled there for operations against the United States. Landing near the mouth of the Mississippi, on December 23, 1814, with a view to the attack of New Orleans, he repulsed a large force of Americans who attempted to stop his march. On the arrival of Sir Edward Pakenham as Commander-in-Chief, Keane—described by Gleig as ' a dashing young officer '—was given command of the 3rd Division. At the subsequent assault of the enemy's works, which resulted in a sanguinary repulse, Pakenham and Gibbs—late 60th, another divisional General—were killed and Keane was badly wounded.

In 1839 he gained a peerage for the capture of Ghuznee in Afghanistan.


## NOTES ON THE SERVICES OF OFFICERS OF THE REGIMENT OTHER THAN THOSE IN THE 5TH BATTALION

*Major-General Thomas Baron.*—In 1778 this officer became a captain in the 3rd Battalion of the 60th and took part in the American War of Independence. He was present at the capture of Savannah, and made Brigade Major to Major-General Prevost, in which capacity he was present at Mark Prevost's brilliant action at Briar Creek and subsequent defence of Savannah. In 1782 Baron was compelled by ill health to go on half-pay and never rejoined the Regiment, although he served for many years in the West India Regiment and was present at various services including the capture of Guadeloupe, for which he received a War Medal. He was promoted to the rank of Major-General in June 1811.

*Lieutenant and Quarter-Master Brennan* served in the Peninsula and was present at Bussaco, Albuera (where he was severely wounded), Vitoria (slightly wounded), the Pyrenees, the Nivelle, the Nive, Orthez, and Toulouse. He was also present at Quatre Bras and Waterloo.

*Lieut.-Colonel Sir John Campbell* was posted to the 2nd Battalion on April 27. 1781, and served in the Regiment in the various grades, until he attained that of Lieut.-Colonel in March 1802. He quitted the Regiment in 1803 but served with the Portuguese Army in the Peninsula and received the Order of the Tower and Sword.

*Major-General Edward Codd.*—In December 1789 this officer was appointed to an ensigncy in the Regiment and served in North America and the West Indies until he reached the rank of Lieut.-Colonel in October 1804. In 1808 he became Lieut.-Colonel of the 2nd Battalion, in which year the battalion embarked for England. In September Colonel Codd sailed with it for Spain in the force under Sir David Baird, destined to join that of Sir John Moore. The battalion being very weak was left in garrison at Coruña, and Colonel Codd was present at the battle of Coruña. In 1819 he was promoted to the rank of Major-General ; but in accordance with the usage of that day his name appeared in the Army List for some few years longer as Colonel of the 2nd Battalion.

*Major-General Gabriel Gordon* was appointed to the 60th as an ensign in 1781, and after distinguished service in the West Indies, particularly at the capture of Tobago in 1793, was in March 1802 promoted to the rank of Lieut.-Colonel of the 4th Battalion. He was present at the capture of Martinique and Guadeloupe and received a War Medal and Clasp.

*Colonel George Mackie* joined the 60th as Major in March 1802, and served in the West Indies. In 1808 he was appointed Lieut.-Colonel of the 3rd Battalion, and was present at the capture of Martinique, for which he received the War Medal. A few years later he performed good service in the repression of a serious insurrection of the negroes, which he effected by tact rather than severity. He was in consequence presented with a piece of plate by the inhabitants, both French and English.

*Frederick Maitland* was made a Captain in the 60th in 1789. Was present at the capture of Tobago, Martinique, St. Lucia, and Guadeloupe, and received the brevet of Lieut.-Colonel. In 1807 he served as Major-General in the attack of the Danish W.I. Islands. and in 1809 commanded a Division of the army under Sir George Beckwith in the capture of Martinique. In 1812 he was a Lieut.-General and Second-in-Command in the Mediterranean. For the capture of Martinique General Maitland received a Medal.

*Major-General Lewis von Mosheim.*—This officer started his

career as a lieutenant in the Wurtemburg Regiment of Foot Guards. In 1794, five years later, he became a Captain in Löwenstein's Riflemen, and was present in command of all the Light Troops of the garrison at the defence of Grave in the last three months of that year. On its surrender he was taken prisoner, but released in the following June, when he came to England and proceeded with his Regiment to the West Indies. In the expedition under Sir Ralph Abercromby he was present as Brigade Major. When Löwenstein's Chasseurs were drafted into our 5th Battalion Von Mosheim became a Major and commanded a detachment at the Island of Trinidad. In 1799, being ordered to join the 6th Battalion in England, he was captured on the voyage, but exchanged a few months later. In 1804 Mosheim became Colonel of this Battalion which he had already commanded as Major for some years.

*Major-General Francis Slater* was ensign in the 60th in 1788 and commanded the Grenadiers of his Regiment at Martinique, St. Lucia, and Guadeloupe. In 1796 he exchanged into the 2nd Life Guards and commanded them in the latter part of the Peninsular War, having assumed the additional name of Rebow.

*Captain Jacob Tonson* served with the Regiment at Tobago, Martinique, Demerara, and St. Vincent, where he was wounded at the head of the Grenadiers of the 3rd Battalion. In 1813 he, as a Brevet Major in the 84th, took command of Robinson's Brigade of the 5th Division during the battle of the Nive.

COLONEL THOMAS BUNBURY.
Lieut.-Colonel, 1824–1835.   Col. Commandant, 1842–1856.
(From a Portrait in the possession of Mr. Hamilton Bunbury.)

## APPENDIX TO VOL. I

THE following notes relate to subjects within the scope of the first volume of these Annals, but contain information acquired since its publication. It is suggested that in the event of a second edition being at any future time needed, this information should be embodied in the text of Vol. I. It may be mentioned that in the writer's own copy of Vol. I. several marginal notes have been made containing still further information, part of which might be utilised in the same way.

### A

Upon the authority of Major Patrick Murray, it was stated in the first volume of these Annals that the original suggestion for the formation of the Royal American Regiment was made by a Mr. Harbot of Berne. On the publication of that volume this statement was traversed by Miss Edith Elliot, who, in a letter to the writer, stated that the credit ought to have been given to her ancestor, General Granville Elliot. Granville Elliot was born in Surrey on October 7, 1713, and having chosen the occupation of a soldier of fortune was at the age of only thirty-two a Lieut.-General of Cavalry in the service of the Elector Palatine of the Rhine. Two years later, in 1747, he entered that of the States General of the Netherlands ; and it was during this phase of his career that he is said to have made the proposal for the raising of the 62nd Royal American Regiment to be trained as light infantry and scouts, with a view to meeting the Red Indian tribes on equal terms. Miss Elliot's statement is probably correct, but the earliest documentary evidence of her ancestor's interest in the subject is dated December 10, 1756, nearly twelve months after the official date of the original existence of the regiment. The following is his letter to William Pitt the Elder, at that time the most powerful Minister in Great Britain :—

'Kew Green, Dec. 10th, 1756.

' Lieut.-General Elliott presents his respects to Mr. Pitt and hopes he will excuse his taking the liberty of troubling him with the inclos'd ; relies upon his Indulgence to an old acquaintance whose best wishes, however insignificant, will always attend him, as they now doe for his speedy recovery.

*Memorial*

' The utility of those kind of troops which the french call " compagnies tranches " and the Germans " free companys " is well known.

' They are made use of entirely for that sort of war, which the french call " La petite Guerre "—such chiefly is the war now carried on in America : a very clear notion of it may be formed from the perusal of a little sketch published not long agoe in france, called " Traité de la petite guerre " par Msieur. de la Croix.

' As it is reported that a considerable body of Troops is to be sent to America ; a certain number (six or more) of these companys, not exceeding 150 men each, would undoubtedly make a very useful part of this Reinforcement.    These sort of soldiers are used to march through woods, lye under or upon Trees, always upon their guard, surprise the enemy, take all the advantages that deep laid stratagems afford, and to live hard, regardless of weather and other hardships.    They commonly have in each company a number of chasseurs, in German, Jagers, who are the best marksmen in Germany, and perhaps in the world ; they make use of a peculiar fire-arm, short and Rifled Barrel.    The king of Prussia has several companys of these chasseurs.    His R.H. the Duke had some of these Compagnies franches in the last war in Flanders.

' Good officers may be had upon advantageous terms ;  There are many officers now in Germany upon half pay who served in these corps during the last war ;  and several in General Cornal's regiment in the Dutch service ;  which regiment was entirely formed of the several compagnies franches when in the service of the States.    A very great advantage may arise from sending such troops over, by their teaching the armed inhabitants of the colonies their stratagems, ambuscades, secret marches and various methods of proceeding in this sort of war, which seems so peculiarly adapted to that country.

' At all events should any inconveniency arise from keeping them in single companies, they will make a useful Body to add to the New Raised Regiment of Royal Americans.

' It is highly probable that a Draught of men used to this service might be made out of the Hessian Corps now in England, and very good chasseurs may be had from that part of Germany they belong to ; supposing it no difficult measure to ask the Landgrave of Hesse Cassel's leave for making a draught of men, and raising a number of chasseurs in his dominions.'

What reply Mr. Pitt sent General Elliot is unknown, but it was owing to the Minister's influence that Count von Elliot shortly afterwards entered the British Army. That the Royal American Regiment was trained upon the lines suggested by him is undeniable.

Elliot now appears to have wished to join the service of his own country. In February 1758 he was appointed to the regiment of Lieut.-General Halkett in the Scottish Brigade of the States Nether-lands and on April 21 became a Colonel in the British service and simultaneously received the rank of Major-General. In August he was on the staff of the secret expedition to St. Malo and commanded a cavalry brigade under the Marquis of Granby at Minden where he received special mention for his gallantry. He died on active service on October 10 the same year.

General Elliot was twice married, first to Jeanne Thérése, Comtesse de Martigny, maid of honour to H.R.H. the Duchess of Loraine, secondly to Elizabeth, daughter of W. D. de Hartham, Colonel of the second troop Horse Grenadier Guards and M.P. for Calne.

General Elliot, who had the title of Comte de Morange, was evidently a man of note. Prince Ferdinand of Brunswick, Com-mander-in-Chief at Minden, who was present at his funeral, spoke of his high character, and he is alluded to in the Stopford Sackville MS. as ' General Elliott whom all men look up to.'

## B

### Governor's Island

#### (NEW YORK, U.S.A.)

Not only the writer but the whole Regiment is greatly indebted to the Rev. Edm. Banks Smith, D.D., Chaplain of Governor's Island, on the history of which he has written a most interesting book emphasising the connection therewith of our Regiment. In the early days of our history, Governor's Island appears to have been its headquarters. A memorial to the 60th has been erected in Dr. Banks Smith's church. After giving a short summary of our history the author proceeds as follows :—

'It is a cause of deep satisfaction to realise that this distinguished regiment, "Celer et Audax" in practice as well as in motto, not only came from our soil in the persons of its recruits but that it gained its growth and training in the island garrison where it remained for a long tour of duty and that by what we may now regard as a most happy occurrence of military routine it was ordered away to the West Indies before the outbreak of hostilities in 1775-6. Thus the 60th were never arrayed against those who were their brethren in blood as well as in sympathy, and the author ventures, at the close of this history of the past, to present his compliments with which he feels he can unofficially join those of the commands stationed on Governor's Island to-day to the 60th of 1756, the King's Royal Rifle Corps of 1913.

'It is not alone in arms that Governor's Island is bound by lasting ties of interest and sympathy to the mother country. The Church and the Army in every land have much in common—the Army to protect and the Church to bless. This garrison has been no exception to the rule. There is, however, a deeper connection than would appear upon the surface as a part of our history. It is that the ministrations of religion here for nearly seventy years carried on at the request of the Army by the venerable Corporation of Trinity Church, have been and are to-day possible because of the royal endowments of the British Crown which constitute the wealth of the parish of Trinity Church. The parish regards it as a privilege to minister to the spiritual needs of the Army. From the point of view of the historian, there is an added interest in reflecting that when in the providence of God the existing close relations between the Army and the old parish began, they not only opened the way in mutual acquaintance and esteem but gave the parish an opportunity, among its other works, to pay from the royal endowments a tribute of appreciation of this very distinguished regiment of the British Army born on Governor's Island.

'Thus remembrance of the Past and loyalty to the Present go hand in hand. The Prince of Wales's feathers still bend over the pulpit of old St. Paul's Chapel. The Coronation of His Majesty King George V. is solemnly observed in the parish church.'

## C

### THE EARL OF LOUDOUN, OUR FIRST COLONEL-IN-CHIEF

In his 'Journal of a Tour to the Hebrides,' Boswell mentions that he and Dr. Johnson dined at Loudoun Castle, in Ayrshire, on October 30, 1773. The story that Lord Loudoun 'jumped for joy' on hearing of the approach of his distinguished visitors may be a slight exaggeration on the part of the writers, but that he received them ' with pleasing courtesy ' may be easily believed. Boswell continues :

' I cannot here refrain from paying a just tribute to the character of John, Earl of Loudoun, who did more service to the county of Ayr in general, as well as to individuals in it, than any man we have ever had. It is painful to think that he met with much ingratitude from persons both in high and low rank ; but such was his temper, such his knowledge of " base mankind," that, as if he had expected no other return, his mind was never soured, and he retained his good humour and benevolence to the last. The tenderness of his heart was proved in 1745–6, when he had an important command in the Highlands, and behaved with a generous humanity to the unfortunate. I cannot figure a more honest politician ; for though his interest in our county was great and generally successful, he not only did not deceive by fallacious promises, but was anxious that people should not deceive themselves by too sanguine expectations. His kind and dutiful attention to his mother was unremitted. At his house was true hospitality ; a plain but a plentiful table ; and every guest being left at perfect freedom, felt himself quite easy and happy. While I live I shall honour the memory of this amiable man.'

## D

EXTRACTS FROM THE JOURNALS OF THE HONOURABLE WILLIAM HERVEY REFERRING TO OUR REGIMENT IN THE SEVEN YEARS' WAR.

| | |
|---|---|
| October 2, 1756. | Colonel Bouquet with a hundred of the Americans at Saratoga. |
| October 3, 1756. | Lieutenant Camel of the Americans killed at Saratoga, where the 1st Battalion then was, by a sergeant's firelock going off at exercise. |
| October 4, 1756. | An officer and thirty of the Americans arrived at Fort Edward. |
| October 5, 1756. | Lord Loudoun arrived at Fort Edward. |
| October 9, 1756. | One complete battalion of the Americans consisting of 1000 at Saratoga. |

October 13, 1756.  Eighty-three wagons with stores arrived escorted by Americans.

November 7, 1756.  Colonel Bouquet marched from Saratoga with the 1st Division of Americans for their winter quarters at New York.

November 10, 1756.  The last Division of the Americans under Haldimand marched for New York.

July 7, 1758.  Ticonderoga.    44th Foot—six companies R.A.R. [1] with 2000 Provincials marched to take possession of the post at the Saw Mill which was found deserted.

July 8, 1758.  (An engineer who was a foreigner and sent out to reconnoitre reported at the Counsel of War : Je voi claire devant moi et s'il y a un retrenchement il vaut que ça soit sous le Canon de Fort.)  Two irregular attacks made without success.  The picquets were driven back before half the army was come to their ground.  The Highland Regiment ran along the front of the 44th marching up.

June 19, 1760.  Amherst advancing from Albany on Montreal —Westward, 42nd, 44th, 46th, 55th, 60th, 77th, 80th, and Provincials.  Northward, 17th, 27th, half Battalion Royals, Provincials.

July 22, 1760.  Light Infantry Brigade (two Battalions 42nd Montgomery, Murray), Oughton, 4th Battalion 60th, Abercrombie), Gage's Brigade, Schyeler (Whiting, Fitch, Wooster, Lyman) —(Corsa, Woodful, Le Roux).

August 7, 1760.  The Grenadiers, six Companies 540.  The Light Infantry Companies, 1st, 42nd, under Haldimand start for the St. Lawrence. Embarcation return 60th 31 officers, 439 men.  Total force, Regulars 287 officers, 5299 men.  Provincials, 287 officers, 4192 men.  Indians, 706 men.  Whale boats, 165. Batteaus, 634.  Guns and Mortars, 35. Cohorns, 10.

---

[1] Royal American Regiment.

MAJOR JOHN RUTHERFORD.
Killed at Ticonderoga, 1758.

September 3, 1760.   Haviland (60th) has taken Isle aux Noix.

September 5, 1760.   Haviland near La Prairie.

September 9, 1760.   Haldimand with L.I. and Grenadiers takes possession of Montreal; 4th Battalion R.A.R. cantonned at St. Henry, Terre-Bonne, La Chenay, St. François, St. Vincent, St. Rose, 445 houses.

February 26, 1762.   Mr. Grant and Mr. Edward Chinn, merchants in the city of Montreal, having been tried by the General Court Martial whereof Major Munster of the 4th R.A. Battalion is president, for having conjunctly abused, assaulted and otherwise ill-treated Ensign Nott of the 4th R.A. Battalion, are found guilty. Grant to be fined £30 and Chinn £20, which sums the Court doth appoint to be disposed of as the General shall direct among the poor and distressed persons in the Government of Montreal. The Court doth further adjudge that Grant and Chinn shall in the most public manner before the Garrison of Montreal severally ask pardon of Ensign Nott in the following words : (' Ensign Nott, I am sorry I have been guilty of assaulting you and do most humbly ask your pardon for it.') The General approves but thinks proper to mitigate the fine from £30 to £20 and from £20 to £13, which sums are to be paid into the hands of the Town Major of Montreal, to be afterwards disposed of as the Court hath appointed. Grant and Chinn will, on Saturday morning next, on the Parade at the time of mounting the Guard, ask pardon of Ensign Nott in the words specified in presence of one Field Officer, two Captains and four Subalterns at least of the Garrison.

    Mr. Forest Oakes, merchant in the City of Montreal was also tried for insulting and giving abusive language to Ensign Nott and for having cast several scurrilous reflections on His Majesty's Army. The Court adjudges

Mr. Oakes to be confined in the custody of the Provost Martial for 14 days for the scurrilous reflections cast by him on the Army, and that before his confinement he shall publickly ask pardon of Mr. Nott in these words : (' Ensign Nott, I ask your pardon for the ill language I was guilty of to you.') The General approves but thinks proper that Mr. Oakes should be released from custody after twenty-four hours upon his finding security for his good behaviour for the remainder of the time of his committment and in consideration that insult and abusive language appear to have been reciprocal between the partys, the General thinks proper to excuse Mr. Oakes from asking the pardon of Ensign Nott.

**March 2, 1762.** Colonel James Prevest to be Colonel of the 1st Battalion R.A. Lieut.-Colonel Marcus Smith of the Royal English Fuziliers to be Colonel of the 4th R.A. Captain Henry Gladwin of the 80th or Regiment of Light Armed Foot to be Major of it.

**April 29, 1762.** War is to be declared to-morrow against Spain, for which purpose the Grenadier and the Light Infantry Companies, colours, drums and fifes of the Battalion will assemble before the Government House at 9 o'clock, when the war will be first declared, and afterwards in the Market Place and on the Parade ; the Artillery to fire 7 guns after each proclamation ; the procession will march from the Government House in the following order :—

A Corporal and six Grenadiers clear the way.

A Gunner properly accoutred and match lighted. A Company of Grenadiers. The music of the 46th Ensigns with the Colours, drums and fifes of the Battalion.

Town Major and Colonel of Militia on horseback.

|  |  |
|---|---|
|  | The Military Officers (if any chosen to go). Captains of Militia, Merchants, Light Infantry Company. |
| May 4, 1762. | The Prisoner Baptist Shorap, convicted of being concerned in taking down his Majesty's Declaration of War against Spain and treating the same with the highest insolence and indignity, is to be flogged to-morrow morning at 10 o'clock at the Cartstale through the town by the executioner; he is to receive a hundred lashes opposite to the Provost's a hundred opposite to the Town Major's, another hundred near Captain Wharton's Quarters, and a hundred at the Market Place directly opposite the place where the Proclamation was put up. The Provost with a Corporal and six men will attend this punishment and see it duly executed. |
| June 7, 1762. | Major Munster of the 4th R.A. to be Lieut.-Colonel. |
| June 25, 1762. | The General expresses his satisfaction of the R. American Battn. and thanks the officers for their care and diligence. N.B. on leaving Montreal. |
| August 26, 1762. | Recruits given to the 4th R. Americans and the 80th are intitulled to a bounty from the King of £5 New York currency over and above what they have received from the Province; and the Provincial Officer who inlisted them a reward of 40s. like money. |
| October 29, 1762. | (On the occasion of the birth of the Prince of Wales.) Three Royal Salutes will be fired this Evening at Sunset; each Salute will be answered by the Musquetry of the Regiment by a rejoicing fire. |
| January 16, 1763. | Louis Mercier inhabitant, for using false weights to receive a hundred lashes on the bare back at the Cartstail by the hands of the common executioners; his beef to be confiscated. The General approves of the confiscation of the beef, is pleased to remit |

February 5, 1763.    the corporal punishment, and orders the beef to be sent to the Recolets, giving 8 lbs. of it to the soldier James Reid of Captain Abercromby's Company.

Baptist Junot for insulting Captain Haultain of the 4th R.A. Battn. to pay five dollars and to suffer 14 days' imprisonment. The General approves but at the desire of Captain Haultain omits the imprisonment.

## E

### DANIEL M'ALPIN

The tradition alluded to in Vol. I. p. 88, that this officer died in a workhouse seems to have been apocryphal.

The following account of his later career may be not without interest. Quitting the 60th in 1765, after a few years he purchased a small property on Saratoga Lake and resided there. On the outbreak of the War of Independence he was peremptorily ordered by the rebels to join their army, and on his firm refusal was made prisoner and confined at Fort Albany, but after a time was allowed out on parole on account of ill-health. In 1777 he raised a small company of loyalists which was, however, shortly afterwards captured by a superior force of the enemy. M'Alpin himself escaped, and lived for a week in a hollow tree, whence he escaped and joined the British army of General Burgoyne. On the surrender of this army at Saratoga M'Alpin contrived to make his way through the American lines into Canada, where he again took a command in the British forces. His health was, however, rapidly breaking down and he died five months afterwards.

## F

### WRECK OF THE BRITISH SLOOP-OF-WAR 'EL ORQUIXO,' REFERRED TO IN VOL. I. pp. 268-9

Off Port Antonio, Jamaica, on November 7, 1805, the sloop *Orquixo* (Captain John Bassett Balderston) was joined by the sloop

*Penguin* (Captain Norris) on a cruise. The *Penguin* coming under the *Orquixo's* stern, Captain Norris hailed her, desiring her commander to come on board, which he did accordingly at about ten or fifteen minutes past two. The *Orquixo's* 1st Lieutenant, Stanton, then hove ship to the larboard tack and set canvases to speak a merchant ship as the captain had ordered, after which he went below for ten minutes, leaving Mr. Coke, the mate of the watch, in charge, and then returned to allow the latter to go to dinner. The ship was then under close reefed topsails and as there was the appearance of a squall to windward these were now lowered on the cap and the canvases hauled up again. About five to ten minutes later the squall struck the ship with such violence that, although previously regarded as a stiff ship, she instantly fell over on her beam ends. Lieutenant Stanton at once gave orders to cut the topsail sheets and a boatswains' mate and some seamen tried to obey but failed from everything being thrown over bodily to leeward and their being deprived of any hold. The boats and booms were also ordered to be cut away, and the captain of the maintop directed to cut away the weather rigging lanyards to right the ship, which he did as far as he could until floated off her side. To add to the confusion, two or three of the after guns, on the larboard side of the quarterdeck broke loose after the ship was upset and went over with a great crash through the leeward side, owing as was supposed to the bolts giving way as the guns were well secured with very good tackling. Directly after all further efforts were rendered useless by the *Orquixo* going down less than half an hour after Captain Balderston left her. The weather was then and for a short time afterwards so very thick that the men left on part of wreck could not be seen from the *Penguin* and only a few survivors were picked up by her boats or picked up in an exhausted state and brought to her by some of her men who jumped overboard and swam to their rescue with the lead and other small lines. Out of 136 souls on board the *Orquixo*, including 3 officers and 30 non-commissioned officers and men of the 6th Battalion 60th Regiment, there were saved Lieutenant Stanton, the Master, boatswain, master's mate, 1 midshipman, 25 petty officers, seamen and boys, Lieutenant Gibson and 2 privates of the 60th and 2 black persons, but 5 of the seamen died on board the *Penguin*, making the total loss 101. Ensigns Johann Voltz and Hassenger of the 60th were drowned with the rest of their detachment. The *Penguin* with the survivors of the disaster arrived at Port Royal, Jamaica, on November 9. The loss of the *Orquixo* was attributed partly to the squall, partly to the guns going through her side.   At a court-martial, of which Captain Vansittart, of the Fort was President, held on board the *Hercule* at Port Royal, Jamaica, on November 11, 1805, Captain Balderston, his officers and ship's company were fully acquitted of all blame in connection with the loss of the *Orquixo*.

# INDEX

PRINTED IN GREAT BRITAIN BY WILLIAM CLOWES AND SONS, LIMITED, LONDON AND BECCLES.

Lightning Source UK Ltd.
Milton Keynes UK
UKHW021346110821
388639UK00004B/247

Printed by Amazon Italia Logistica S.r.l.
Torrazza Piemonte (TO), Italy

That's it. That's a great question to finish with. I hope you enjoyed this book, and I hope you got most of the answers right.

I also hope you learnt some new facts about the club, and if you saw anything wrong, or have a general comment, please visit the glowwormpress.com website.

Thanks for reading, and if you did enjoy the book, would you be so kind as leave a positive review on Amazon.

A100. Ex-player Alan Pardew scored a memorable winning goal in a 4-3 victory over Liverpool in the 1990 FA Cup Semi-Final at Villa Park on 8th April 1990. It was a fitting finale to one of the most thrilling matches you could ever wish to see.

A101. Kayla, an American bald eagle, flies from one end of the pitch to the other at the start of every home game. She has been making appearances at Selhurst since 2010.

Here is the final set of answers.

A91. Cheikhou Kouyate transferred from West Ham United to join Crystal Palace in August 2018 for a reported fee of £9.5 million.

A92. Townsend wears the number 10 shirt.

A93. The legendary Jim Cannon was the only Scottish player to make the club's Centenary XI.

A94. Dean Kiely is the club's goalkeeping coach.

A95. Andy Jonson scored a massive 74 league goals for the club in 140 appearances.

A96. Italian manager Dario Gradi was the club's first foreign manager, taking over the club in 1981.

A97. Jordan Ayew was born in Marseille in France, but has represented Ghana over 50 times at international level.

A98. In 1998 Palace became the first English team to sign two Chinese players when they signed Fan Zhiyi and Sun Jihai.

A99. Crystal Palace was originally nicknamed the Glaziers. The change of nickname from the Glaziers to the Eagles was part of Malcolm Allison's overhaul of the club image in the early 1970s, as he changed the colour of the club's shirts, the badge and the nickname.

c. Kylie

A. Dario Gradi
B. Attilio Lombardo
C. Roberto Mancini

97. Where was Jordan Ayew born?
   A. France
   B. Ghana
   C. Ivory Coast

98. In 1998 Palace became the first English team to do what?
   A. Sign two Chinese players
   B. To have the same manager take over the club for the third time
   C. Fall into administration

99. What was Crystal Palace's original nickname?
   A. The Crystals
   B. The Diamonds
   C. The Glaziers

100. Which ex-player scored the winning goal against Liverpool in the 1990 FA Cup semi-final?
   A. Steve Coppell
   B. Dougie Freedman
   C. Alan Pardew

101. What is the name of the eagle that you can see at home games?
   A. Kayla
   B. Kanye

Here is the final set of questions. Good luck.

91. From which club did Cheikhou Kouyate transfer from to join Palace in 2018?
    A. West Bromwich Albion
    B. West Ham United
    C. Wolverhampton Wanderers

92. What is Andros Townsend's shirt number?
    A. 10
    B. 11
    C. 12

93. Who was the only Scot to be included in the club's Centenary XI?
    A. Jim Cannon
    B. Dougie Freedman
    C. George Wood

94. Who is the club's goalkeeping coach?
    A. Dean Kiely
    B. Nigel Martyn
    C. George Wood

95. How many goals did Andrew Johnson score for the club in the League?
    A. 74
    B. 75
    C. 76

96. Who was the club's first foreign manager?

A89. Palace based their badge on Portuguese side Benfica, with an eagle clutching a football.

A90. Legendary goalkeeper Nigel Martyn was placed in the club's Centenary XI.

Here are the answers to the last set of questions.

A81. Billy Davies was the club's first internationally capped player. He played for Wales against Scotland in 1908.

A82. Gregg Berhalter was the first Palace player to play in the World Cup Finals.

A83. He played for the United States during the 2002 World Cup Final tournament held in South Korea and Japan.

A84. Archie Needham was the first player to reach 100 games for Crystal Palace.

A85. Attilio Lombardo played for the club between 1997 and 1999 and he was nicknamed the 'Bald Eagle' because of his lack of hair and the massive influence he had on the pitch.

A86. Wilfried Zaha has been voted Young Player of the Year twice; he won it after the 2011-12 and 2012-13 seasons.

A87. Julian Speroni has been voted the club's Player of the Year four times in total.

A88. Puma is the official kit supplier. It would be fair to say Crystal Palace have had many kit manufacturers over the years.

A. 1
B. 2
C. 3

87.     How many times has Julian Speroni won the club's Player of the Year award?
    A. 2
    B. 3
    C. 4

88.     Who is the official kit supplier?
    A. Nike
    B. Puma
    C. Umbro

89.     After which club did Crystal Palace base their club badge?
    A. Benfica
    B. Sporting Lisbon
    C. Valencia

90.     Who did Crystal Palace place in goal for the Club's Centenary XI?
    A. Paul Barron
    B. John Burridge
    C. Nigel Martyn

Here is the next set of questions.

81.     Who is the club's first internationally capped player?
    A. Kenny Samson
    B. Horace Colclough
    C. Billy Davies

82.     Who was the first Crystal Palace player to play in the World Cup Finals?
    A. Gregg Berhalter
    B. Andrew Johnson
    C. Geoff Thomas

83.     Which country did he play for?
    A. England
    B. United States
    C. Wales

84.     Who was the first Crystal Palace player to reach 100 appearances?
    A. Harry Collyer
    B. Wilf Innerd
    C. Archie Needham

85.     Which Crystal Palace player was nicknamed the 'Bald Eagle'?
    A. Andy Johnson
    B. Attilio Lombardo
    C. Ian Wright

86.     How many times has Wilfried Zaha win the club's Young Player of the Year Award?

A80. Crystal Palace's first manager was John Robson; he was manager between 1905 and 1907.

Here are the answers to the latest set of questions.

A71. Crystal Palace's first and only European opponents were Turkish side Samsunspor.

A72. Crystal Palace lost both legs against Samsunspor 2-0.

A73. Crystal Palace's only time in Europe came in July 1998.

A74. The club's record attendance for their solitary European appearance was 11,758.

A75. The highest number of people who have ever watched a Palace game is 82,025. This was for the 2013 Football League Play Off Final against Watford at Wembley Stadium on 27th May 2013.

A76. Palace scored a whopping 110 goals in a season, easily the most they have ever scored in one single season.

A77. Palace scored 110 goals during the 1960-61 season in Division Four.

A78. Crystal Palace's lowest goal tally in the Premier League is just 33 goals, which came during the 2013-14 season.

A79. Ray Lewington started the 2019-20 season as assistant manager.

C. 110

77. During which season?
   A. 1959-60
   B. 1960-61
   C. 1962-63

78.     What is the fewest number of goals the
   club has scored in a Premier League season?
   A. 32
   B. 33
   C. 34

79.     Who started the 2019-20 season as
   assistant manager?
   A. Ray Lewington
   B. Dave Reddington
   C. Ed Richmond

80.     Who was the club's first ever manager?
   A. Edmund Goodman
   B. Alex Maley
   C. John Robson

71. Who were the club's first European
    opponents?
    A. Antalyaspor
    B. Samsunspor
    C. Trabzonspor

72.      What was the score?
    A. 0-2
    B. 1-2
    C. 1-3

73.      What was the year?
    A. 1996
    B. 1997
    C. 1998

74.      What was the club's record attendance
    in Europe?
    A. 11,758
    B. 12,758
    C. 13,758

75.      What is the highest number of
    spectators Crystal Palace has played in front
    of?
    A. 80,250
    B. 81,520
    C. 82,025

76.      What is the highest number of goals the
    club has scored in a season?
    A. 100
    B. 105

Here are the answers to the last set of questions.

A61. The traditional colour of the home kit is red and blue stripes; it changed from claret and blue stripes in 1973.

A62. The away colour has changed regularly over the years, although it is generally agreed that the traditional away kit is mainly yellow. Or white!

A63. International gaming company ManBetX is the official club sponsor.

A64. Crystal Palace's first kit sponsor was Red Rose, with a deal signed in time for the 1983-84 season.

A65. Virgin Atlantic once sponsored Crystal Palace's kit, from 1988 to 1991.

A66. Steve Parish is currently the club chairman.

A67. Palace and Welsh legend Chris Coleman was included in the club's Centenary XI.

A68. Tony Collins was the club's first black player; signing for the club way back in 1957.

A69. Crystal Palace played their first ever league match against Merthyr on the 28th August 1920.

A70. Roy Hodgson started the 2019-20 season as manager. He was appointed to the role in September 2017.

C. Steve Parish

67. Which Welsh legend was included in Palace's Centenary XI?
   A. Chris Coleman
   B. Peter Nicholas
   C. Ian Rush

68. Who was the club's first black player?
   A. Mark Bright
   B. Tony Collins
   C. John Salako

69. Who was the club's first match in the league against?
   A. Loughborough
   B. Merthyr
   C. New Brompton

70. Who started the 2019-20 season as manager?
   A. Sam Allardyce
   B. Roy Hodgson
   C. Tony Pulis

Let's give you some easier questions.

61. What is the traditional colour of the home shirt?
    A. Red and Black stripes
    B. Red and Blue stripes
    C. Red and White stripes

62. What is the traditional colour of the away shirt?
    A. Blue
    B. Green
    C. Yellow

63. Who is the current club sponsor?
    A. Betfair
    B. ManBetX
    C. Mansion

64. Who was the first shirt sponsor?
    A. AVR
    B. Red Rose
    C. Top Score

65. Which of these airlines have once sponsored the club?
    A. Emirates
    B. Flybe
    C. Virgin Atlantic

66. Who is currently the club chairman?
    A. Jeremy Hosking
    B. Martin Long

A58. Crystal Palace has no club motto.

A59. Crystal Palace has a rivalry with Millwall, but a bitter rivalry with Brighton and Hove Athletic. The rivalry started when Palace were relegated to the Third Division in 1974 and the clubs started playing one another in the League. Both clubs saw themselves as the biggest clubs in the division and a number of incidents strengthened the rivalry. One such incident was at a controversial cup tie in 1976 involving a disputed re-taken penalty. The Brighton manager at the time Alan Mullery was enraged at the end of the match, and after having a cup of coffee thrown at him, he reacted angrily swearing and making V signs at the Palace fans, and a deep antagonism was reinforced.

A60. "Walking down the Holmesdale Road" is a famous Palace song that is regularly sung from the stands.

Here are the answers to the last block of questions.

A51. Steve Coppell can be considered to be Palace's most successful manager. Under his management the club won 2 play-off finals, were runners up in the FA Cup, finished 3rd in the League and also won the Full Members Cup.

A52. Edmund Goodman is Palace's longest serving manager; he was at the club for 18 years, from 1907-1925.

A53. Steve Coppell is the club's longest serving post war manager; he has been manager of the club four times with a combined total of 12 years.

A54. Riihilahti was born in Helsinki in Finland. He joined Palace in 2001 and quickly became a fans favourite. Riihilahti was so loved by the Palace faithful that a Finnish flag with the legend 'AKI 15' across the centre was hung behind one of the Selhurst Park goals for an entire Premier League season.

A55. Jonathon Rogers is the editor of the Matchday Programme.

A56. Five Year Plan is a famous Crystal Palace fanzine. It is now a website that also offers popular podcasts.

A57. Of course there is an eagle on the club's crest, hence the nickname The Eagles.

56. Which of these is a Crystal Palace fanzine?
   A. Palace Centre
   B. Five Year Plan
   C. Support Selhurst

57. What animal is on the club crest?
   A. Eagle
   B. Hawk
   C. Ostrich

58. What is the club's motto?
   A. Never give up
   B. Nothing but the best
   C. There is no motto

59. Who is considered as Crystal Palace's main rivals?
   A. Brighton & Hove Albion
   B. Charlton Athletic
   C. Millwall

60. What could be regarded as the club's most well-known song?
   A. Red side of London
   B. The Selhurst March
   C. Walking down the Holmesdale Road

I hope you're learning some new facts about the Eagles.

51. Who is Crystal Palace's most successful manager?
   A. Steve Bruce
   B. Steve Coppell
   C. Alan Smith

52. Who is the club's longest serving manager of all time?
   A. Steve Coppell
   B. Edmund Goodman
   C. Bert Head

53. Who is the club's longest serving post war manager?
   A. Steve Bruce
   B. Steve Coppell
   C. Alan Smith

54. What nationality was cult hero Aki Riihilahti?
   A. Danish
   B. Finnish
   C. Norwegian

55. Who is the editor of the match day programme?
   A. Martin Long
   B. Jonathon Rogers
   C. Steve Parish

October 2007. He went on to play just four times for the club. The last we heard he was still a professional footballer in Belgium.

A49. Jack Little is the oldest player to ever put on a Crystal Palace shirt, at the staggering age of 41 years and 68 days when he turned out against Gillingham on 3rd April 1926. It's a record that is unlikely to ever be beaten.

A50. Kevin Phillips is Palace's oldest ever goal scorer for the club, scoring a penalty at the ripe old age of 39 years and 306 days after coming on as a substitute against Watford in the Play-Off Final on 27th May 2013. It was a crucial goal in a crucial match.

Here are the answers to the last set of questions.

A41. The highest transfer fee paid by the club is £27 million to Liverpool for a Belgian forward.

A42. This amount of money was spent on Christian Benteke in August 2016, which smashed the previous record of £13 million paid for Andros Townsend just weeks before.

A43. The record transfer fee received is £45 million.

A44. This huge fee was received for Aaron Wan-Bissaka who was sold to Manchester United in June 2019. This beat the previous record of £25 million received for Yannick Bolasie from Everton in August 2016.

A45. Harry Colcough was the first Palace player to play for England, and his first and only appearance came in 1914.

A46. Mile Jedinak won 79 caps in total for Australia, playing 37 times for his country whilst a player at Palace.

A47. Welsh Wizard Ian Walsh scored the most international goals whilst a Palace player scoring seven goals for his country.

A48. John Bostock is the youngest player to ever represent the club. He was aged 15 years, 287 days when he made his debut against Watford on 29th

A. Mile Jedinak
B. Aki Riihilati
C. Paddy Mulligan

47. Who has scored the most international goals whilst a Crystal Palace player?
   A. Chris Coleman
   B. Ian Walsh
   C. Fan Zhiyi

48. Who is the youngest player ever to represent the club?
   A. John Bostock
   B. Dwight Gayle
   C. Sean Scannell

49. Who is the oldest player ever to represent the club?
   A. Jack Little
   B. Kevin Phillips
   C. Julian Speroni

50. Who is the club's oldest ever goal scorer?
   A. Andrew Johnson
   B. Kevin Phillips
   C. Gareth Southgate

I hope you're having fun, and getting most of the answers right.

41. What is the record transfer fee paid?
    A. £21 million
    B. £24 million
    C. £27 million

42. Who was the record transfer fee paid for?
    A. Christian Benteke
    B. Mamadou Sakho
    C. Andros Townsend

43. What is the record transfer fee received?
    A. £25 million
    B. £35 million
    C. £45 million

44. Who was the record transfer fee received for?
    A. Aaron Wan-Bissaka
    B. Yannick Bolasie
    C. Dwight Gayle

45. Who was the first Crystal Palace player to play for England?
    A. Jack Alderson
    B. Harry Colclough
    C. Billy Turner

46. Who has won the most international caps whilst a Crystal Palace player?

A39. Mile Jedinak was the last captain to lift a trophy for Palace; he lifted the trophy after winning the play offs in 2013.

A40. Luka Milivojevic started the 2019-20 season as club captain.

Here are the answers to the last block of questions.

A31. Peter Simpson has scored the most hat-tricks for the club, scoring 20 in total for the club.

A32. In 1990 Crystal Palace reached the final of the FA Cup losing 1-0 to Manchester United in a replay. The club also got to the Final in 2016 and lost 2-1 to Manchester United. So near, yet so far.

A33. Crystal Palace won the Full Members Cup, known as the Zenith Data Systems Cup for sponsorship reasons, on the 7th April 1991.

A34. The Full Members Cup win was by the convincing score line of 4-1 with three of the goals being scored in extra time.

A35. Palace beat Everton 4-1 in the final at Wembley in front of 52,460 supporters.

A36. The Eagles have won the Second Division twice, in 1978-79 and 1993-94.

A37. Crystal Palace finished 3rd in the top flight at the end of the 1990-91 season, the highest they have ever finished.

A38. Gareth Southgate was the last captain to lift a league trophy, leading the club to the 1993-94 First Division title.

B. 2

C. 3

37. What's the highest position Palace have finished in the top division?
   A. 3rd
   B. 5th
   C. 7th

38. Who was the last captain to lift a League trophy?
   A. Andy Roberts
   B. Gareth Southgate
   C. Geoff Thomas

39. Who was the last captain to lift a trophy for the club?
   A. Mile Jedinak
   B. Paddy McCarthy
   C. James Tonkins

40. Who started the 2019-20 season as club captain?
   A. Gary Cahill
   B. James McCarthy
   C. Luka Milivojevic

Here is the next set of questions.

31. Who has scored the most hat tricks for Crystal Palace?
   A. Kevin Phillips
   B. Peter Simpson
   C. Ian Wright

32. What is the furthest the club has reached in the FA Cup?
   A. Quarter Finals
   B. Semi-Finals
   C. Final

33. When did the club win the Full Members Cup?
   A. 1989
   B. 1990
   C. 1991

34. What was the score?
   A. 3-1
   B. 4-1
   C. 5-1

35. Who did they beat?
   A. Everton
   B. Liverpool
   C. Ipswich

36. How many times have Crystal Palace won Division 2?
   A. 1

A30. @CPFC is the club's official twitter account. It tweets multiple times daily and the account now has close to a million followers.

Here are the answers to the last block of questions.

A21. Crystal Palace spent their early years playing in the Southern League, before joining the Football League in 1920.

A22. The club incredibly won the League in its first season, becoming inaugural champions of the newly formed Third Division in 1920-21.

A23. Palace finished a very credible 12th at the end of the 2018-19 season.

A24. Crystal Palace's record win was 13-1 in a 2014-15 pre-season friendly in Austria against GAK Graz. However, the club's biggest victory in the league is 9-0.

A25. This convincing 9-0 victory came against Barrow.

A26. Crystal Palace beat Barrow 9-0 on the 10th October 1959, during the 1959-60 season.

A27. Similarly Crystal Palace's biggest defeat is also 9-0.

A28. This crushing defeat came on the 12th December 1989 during the 1989-90 season away to a very strong Liverpool side at Anfield.

A29. cpfc.co.uk is the club's official website address.

B. 1960-61

C. 1961-62

27.       What is the club's record league defeat?
   A. 8-0
   B. 9-0
   C. 10-0

28.       Who against?
   A. Burnley
   B. Aston Villa
   C. Liverpool

29. What is the club's official website address?
   A. cpfc.co.uk
   B. crystalpalace.com
   C. crystalpalace.co.uk

30. What is the club's official twitter account?
   A. @CPFC
   B. @CrystalPalace
   C. @PalaceOfficial

Now we move onto some questions about the club's records.

21. When did the club join the Football League?
    A. 1920
    B. 1922
    C. 1924

22. What position did the club finish in its first season in the Football League?
    A. 1st
    B. 7th
    C. 13th

23. What position did the club finish at the end of the 2018-19 season?
    A. 12th
    B. 14th
    C. 16th

24. What is the club's record win in the league?
    A. 7-0
    B. 8-0
    C. 9-0

25. Who did they beat?
    A. Barrow
    B. Clapham
    C. Luton Town

26. In which season?
    A. 1959-60

Here are the answers to the last block of questions.

A11. During the 1960-61 season in Division Four, Crystal Palace scored a record 110 goals, the most they have ever scored in one season.

A12. During the 1978-79 season in Division Two, Palace only conceded 24 goals, a fine statistic for any team.

A13. Peter Simpson has scored the most penalties for the club; he scored 20 in all competitions.

A14. The players run out to the Dave Clark Five song "Glad All Over".

A15. Selhurst Park is on Whitehorse Lane.

A16. The Arthur Wait Stand has the largest capacity at the stadium.

A17. The home end of the ground is called the Holmesdale Road Stand; and it has a capacity of just over 8,000.

A18. The club's record attendance was 51,801 on 11th May 1979 against Burnley.

A19. Crystal Palace's training ground is situated on Copers Cope Road in Beckenham.

A20. The size of the pitch at Selhurst Park is 110 metres long x 74 metres wide.

A. Arthur Wait Stand
B. Main Stand
C. Holmesdale Road Stand

17. What is the home end of the ground known as?
  A. Arthur Wait Stand
  B. Holmesdale Road Stand
  C. Whitehorse Lane Stand

18.    What is the club's record attendance?
  A. 51,800
  B. 51,801
  C. 51,802

19. Where is Crystal Palace's training ground?
  A. Beckenham
  B. Biggin Hill
  C. Bromley

20.    What is the size of the pitch?
  A. 110m x 74m
  B. 112m x 75m
  C. 114m x 76m

OK, back to the questions.

11. What is the highest number of goals that Crystal Palace has scored in a league season?
    A. 105
    B. 110
    C. 115

12. What is the fewest number of goals that Crystal Palace has conceded in a league season?
    A. 24
    B. 25
    C. 26

13. Who has scored the most penalties for the club?
    A. Dougie Freedman
    B. Andy Johnson
    C. Peter Simpson

14. What song do the players run out to?
    A. 'Glad All Over'
    B. 'We Love You'
    C. 'You Are My Palace'

15. What is the name of the road the ground is on?
    A. Black Horse Street
    B. Grey Horse Road
    C. Whitehorse Lane

16. Which stand has the biggest capacity?

A7. Jim Cannon has made the most appearances for the club in the league, 571 in total. Legend.

A8. Peter Simpson is the club's record goal scorer with 165 goals in 195 games, which is a very impressive strike rate.

A9. Keith Smith is the club's fastest ever goal scorer, scoring after just six seconds against Derby County on 12th December 1964.

A10. The Clifton Arms is a well known pub near the ground. Be prepared to queue for a pint though!

OK, so here are the answers to the first ten questions. Expect to get seven or more right, but don't get too cocky, as the questions do get harder.

A1. Crystal Palace was founded on 10th September 1905. The club was founded at the famous Crystal Palace Exhibition building by the owners of the FA Cup Final stadium, who wanted their new team to play at the historic venue.

A2. Crystal Palace adopted the nickname of The Eagles in 1973. A question later in the book will ask what the club's previous nickname was.

A3. The club plays their home games at Selhurst Park. It has been their home since 1924.

A4. Selhurst Park is quite a modest stadium by modern standards, and one of the smaller venues for a top flight club with a current maximum capacity of 25,486; but when it is full it feels like there are a lot more people inside.

A5. The club mascot is Pete the Eagle. For reasons that have never been clear, Pete wears sunglasses. There is also a female mascot called Alice the Eagle. Give yourself a bonus point if you knew that.

A6. Legendary centre back Jim Cannon has made the most appearances for the club with a total of 660 during his time at the club from 1973 to 1988.

7. Who has made the most *League* appearances for the club?
   A. Jim Cannon
   B. John Jackson
   C. Julian Speroni

8. Who is the club's record goal scorer?
   A. Dougie Freedman
   B. Peter Simpson
   C. Ian Wright

9. Who is the fastest ever goal scorer for the club?
   A. Andy Johnson
   B. Kevin Phillips
   C. Keith Smith

10. Which of these is a well known pub near the ground?
   A. Clifton Arms
   B. The Eagles Nest
   C. Palace Arms

Let's start with some relatively easy questions.

1. When were Crystal Palace founded?
   A. 1904
   B. 1905
   C. 1906

2. What is Crystal Palace's nickname?
   A. The Buzzards
   B. The Eagles
   C. The Hawks

3. Where do Crystal Palace play their home games?
   A. Sandhurst Park
   B. Selhurst Park
   C. St James Park

4. What is the stadium's capacity?
   A. 25,486
   B. 25,648
   C. 25,864

5. Who or what is the club mascot?
   A. Eddie the Eagle
   B. Kayla the Eagle
   C. Pete the Eagle

6. Who has made the most appearances for the club in total?
   A. Jim Cannon
   B. Dougie Freedman
   C. Terry Long

# FOREWORD

When I was asked to write a foreword to this book I was flattered.

I have known the author Chris Carpenter for many years and his knowledge of facts and figures is phenomenal.

His love for football and his skill in writing quiz books make him the ideal man to pay homage to my great love Crystal Palace Football Club.

This book came about as a result of a challenge on a golf course.

I do hope you enjoy the book.


Derek Seddon

# 2019-20 Season Edition

# Crystal Pal Quiz Book

C000115222

## 101 Questions To Test Your Knowledge of Crystal Palace Football Club

Published by Glowworm Press
7 Nuffield Way
Abingdon OX14 1RL

By Chris Carpenter

## Crystal Palace Football Club

This book contains one hundred and one informative and entertaining trivia questions with multiple choice answers. Some questions are easy, some are more challenging, yet this entertaining book will test your knowledge of your club. The questions cover a wide range of topics associated with Crystal Palace Football Club for you to test yourself.

You will be quizzed on players, legends, managers, opponents, transfer deals, trophies, records, honours, fixtures, terrace songs and much more, guaranteeing you both an educational experience and hours of fun. The **Crystal Palace FC** Quiz Book will provide the ultimate in entertainment for Crystal Palace fans of all ages.

# RESULTS

OF

# RAILWAY EXTENSION.

*A PAPER READ BEFORE THE STATISTICAL SOCIETY*

*OF LONDON, BY R. DUDLEY BAXTER, M. A.,*

*IN NOVEMBER,* 1866.

Four Hundred and Forty Millions of Government Subsidy
to the Cotton Railroads
in India.

# RAILWAY EXTENSION AND ITS RESULTS.

·I.—INTRODUCTION.

If a Roman Emperor, in the most prosperous age of the empire, had commanded a history to be written of that wonderful system of roads which consolidated the Roman power, and carried her laws and customs to the boundaries of the accessible world, it would have afforded a just subject for national pride. The invention and perfecting of the art of road making, its sagacious adoption by the State, its engineering triumphs, its splendid roads through Italy, through Gaul, through Spain, through Britain, through Germany, through Macedonia, through Asia Minor, through the chief islands of the Mediterranean, and through Northern Africa; all these would have been recounted as proofs of Roman energy and magnificence, and as introducing a new instrument of civilization, and creating a new epoch in the history of mankind.

A similar triumph may fairly be claimed by Great Britain. The Romans were the great road-makers of the ancient world—the English are the great railroad makers of the modern world. The tramway was an English invention, the locomotive was the production of English genius, and the first railways were constructed and carried to success in England. We have covered with railroads the fairest districts of the United Kingdom, and developed railways in our colonies of Canada and India. But we have done much more than this, we have introduced them into almost every civilized country. Belgian railways were planned by George Stephenson. The great French system received an important impulse from Locke. In Holland, in Italy, in Spain, in Portugal, in Norway, in Denmark, in Russia, in Egypt, in Turkey, in Asia Minor, in Algeria, in the West Indies, and in South America, Englishmen have led the way in railway enterprise and construction. To this day, wherever an under- taking of more than ordinary difficulty presents itself, the aid is invoked of English engineers, English contractors, English navvies, and English shareholders; and a large portion of the rails with which the line is laid, and the engines and rolling stock with which it is worked, are brought from England.

To Englishmen the annals of railways must always be of the highest interest, and I trust that the brief inquiry upon which I am about to en- ter will not be deemed a waste of labor. I propose to examine into the extension of railways at home and abroad, to show the rate at which it is proceeding, the expenditure which it has cost, and its vast commercial

results.  The practical questions will follow whether the construction of
railways in the United Kingdom has reached its proper limit?  Are we
over-railroaded, as some assert, so that railways ought to be discouraged?
Or are we under-railroaded, so that fresh railways ought to be invited?
Are other nations passing us in the race of railway development?  And,
lastly, can any improvement be introduced into our railway legislation?

## II.—RAILWAYS IN THE UNITED KINGDOM.

So far as roads are concerned, the dark ages may be said to have lasted
from the evacuation of Britain by the Romans in 448 to the beginning
of the last century.  During the whole of that period nothing could be
more barbarous or impassable than English highways.  The Sçotch re-
bellions first drew attention to the necessity of good roads.  The first step
was to establish turnpikes, with their attendant wagons and stage coaches,
superseding the long strings of pack-horses which, up to that time, had
been the principal means of transport.  The second step was to render
navigable the rivers which passed through the chief seats of industry.
The third, which commenced later in the century, was to imitate the
rivers by canals, and to construct through the north and centre of England a net-work of 2,600 miles of water communication at an outlay of
£50,000,000 sterling.  But roads and canals combined were insufficient
for the trade of Lancashire and Yorkshire, and bitter complaints were
made of expense and delay in the transmission of their goods.

The desired improvement came from the mining districts.  Since the
year 1700 it had been the custom to use wooden rails for the passage of
the trucks.  About the year 1800 Mr. Outram, in Derbyshire, laid down
iron rails upon stone sleepers, and the roads so constructed took from him
the name of Outram's Ways or Tramways.  About the year 1814 the ingenuity of mining engineers developed the stationary steam engine into a
rude locomotive, capable of drawing heavy loads at the rate of four or
five miles an hour.  It was proposed to construct a public railway on this
principle between Stockton and Darlington.  After much delay the line
was opened by George Stephenson in 1825, and the experiment was successful as a goods line—unsuccessful, from its slowness, as a passenger
line.  The next experiment was the Manchester and Liverpool railway,
projected as a goods line to accommodate the increasing trade of those
two places, which was crippled by the high rates of the canal and navigation.  Before the railway was completed, another great improvement
had taken place in the construction of locomotives by the discovery of
the multitubular boiler, which immensely increased the volume of steam
and the speed attainable.

The opening of the Manchester and Liverpool railway on 15th September, 1830, was the formal commencement of the railway era.  On that
day the public saw for the first time immense trains of carriages, loaded
with passengers, conveyed at a rate of more than fifteen miles an hour, a
speeed which was largely exceeded in subsequent trials.  The desidera-
tum was at length obtained, viz: the conveyance of large masses of passengers and goods with ease and rapidity ; and it was seen that the discovery must revolutionize the whole system of inland communication.

The public feeling was strangely excited.  Commercial men and men
of enterprise were enthusiastic in favor of the new railways, and eager

for their introduction all over the country. But the vested interests of roads and canals, and landed proprietors who feared that their estates would be injured, together with the great body of the public, were violently prejudiced against them. Railways had to fight their way against the most strenuous opposition. I quote from the " Life of Robert Stephenson," the engineer of the London and Birmingham line :

"In every parish through which Robert Stephenson passed he was eyed with suspicion by the inhabitants, and not seldom menaced by violence. The aristocracy regarded the irruption as an interference with territorial rights. The humbler classes were not less exasperated, as they feared the railway movement would injure those industrial interests by which they lived. In London, journalists and pamphleteers distributed criticisms, which were manifestly absurd, and prophecies which time has signally falsified."
—Vol. i., p. 169.

The city of Northampton was so vehement in its opposition that the line was diverted to a distance of five miles, through the Kilsby Tunnel, to the permanent injury both of the city and railway. The bill was thrown out in Parliament, and only passed in the following session by the most lavish expenditure in buying off opposition.

Other lines were soon obtained in spite of the same vehement hostility. The Grand Junction railway from Liverpool to Birmingham was passed in 1833. The Eastern Counties railway was sanctioned in 1834. It was launched as a 15 per cent. line. It is said that a wealthy banker in the eastern counties made a will, leaving considerable property to trustees to be expended in parliamentary opposition to railways. The Great Western was thrown out in 1834, but passed in 1835. The London and Southampton, now the London and Southwestern, was proposed in 1832, but was not sanctioned till 1834.

In 1836 came the first railway mania. Up to this time the difficulty had been to pass any bill at all; now competing schemes began to be brought before Parliament. Brighton was fought for by no less than five companies, at the total expenditure of £200,000. The Southeastern obtained its act after a severe contest with the Mid Kent and Central Kent. Twenty-nine bills were passed by Parliament authorizing the construction of 994 miles of railway. In the autumn the mania raged with the greatest violence. " There is scarcely," said the Edinburgh Review, " a practicable line between two considerable places, however remote, that has not been occupied by a company; frequently two, three, or four rival lines have started simultaneously." The winter brought a crash, and the shares of the best companies became almost unsaleable.

In 1845 most of the great lines had proved a success. The London and Birmingham was paying 10 per cent., the Grand Junction 11 per cent., the Stockton and Darlington 15 per cent., and railway shares were on an average at 100 per cent. premium. The railway mania broke out with redoubled violence; railways appeared an El Dorado. The number of miles then open was 2,148. The number of miles sanctioned by Parliament in the three following sessions was:

|  |  |
|---|---|
| 1845........................................................ | 2,700 |
| 1846........................................................ | 4,538 |
| 1847........................................................ | 1,354 |
| Total........................................................ | 8,592 |

**Had** all these lines been constructed, we should have had in 1852 more than 10,700 miles of railway, a number which was not actually reached till 1861, or nine years later. But the collapse in 1846 was so severe that an act was passed for the purpose of facilitating the dissolution of companies, and a large number of lines were abandoned, amounting, it is said, to 2,800 miles.

Railway extension was now menaced with a new danger. The effect of the panic was so great, and the losses on shares so severe, that the confidence of the public was destroyed. Besides this, as the new lines were opened, the dividends gradually decreased till the percentage of profit on capital had gone down from 5½ per cent. in 1845 to 3½ in 1849 and 3½ in 1850, leaving scarcely anything for ordinary shareholders. As a consequence, shareholders' lines were at an end. But since 1846 a new custom had been gaining ground of the amalgamation of smaller into larger companies. I may instance the North Eastern Company, which consists of twenty-five originally independent railways. In this manner eleven powerful companies had been formed, which divided the greater part of England between them. The competition between these companies for the possession of the country was very great, and by amalgamations, leases, guarantees, and preference stocks, they financed a large number of lines which otherwise could not be made. In this manner the construction of railways between 1850 and 1858 progressed at the rate of nearly 400 miles a year.

But towards the end of 1858 the great companies had exhausted their funds and ardor, and proposed terms of peace. The technical phrase was "that the companies required rest." Again it seemed probable that railway extension would be checked. But a new state of things arose. Twenty years of railway construction had brought forward many great contractors, who made a business of financing and carrying through lines which they thought profitable. The system had grown up gradually under the wing of the companies, and it now came to the front, aided by a great improvement in the value of railway property, on which the percentage of profits to capital expended had gradually risen from 3½ per cent. in 1850 to 4½ in 1860. The companies also found it their interest to make quiet extensions when required by the traffic of the country. Thus railway construction was continued in the accelerated ratio of more than 500 miles a year. The following table gives a summary of the rate of progress from 1845 to 1865 :—

UNITED KINGDOM—MILES CONSTRUCTED.

| Year. | Miles Opened. | Average Number. Opened per An. |
|---|---|---|
| 1834 ...... about | 200 | 133 |
| 1840 ...... " | 1,200 | |
| | | 240 |
| 1845 ...... | 2,440 | |
| | | 812 |
| 1850 ...... | 6,500 | |
| | | 367 |
| 1855 ...... | 8,335 | |
| | | 425 |
| 1860 ...... | 10,434 | |
| 1865 ...... | 13,289 | 571 |

During the same year the percentage of profits to capital expended were as follows:—

| | Per cent. | | Per cent. |
|---|---|---|---|
| 1845 | 5.48 | 1860 | 4 39 |
| 1850 | 3.31 | 1865 | 4.46 |
| 1855 | 3.90 | | |

The latter table, which is abridged from an annual statement in *Herepath's Journal*, scarcely gives an idea of the gradual manner in which the dividends sunk from their highest point in 1845 to their lowest in 1850, and of their equally gradual recovery from 1850 to 1860 and 1865. The main results of the two tables are, first, the close connection between the profit of one period and the average number of miles constructed in the next five years; and, second, the fact that the construction of railways in the United Kingdom has been steadily increasing since 1855, and is now more than 500 miles per annum.

The number of miles authorized by Parliament during the last six years is stated in the *Railway Times* to be as follows:—

| Year. | Miles. | Year. | Miles. |
|---|---|---|---|
| 1861 | 1,332 | 1864 | 1,329 |
| 1862 | 809 | 1865 | 1,996 |
| 1863 | 795 | 1866 | 1,062 |
| | | | 7,323 |
| Average | | | 1,220 |

Hence the miles authorized by Parliament for the last six years have been double the number constructed; and there must be about 3,500 miles not begun or not completed—a number sufficient to occupy us for fully seven years, at our present rate of construction.

Such is a brief summary of the history of 'railway extension in Great Britain and Ireland. It may be thrown into five periods:

1. The period of experiment, from 1820 to 1830.
2. The period of infancy, from 1830 to 1845.
3. The period of mania, from 1845 to 1848.
4. The period of competition by great companies, from 1848 to 1859.
5. The period of contractors' lines and companies' extensions, from 1859 to 1865.

### III.—DISTRIBUTION OF RAILWAYS IN THE UNITED KINGDOM.

The returns of the Board of Trade to the end of 1865 give the following distribution of the 13,289 miles then open:—

| | Double Lines. | Single Lines. | Total Miles Open. |
|---|---|---|---|
| England and Wales | 6,081 | 3,170 | 9,251 |
| Scotland | 946 | 1,254 | 2,200 |
| Ireland | 476 | 1,362 | 1,838 |
| | 7,503 | 5,786 | 13,289 |

Hence there is a considerable preponderance of double lines over single lines in England, and of single lines over double in Scotland and Ireland.

The following table shows which country has the greatest length of railways in proportion to its area :—

| | Area in Square Miles. | Railway Mileage. | Square Miles. per Mile of Railway |
|---|---|---|---|
| England and Wales | 57,812 | 9,251 | 6.25 |
| Scotland | 30,715 | 2,200 | 14. |
| Ireland | 32,512 | 1,838 | 17.7 |

*So that England and Wales have a mile of railway for every six and a half square miles of country, being the highest proportion in the world,* while Scotland has less than half that accommodation, and Ireland little more than one-third.

The following table shows which country has the greatest length of railway in proportion to population :—

| | Population in 1860. | Railway Mileage. | Population per Mile of Railway. |
|---|---|---|---|
| England and Wales | 20,228,497 | 9,251 | 2,186 |
| Scotland | 3,096,308 | 2,200 | 1,409 |
| Ireland | 5,850,309 | 1,838 | 3,182 |

So that Scotland, a thinly inhabited country, has the greatest railway mileage in proportion to her population, and we shall afterwards find that she stands at the head of all European countries in this respect.

The manner in which this railway mileage is distributed through England deserves some attention. A railway map will show that the general direction of English lines is towards the metropolis. London is a centre to which nearly all the main lines converge. Every large town is, in its degree, a centre of railway convergence. For example, look at the lines radiating from Leeds, from Hull, from Birmingham, or from Bristol. But all those lesser stars revolve, so to speak, round the metropolis as a central sun.

A great deal may be learned of the character and political state of a country from the convergence of its railway lines. Centralizing France concentrates them all on Paris. Spain, another nation of the Latin race, directs her railways on Madrid. Italy shows her past deficiency of unity, and want of a capital, by her straggling and centerless railroads. Belgium is evidently a collection of co-equal cities without any preponderating focus. Germany betrays her territorial divisions by the multitude of her railway centres. Austria, on the contrary, shows her unity by the convergence of her lines on Vienna. The United States of America prove their federal independence by the number of their centres of radiation.

The national character of the English nation may be traced in the same way. Though our railways point towards London, they have also another point of convergence—towards Manchester and the great port of Liverpool. The London and Northwestern, the Great Northern (by the Manchester, Sheffield and Lincolnshire line), the Great Western and the Midland run to Manchester and Liverpool from the south. The Manchester, Sheffield and Lincolnshire railway, the London and Northwestern Yorkshire and Carlisle lines, and the network of the Lancashire and Yorkshire Company converge on them from the east and north. The London and Northwestern Welsh railways and the Mid Wales and South Wales lines communicate with them from the west. Thus our railway system shows that Manchester and Liverpool are the manufacturing and commercial capitals of the country, as London is its monetary and political metropolis, and that the French centralization into a single great city does not exist in England.

It remains to describe the great systems into which the English railways have been amalgamated. There are in England twelve great companies, with more than £14,000,000 each of capital, which in the aggregate comprises nearly seven-eighths of our total mileage and capital. They divide the country into twelve railway kingdoms, generally well defined, but sometimes intermingled in the·most intricate manner. They may be classified into the following seven districts:—

|  | Miles Open. | Capital Expended. |
|---|---|---|
| 1. *Northwestern District*—London and Northwestern Railway....... ......................... | 1,306 | £53,210,000 |
| 2. *Midland District*—Midland Railway.... | 677 | 26,103,000 |
| 3. *Northeastern District*—Great Northern Railway........ | 422 | 18,200,000 |
| Northeastern Railway... .................. | 1,121 | 41,158,000 |
| 4. *Mersey to Humber District*—Lancashire and Yorkshire Railway ................................. | 403 | 21,114,000 |
| Manchester, Sheffield and Lincolnshire Railway...... | 246 | 14,113,000 |
| 5. *Eastern District*—Great Eastern Railway................. | 709 | 23,574,000 |
| 6. *Southeastern District*—Southeastern Railway........... | 319 | 18,626,000 |
| London, Chatham and Dover Railway | 175 | 14,768,000 |
| London and Brighton Railway.......... | 294 | 14,561,000 |
| 7. *Southwestern District*—London and Southwestern Railway................. ...................................... | 500 | 16,364,000 |
| Great Western Railway................... .................... | 1,292 | 47,630,000 |
| Total............................................ | 7,564 | £309,421,000 |

In Scotland there are three great companies :—

|  | Miles Open. | Capital Expended. |
|---|---|---|
| 1. *Southeast Coast*—North British Railway.................. | 732 | £17,802,000 |
| 2. *Central District*—Caledonian Railway.......... .......... | 561 | 14,797,000 |
| 3. *Southwest Coast*—Glasgow and Southwestern.......... | 249 | 5,603,000 |
| Total.............. ................... .............. | 1,542 | £38,202,000 |

which include three-fourths of the whole mileage and capital of Scotch railways.

In Ireland there are only two large companies:—

|  | Miles Open. | Capital Expended. |
|---|---|---|
| 1. *Southwestern District*—Great Southern and Western .......... | 420 | £5,712,000 |
| 2. *Midland District*—Midland Great Western............. .......... | 260 | 3,625,000 |
| Total ................................................ | 680 | £9,337,000 |

which embrace rather more than two-fifths of the capital and mileage.

The above figures are taken from *Herepath's Railway Journal*, made up very nearly to the present time.

The following table shows the average gross receipts and net profits, for three years, for the United Kingdom, and also the dividends paid on ordinary stock in the above great companies, except the London, Chatham and Dover:—

AVERAGE RECEIPTS AND DIVIDENDS PER CENT.

|  | 1857. | 1861. | 1865. |
|---|---|---|---|
| Gross receipts ................................................. ................ | 7.87 | 8.27 | 8.57 |
| Net profits ...................................... .......................... | 4.19 | 4.30 | 4.46 |

| Dividends of Great Companies : | 1857. | 1861. | 1865. |
|---|---|---|---|
| 12 English | 4.00 | 4.45 | 4.65 |
| 3 Scotch | 4.55 | 4.90 | 5.70 |
| 2 Irish | 5.00 | 5.00 | 3.56 |
| Average dividends | 4.51 | 4.78 | 4.64 |

## IV.—Cost of Railways in the United Kingdom.

The total capital authorized and expended, up to the end of 1865, is given in the Board of Trade Returns, as follows, including the companies estimated for, which have not made a return :—

CAPITAL AUTHORIZED.

| | |
|---|---|
| Shares | £434,457,000 |
| Loans | 143,968,000 |
| Total | £578,425,000 |

CAPITAL EXPENDED.

*Debenture Capital :*

| | | |
|---|---|---|
| Stock | £13,312,000 | |
| Mortgages | 98,059,000 | —111,871,000 |
| Preference capital | | 124,517,000 |
| Ordinary capital | | 220,033,000 |
| | | £456,421,000 |

Hence the following conclusions:—

1. *The capital expended is more than half as large as the national debt.*

2. The debenture and preference capital, which are practically first and second mortgages of railway property, amounted in 1865 to more than half the whole capital expended.

So that railway property is virtually mortgaged to the debenture and preference capitalist for about half its value.

The preference capital has for some years been steadily increasing, while the ordinary capital has remained almost stationary. During 1865 the preference capital increased by £19,615,000, while the ordinary capital only increased by £4,650,000. As the old companies almost always increase their capital by preference stock, I anticipate that in seven or eight years the debenture and preference capital will have risen to two-thirds of the capital expended.

3. The unissued or unpaid capital was, in 1864, £95,000,000. This increased largely in 1865, by the great number of miles authorized in that year, and in the return for that year is £122,000,000.

The expenditure was, in 1864, divided between the three kingdoms in the following proportions, including non-returning companies:—

| | Capital Expended. | Cost per Mile of Railway. |
|---|---|---|
| England and Wales | £379,000,000 | £41,033 |
| Scotland | 50,206,000 | 22,820 |
| Ireland | 26,394,000 | 14,360 |

Thus Ireland has made her railways for one-third the cost, and Scotland for little more than half the cost of the English railways—a result which might be partly expected from their larger poportions of single lines, the greater cheapness of land, and in Ireland the lower wages of labor.

But the English expenditure is the highest in the world, and has given rise to severe remarks on the wastefulness of the English system. Let us examine the causes of expense.

1. The English expenditure includes, on a probable estimate, no less than £40,000,000 sterling absorbed by metropolitan railways and termini. This of itself is £4,500 per mile on the 8,890 miles constructed.

It also includes very large sums for termini in Manchester, Liverpool, Leeds, Sheffield, Birmingham and other great towns, far beyond what is paid in continental cities.

2. The English expenditure also includes considerable capital for docks, as at Grimsby, where £1,000,000 was laid out by the Manchester, Sheffield and Lincolnshire Company; and at Hartlepool, where £1,250,000 was spent by a company now merged in the Northeastern.

It also includes in many instances capital expended on steamers and capital for the purchase of canals.

3. The counties whose trade and population is greatest, and which are most thickly studded with railways, as Lancashire, Yorkshire, and Glamorgan, are exceedingly hilly, and necessitate heavy embankments, cuttings and tunnels, which enormously increase the cost of construction. The Lancashire and Yorkshire Railway has cost £52,400 per mile for the whole of its 403 miles. Had those counties been as flat as Belgium the company might probably have saved something like £20,000 per mile, or £8,000,000 sterling. The Manchester, Sheffield and Lincolnshire Company, even after deducting £1,000,000 for the docks of Grimsby, have spent £53,000 per mile. A flat country might have saved them a similar sum per mile, or £5,000,000 sterling.

4. England, as the inventor of railways, had to buy experience in their construction. Other nations have profited by it. There is no doubt that our present system of lines could now be made at very much less than their original cost. In addition we have paid for experiments, such as the broad guage and atmospheric railway.

5. The great preponderance of double lines over single (6,081 miles against 3,170) has largely increased the expense as compared with the single lines, which predominate in other countries.

6. The price of land in a thickly populated country like England must necessarily be higher than in the more thinly inhabited continental countries. But beyond this, English landowners, in the first vehement opposition to railways, acquired the habit of being bought off at high prices, and of exacting immense sums for imaginary damages. The first Eastern Counties line was said to have paid £12,000 per mile for land through an agricultural country, being about ten times its real value. This habit of exaction has been perpetuated to our own day. As an every day instance I may mention that. only a few months ago a gentleman of great wealth was selling to a railway company, which he had supported in Parliament, thirty acres of grass land, of which the admitted agricultural value was £100 an acre, and three acres of limestone, of which the proved value to a quarryman was £300 an acre. There was no residential damage, and the railway skirted the outside of the estate. The price of the whole in an auction room would have been about £4,000. The proprietor's agent, supported by a troop of eminent valuers, demanded £25,000.

7. Parliamentary expenses are an item of English expenditure not occurring in countries where the concession of railways is the province of a department of the government. But in those countries there is almost always a "promoter's fund" and secret service fund, which often attain very large dimensions. Which is the preferable alternative? Besides, those who object to parliamentary committees must be prepared to give us a practicable substitute, which will suit the habits and feelings of the British nation. Now, a free nation must have liberty to bring forward schemes for the public accommodation, and to have them decided by some public tribunal, after full investigation and hearing all parties. There must be witnesses, and, where millions of money are at stake, there must be the power of being represented by the ablest advocates. Commissions appointed by the Board of Trade, or of any other department, would be just as expensive. The expense of parliamentary committees is the price we pay for free trade in railways, and for our present amount of railway development.

I believe that these causes will fully account for the higher cost of English railways, and, except as regards the cost of land, I think that no valid or practical objection can be taken to them. There is certainly the consolation of knowing that in return for our money we have a more efficient system of railways than any other country.

## V.—TRAFFIC AND BENEFIT OF RAILWAYS IN THE UNITED KINGDOM.

In order to appreciate the wonderful increase of traffic which has resulted from railways, it is necessary to know the traffic of the kingdom before their introduction.

Previous to the opening of the great trunk lines in 1835, passengers were conveyed by mail and stage coaches, a system which had reached a high degree of perfection. Mr. Porter, in his "Progress of the Nation," has calculated, from the stage-coach license returns, the total number of miles traveled by passengers during 1834 as 358,290,000, which represented 30,000,000 persons traveling 12 miles each. The fares were very high, being by the mails 6d. a mile inside and 4d. outside, exclusive of coachmen and guards, and rather less on the stage coaches. Including coachmen and guards, the average fares paid may be taken above 5d. per mile. Hence the 30,000,000 passengers paid a total of £6,250,000.

Goods were conveyed by water or by road.

Water communication had been developed with great perseverance, and was nearly as follows:—

| | Miles. | Miles. |
|---|---|---|
| Canals—England | | 2,600 |
| Scotland | 225 | 225 |
| Ireland | 275—3,100 | 275—3,100 |
| Navigations | 900 | 900 |
| Total | 4,000 | 4,000 |

Being one mile to every thirty square miles of country.

Canal companies always regarded with great jealousy any attempt to ascertain the amount of their traffic, and the only calculation I can find is in Smiles' "Life of Brindley," [p. 464,] where it is estimated at 20,000,000 tons annually. The rates charged by canal carriers were, for the great bulk of general goods, about 4d. per ton per mile. Thus, Lon-

The effect of railways was very remarkable. It might reasonably be supposed that the new means of communication would have supplanted and destroyed the old. Singular to relate, no diminution has taken place either in the road or canal traffic. As fast as coaches were run off the main roads they were put on the side roads, or re-appeared in the shape of omnibuses. At the present moment there is probably a larger mileage of road passenger traffic than in 1834. The railway traffic is new and additional traffic. But railways reduced the fares very materially. For instance, the journey from Doncaster to London by mail used to cost £5 inside and £3 outside, (exclusive of food,) for 156 miles, performed in twenty hours. The railway fares are now 27s. 6d. first class, and 21s. second class for the same distance, performed in four hours. The average fares now paid by first, second, and third class passengers are 1¼d. per mile, against an average of 5d. in the coaching days, being little more than one-fourth of the former amounts.

On canals the effect of railway competition was also to lower the rates to one-fourth of the former charges. In consequence the canal tonnage actually increased, and is now considerably larger than it was before the competition of railways. Hence the railway goods traffic, like its passenger traffic, is entirely a new traffic. The saving in cost is also very great; goods are carried by rail at an average of 1½d. per ton, or 40 per cent. of the old canal rates.

Now observe the growth of this new railway traffic. The following table from the Parliamentery returns (except for 1865) shows the receipts from passenger and goods traffic on railways in the following years:—

INCREASE OF TRAFFIC.

| | Total Receipts. | Average Annual Increase. | Average of whole 22 years. |
|---|---|---|---|
| 1843 | £4,535,000 | £1,070,000 | |
| 1848 | 9,933,000 | 1,653,000 | |
| 1855 | 21,507,000 | 1,252,000 | £1,423,000 |
| 1860 | 27,766,000 | 1,619,000 | |
| 1865 | 35,890,000 | | |

Thus the average annual increase for the whole twenty-two years was £1,423,000 per annum; and the increase was largest in the latest years. The traffic in 1864 and 1865 was thus made up :—

|  | 1864. | 1865. |
|---|---|---|
| Passengers | £15,684,000 | £16,572,000 |
| Goods | 18,331,000 | 19,318,000 |
| Total receipts | £34,015,000 | £35,890,000 |

And the things carried were, exclusive of carriages and animals,—

|  | 1864 | 1865 |
|---|---|---|
| Passengers | 229,272,000 | 251,863,000 |
| Goods, tons | 110,400,000 | 114,593,000 |

Being six times as many as before the introduction of railways.
The increase was extraordinary.

|  | 1864 over 1863. | 1865 over 1864. |
|---|---|---|
| Increase in passenger receipts | £1,163,000 | £888,000 |
| " goods " | 1,696,000 | 986,000 |
|  | £2,859,000 | £1,874,000 |

So that the increase in 1864 was just double the average annual increase. The increase in things carried was :

|  | 1864 over 1863. | 1865 over 1864. |
|---|---|---|
| Increase in number of passengers | 24,637,000 | 22,590,000 |
| " tons of goods | 9,800,000 | 4,233,000 |

An increase in 1864 equal to five-sixths of the whole number of passengers in 1834, and to five-twelfths of the total goods tonnage in 1834; a wonderful proof of the capabilities and benefits of the railway system.

Now let us examine the saving to the country. Had the railway traffic of 1865 been conveyed by canal and road at the pre-railway rates, it would have cost three times as much. Instead of £36,000,000 it would have cost £108,000,000. *Hence there is a saving of £72,000,000 a year, or more than the whole taxation of the United Kingdom.*

But the real benefit is far beyond even this vast saving. If the traffic had been already in existence it would have been cheapened to this extent. *But it was not previously in existence; it was a new traffic, created by railways, and impossible without railways.* To create such a traffic, or to furnish the machinery by which alone it could exist, is a far higher merit than to cheapen an existing traffic, and has had far greater influence on the prosperity of the nation.

Look at the effects on commerce. Before 1833 the exports and imports were almost stationary. Since that time they have increased as follows :—

INCREASE OF EXPORTS AND IMPORTS.

| One Year. | Total Exports and Imports | Per cent. Increase. | Per cent. per annum Increase. |
|---|---|---|---|
| 1833 | £85,500,000 | 36 | 4 |
| 1842 | 116,000,000 | 47 | ʊ |
| 1850 | 171,000,000 | 52 | 10.4 |
| 1855 | 260,000,000 | 44 | ᴏ |
| 1860 | 375,000,000 | 30 |  |
| 1865 | 490,000,000 |  |  |

I am far from attributing the whole of this increase to railways. Free trade, steamboats, the improvements in machinery, and other causes contributed powerfully to accelerate its progress. But I wish to call attention to two facts.

1. This increase could not have taken place without railways. It would have been physically impossible to convey the quantity of goods, still less to do so with the necessary rapidity.

Mr Francis, in his "History of Railways," draws a striking picture of the obstacles to commerce in 1824, from the want of means of conveyance :

"Although the wealth and importance of Manchester and Liverpool had immensely increased, there was no increase in the carriage power between the two places. The canal companies enjoyed a virtual monopoly. Their agents were despotic in their treatment of the great houses which supported them. The charges, though high, were submitted to, but the time lost was unbearable. Although the facilities of transit were manifestly deficient, although the barges got aground, although for ten days during summer the canals were stopped by draught, and in severe winters frozen up for weeks, yet the agents established a rotation by which they sent as much or as little as suited them, and shipped it how or when they pleased. They held levees attended by crowds, who almost implored them to forward their goods. The effects were disastrous ; mills stood still for want of material ; machines were stopped for lack of food. Another feature was the extreme slowness of communication. The average time of one company between Liverpool and Manchester was four days, and of another thirty-six hours ; and the goods, although conveyed across the Atlantic in twenty-one days, were often kept six weeks in the docks and warehouses of Liverpool before they could be conveyed to Manchester. 'I took so much for you yesterday, and I can only take so much to-day,' was the reply when an urgent demand was made. The exchange of Liverpool resounded with merchants' complaints ; the counting-houses of Manchester re-echoed the murmurs of manufacturers."—Vol. i., p. 77 and 78.

This intolerable tyranny produced the Manchester and Liverpool railway, and gave the greatest impetus to railway development.

2. *The increase of imports and exports was in strict proportion to the development of railways.* The following table shows the miles of railway and navigation opened, and the total exports and imports. It must be remembered that there are about 4,000 miles of navigation, and that the exports and imports had been for some time stationary before 1833 :—

PROPORTION OF EXPORTS AND IMPORTS TO RAILWAYS AND NAVIGATION.

| Year. | Miles of railway and navigation. | Total exports and imports. | Exports and imp'ts per mile. |
|---|---|---|---|
| 1833 | 4,000 | £85,500,000 | £21,375 |
| 1840 | 5,200 | 119,000,000 | 22,884 |
| 1845 | 6,441 | 135.000,000 | 20,959 |
| 1850 | 10,733 | 171,800,000 | 16,006 |
| 1855 | 12,334 | 260,234,000 | 21,098 |
| 1860 | 14,433 | 375,052,000 | 52,985 |
| 1865 | 17,289 | 490,000,000 | 28,341 |

Here the increase in exports and imports keeps pace with railway development from 1833 to 1845, falls below it during the enormous multiplication of railways and the railway distress from 1845 to 1850, rises again to the former level in 1855, and outstrips it after that year, aided by the lowering of fares and the greater facilities for through booking and interchange of traffic. I cannot think that this correspondence within the two increases is accidental, especially as I shall show that it exists also in France.

But, it may be said, how do exports and imports depend on the development of the railway system? I answer, because they depend on the goods traffic; and the goods traffic increases visibly with the increase of railway mileage and the perfecting of railway facilities. Goods traffic means raw material and food brought from ports, or mines, or farms, to the producing population, and manufactured articles carried back from the producers to the inland or foreign consumers. The exports and imports bear a variable but appreciable proportion to the inland traffic. Every mineral railway clearly increases them; every agricultural railway increases them less clearly but not less certainly. *Hence I claim it as an axiom, that the commerce of a country increases in direct proportion to the improvement of its railway system, and that railway development is one of the most powerful and evident causes of the increase of commerce.*

Now, let us turn to the benefits which railways have conferred on the working classes. For many years before 1830 great distress had prevailed through the country. Mr. Molesworth, in his "History of the Reform Bill," says that it existed in every class of the community. "Agricultural laborers were found starved to death. In vain did landlords abate their rents and clergymen their tithes; wages continued to fall till they did not suffice to support existence." Innumerable petitions were presented from every county in England, stating that the distress "was weighing down the landholder and the manufacturer, the shipowner and the miner, the employer and the laborer." Trade and commerce were standing still while population was rapidly increasing at nearly the same rate as during the most busy and prosperous period of the French war. The increase from 1801 to 1861 is given in the census:—

ENGLAND AND WALES.

| Year. | Population. | Inc. per ct. for 10 years. | Year. | Population. | Inc. per ct. for 10 years. |
|---|---|---|---|---|---|
| 1801...... | 8,892,536 | 11 | 1841 | 15,914,148 | 14 |
| 1811...... | 10,164,256 | 14 | 1851 | 17,927,609 | 13 |
| 1821...... | 12,000,236 | 18 | 1861............... 20,066,224 | | 12 |
| 1831...... | 13,896,797 | 16 | | | |

The increase during the ten years from 1821 to 1831, which included so much distress, was no less than 16 per cent., distributed pretty uniformly between the agricultural and manufacturing counties, and in itself almost a sufficient cause for the distress. But what has happened since? Increased facilities of transit led to increased trade; increased trade gave greater employment and improved wages; the diminution in the cost of transit and the repeal of fiscal duties cheapened provisions; and the immense flood of commerce which set in since 1850 has raised the incomes and the prosperity of the working classes to an unprecedented height. Railways were the first cause of this great change, and are entitled to share largely with free trade the glory of its subsequent increase and of the national benefit. But one portion of the result is entirely their own. Free trade benefited the manufacturing populations, but had little to do with the agriculturists. Yet the distress in the rural districts was as great or greater, than in the towns, and this under a system of the most rigid protection. How did the country population attain their present prosperity? Simply by the emigration to the towns or colonies of the redundant laborers. This emigration was scarcely possible till the construction of railways. Up to that time the farm laborer was unable to migrate; from that time he became a migratory animal. The increase of population in agricultural counties stopped, or was changed into a decrease, and the laborers ceased to be too numerous for the work. To this cause is principally owing the sufficiency of employment and wages throughout the agricultural portion of the kingdom. If I may venture on a comparison, England was, in 1830, like a wide-spreading plain flooded with stagnant waters, which were the cause of malaria and distress. Railways were a grand system of drainage, carrying away to the running streams, or to the ocean, the redundant moisture, and restoring the country to fertility and prosperity.

## VI.—RAILWAYS IN FRANCE.

In turning from England to France we enter a country completely different in its railway organization. In England everything is left to individual enterprise and independent companies. In France nothing can be done without the aid of the government. They tried the English system, and failed, just as they tried parliamentary government and failed. The independent railway companies broke down, and it was found absolutely necessary to change to a *regime* of government guarantees and government surveillance, suited to the genius of the French people, and under which they regained confidence and prosperity.

Before the introduction of railways France possessed an extensive system of water communication, which is now of the following extent:—

|  | Miles. |
|---|---|
| Navigable rivers | 4,820 |
| Canals | 2,880 |
| Total | 7,700 |

by which goods were conveyed at very reasonable rates, varying from 1d. to 2d. per ton per mile, or about half the English charges. But the delays were very great; three or four months for a transit of 150 miles was quite usual. And the canals paid scarcely 1 per cent. dividend, while their English cotemporaries were paying 5 to 20 per cent. .

Communication by road was also cheaper, but slower, than in England. The passengers paid from 1½d. to 3d. per mile, instead of the 3d. to 6d. paid in England. But they only traveled five to six miles an hour instead of the English eight to ten. Goods paid by road about 3d. per ton per mile for ordinary conveyance, and 6d. for quick despatch, being less than half the English charges. The distances in France were greater than in England, the commerce was less, and labor and food were cheaper; thus fully accounting for the difference.

Tramways were introduced into France in 1823, by the construction of a line of eleven miles from the coal mines of St. Etienne, and this was followed by two much longer lines of a similar character, which were opened by sections between 1830 and 1834. They are dignified in French books with the title of railways, but they were really nothing but horse tramways, and were sometimes even worked by oxen.

The success of the Manchester and Liverpool railway provoked some real though short railways in France, especially those from Paris to St. Germain and to Versailles. But in 1837 only 85 miles had been opened, against nearly 500 in England. In 1837 and 1838 the French Chambers

2

threw out a scheme of their government for the construction by the State of an extensive system of railways, but granted concessions to private companies for lines to Rouen, Havre, Dieppe, Orleans, and Dunkerque. These lines were abandoned for a time, in 1839, for want of funds.

In this emergency Mr. Locke, the great English engineer, restored the fortunes of French railways. Assisted by the London and Southwestern company, and Mr. Brassey, and with subventions from the French Government, and subscriptions from English shareholders, and a powerful corps of English navvies, he recommenced, carried through the line from Paris to Rouen, and from Rouen to Havre, and fairly gave the start to railway enterprise in France.

In 1842 a new law was passed, by which the State undertook the earthworks, masonry, and stations, and one-third of the price of land; the departments were bound to pay by instalments the remaining two-thirds of the land; and the companies had only to lay down rails, maintain the permanent way, and find and work the rolling stock. *It was intended that three-fifths of the total cost should be borne by the State and departments, and two-fifths by the companies.* Under this system of subventions, a number of concessions were made, the shares rose to 50 per cent. premium, and in 1848 a total of 1,092 miles had been opened. The revolution of 1848 was a terrible shock to their credit, and shares went down to half their value. Many lines became bankrupt and were sequestrated, and for three years fresh concessions were entirely stopped. But the concessions already made were slowly completed, and by the end of 1851 France had opened 2,124 miles against 6,889 opened in the United Kingdom.

In 1852 the Emperor took French railways in hand, and by a system of great wisdom, singularly adapted to the French people, he put an end to the previously feeble management, and launched into a bold course of railway development. The French public shrank from shares without a guarantee; he gave a State guarantee of 4 or 5 per cent. interest. The French public preferred debentures to shares; he authorized an enormous issue of debentures. The companies complained of the shortness of their concessions; he prolonged them to a uniform period of ninety-nine years. At the same time he provided for the interest of the State by a rigid system of government regulation and audit. And, lastly, coming to the conclusion that small companies were weak and useless, he amalgamated them into six great companies, each with a large and distinct territory; and able, by their magnitude, to inspire confidence in the public, and aid the government in the construction of fresh railways. This vigorous policy was very soon successful. Capital flowed in rapidly, construction proceeded with rapidity, and between the end of 1851 and 1857 the length of the railways opened was increased from 2,124 miles to 4,475, or more than doubled. England at that time had opened 9,037 miles.

France was now exceedingly prosperous. *Her exports and imports had increased from £102,000,000 in 1850, to £213,000,000 in 1857, or more than 100 per cent in seven years.* The six great companies were paying dividends which averaged 10 per cent.; *and the government guarantee had never been needed.* Railways united all the great towns and ports, and met the most pressing commercial wants. But the Emperor was not satisfied. France, with double the territory of England, had only half the railway accommodation, and wide districts between all the trunk lines were totally unprovided with railways. The government engineers of the *ponts et chaussees* were prepared with plans and estimates for 5,000 miles of lines, which had been inquired into, and officially declared to be *d'utilite publique, i. e.*, a public necessity. The country districts clamored for these lines. But how were they to be made? The public were not prepared to subscribe for them, the government could not undertake them, and the great companies were too well satisfied with their 10 per cent. dividend to wish to endanger it by unremunerative branches.

The plan of the Emperor was intricate but masterly. He said to the companies: "You must make these lines. The 4,520 miles of railway already made shall be a separate system for the present, under the name of *Ancien Reseau*, the old lines. You no longer require the guarantee of the State for these lines. But I will give you an extension of the ninety-nine years of your concessions, by allowing them to commence at later dates; beginning with 1852 for the Northern Company, and at various dates for the rest, up to 1862, for the Southern Company. I also engage that £9,000,000 sterling of the net revenue of these old lines shall for ever be divisable among the shareholders, without being liable for any deficit of the extension lines, an amount which will give you a clear and undefeasible dividend of 6 to 8 per cent.; with a strong probability— almost a certainty—of getting much more from surplus traffic."

"Next the new lines, 5,128 miles in length, shall be a separate system, under the name of *Nouveau Reseau*, or extension lines. Their estimated cost is £124,000,000, and you, the companies, may raise this sum by debentures, on which the government will guarantee 4 per cent. interest, and .65 sinking fund for the paying them off in fifty years. Any extra cost you must pay yourselves."

These, in their briefest possible form, are the terms on which the Emperor imposed an average of nearly 1,000 miles per company on the six great companies of France. They were accepted with considerable reluctance. Their effect has heen to lower the value of the shares of the great companies, for the bargain is considered disadvantageous. The companies cannot borrow at less than 5.75, so losing 1.10 per cent. per annum on every debenture; and as the lines cost more than the £124,000,000, the overplus has been raised by the companies by debentures, for which they alone are responsible. But on the other hand, they get an immense amount of fresh traffic over their old lines, which must ultimately more than repay this loss. English railways would be thankful if their extensions cost them so little.

In the following years other lines were added, with similar guarantees and with considerable subventions from the State, and in 1863 an additional series of lines, 1,974 miles in length, were imposed on similar terms, but with some modifications of the conventions with two of the weakest companies.

Besides the government lines, the Emperor encouraged to the utmost the efforts of the departments, and in July, 1865, a law was passed respecting *chemins de fer d'interet local*, which authorized departments and communes to undertake the construction of local railways at their own expense, or to aid concessionaires with subventions to the extent of one-fourth, one-third, or in some cases one-half the expense, not exceeding £240,000.

Not content with passing this law; the minister of public works, in the very next month wrote to the prefets of the 88 departments of France, to acquaint them fully with its provisions, and to invite them to communicate with their councils general, and deliberate upon the subject. The result was that sixteen councils requested their prefets to make surveys and inquiries to ascertain what lines would be advisable. 32 departments authorized their prefets to prepare special plans, and even to make provisional agreements with the companies to carry out lines, subject to confirmation by the councils. Two of these made immediate votes, viz., the department of Ain, £56,000, and Herault, £260,000, for lines which they approved. A third, the department of Calvados, voted subventions amounting to £1,000 per mile for one line, and £2,000 per mile for another line. Besides, these five departments put railroads into immediate execution by contracts with independent companies. Among these were:

|  | Subvention. |
|---|---|
| Saone et Loire | £14,000 |
| " (besides the land) | 40,000 |
| Manche (with an English company, and including land) | 40,000 |
| Rhone | 240,000 |
| Tarn | 171,000 |

By these measures the Emperor has brought up the concessions to the following total :—

|  | Miles. |
|---|---|
| *Ancien Reseau*, or old lines | 5,027 |
| *Nouveau* " or extension lines | 7,565 |
|  | 12,592 |

Being very nearly the length of our constructed lines in 1864.

But of this mileage there has been constructed up to the present time only ...... 8,134

Leaving still unconstructed ...... 4,458

being one-third of the whole concessions. Of this, 1,800 miles are now being constructed, and 1,600 miles are expected to be opened by the end of 1867.

Hence the lines constructed in France up to and including 1865, are 8,134 miles, or about the same length as the lines constructed in the United Kingdom to the end of 1865; so that France is ten years behind England in actual length of railways constructed, and at least fifteen years behind England if her larger territory and population are taken into account; and I must add that France would have been very much farther behind had it not been for the vigorous impulse and the wise measures of the Emperor Napoleon.

The progress of completion from 1837 to the present time is shown in the following table :—

| Year. | MILES CONSTRUCTED. Miles open. | Average annual Increase. |
|---|---|---|
| 1837 | 85 | 84 |
| 1840 | 338 | 34 |
| 1845 | 508 | 259 |
| 1850 | 1,807 | 301 |
| 1855 | 3,315 | 454 |
| 1860 | 5,586 | 509 |
| 1865 | 8,134 | |

This shows the insignificant rate of progress up to 1845, and the larger but still slow progress up to 1855. From that time the effect of the Emperor's policy becomes visible in the increased rate of progression. It is expected that between 1852 and 1872 more than 9,500 miles will have been opened, quadrupling the number constructed in the previous twenty years, and contributing in the highest degree to the prosperity and wealth of the French nation.

Railway history in France may be briefly summed up in four periods:

1. The period of independent companies, from 1831 to 1841.

2. The period of joint partnership of the State and the companies, from 1842 to 1851.

3. The period of Imperial amalgamations and guarantees, from 1852 to 1857.

4. The period of guaranteed extension lines from 1858 to the present time.

## VII.—Cost and Results of French Railways.

The French system of railway organization is worthy of attentive study. It is in many points novel to an Englishman; it is often characterized by remarkable talent; and some of its regulations are very instructive and worthy of imitation.

In extent the French lines are far inferior to the English, whether judged by the area or population of the two countries.

COMPARISON BY AREA.

| Country. | Area in Square Miles. | Railway Mileage. 1865. | Square Miles per Mile of Railway. |
|---|---|---|---|
| United Kingdom | 120,927 | 13,289 | 9 |
| France | 211,852 | 8,134 | 26 |

COMPARISON BY POPULATION.

| Country. | Population, 1861. | Railway Mileage. 1865. | Population per Mile of Railway. |
|---|---|---|---|
| United Kingdom | 29,321,000 | 13,289 | 2,206 |
| France | 37,382,000 | 8,134 | 4,595 |

Hence, measured by area, France has only one-third of the railway accommodation, and measured by population only one-half of the railway accommodation of the United Kingdom.

The capital authorized and expended to the 31st December, 1865, was as follows:—

CAPITAL AUTHORIZED.

*Ancien Reseau,* or old lines............................................. £151,000,000
*Nouveau* " or extension lines................................... 209,000,000—£360,000,000
Including 64,000,000 subventions.

CAPITAL EXPENDED, 1866.

Debentures................................................................. £178,700,000
Shares...................................................................... 54,800,000
Subventions ............................................................. 27,500,000—£261,000,000

So that the French companies borrow more than three times the amount of their share capital; reversing the English rule, of borrowing only one-third of the share capital. But if we consider preference capital as a second mortgage; the English practice is to borrow an amount equal to the ordinary share capital. This, however, is still a long way from them the French regulations.

The capital not paid up is nearly £100,000,000. Of this nearly one-half will be required in the next three years for lines approaching completion.

The cost per mile of French railways is as follows:—

| | |
|---|---|
| *Ancien Reseau*......................................................... | $30,650 |
| *Nouveau "*............................................................. | 27,350 |

As the *Nouveau reseau* is almost entirely composed of single lines, this does not show very great cheapness of construction. We are making our country lines much cheaper, particularly in Ireland and Scotland.

The effect of railway competition with canals was the same as in England. The canal rates were reduced to one-third of their former amount, and the canal traffic has increased instead of diminishing. The average railway fares and rates are stated by M. Flachat, in his works on railways, to be 6 to 7 centimes for each passenger per kilometre, being 1d. to 1⅒th per mile; as compared with 1½d. per mile, the average on English railways.

The increase of traffic since 1850 is stated in the following as follows:—

INCREASE OF TRAIN TRAFFIC.

| Year. | Total Receipts. | Average Annual Increase | Average Annual Increase for Fifteen Years. |
|---|---|---|---|
| 1850 ...... | £3,824,400 | | |
| | | £1,307,000 | |
| 1855 ...... | 10,358,000 | | |
| | | 1,217,000 | £1,238,400 |
| 1860 ...... | 16,443,000 | | |
| | | 1,192,000 | |
| 1865 ...... | 22,400,000 | | |

Thus the increase has been more equable than in England, but smaller in amount, showing an average of £1,238,400, as against £1,423,000 for England. But I see it stated in the railway papers that the first nine months of 1866 show much more than the usual increase.

M. Flachat gives a calculation of the savings to the nation by railway conveyance, which he makes a minimum of £40,000,000 a year. But it is based on the supposition that all the then traffic would have been carried by roads, which is obviously untenable. Probably £25,000,000 to £30,000,000 is a safer estimate. A writer in the Dictionnaire du Commerce goes into elaborate calculations of the money saving arising out of the greater rapidity of railways, and values it at £8,000,000 on the basis that the time of a French citizen is worth 5d. an hour. I give the passage entire:—

" In France, the number of kilomètres travelled by passengers in 1856 was 2,200,000,000. In travelling this distance they would have spent 290,000,000 hours while they have only been 50,600,000 hours on the railway. The saving in time of travelling by railway has therefore been 240,000,000 hours, which, at the moderate price of 5d. per hour, represent an economy of £29,000,000. Besides this the time lost in stoppages at small inns (auberges) used to exceed that spent in travelling, and hence on this head alone we may calculate a saving of more than 100,000,000 hours. But even if we should reduce this valuation to 80,000,000 or still lower to 60,000,000 hours, there cannot however be any doubt that the saving to the traveller in the matter of time exceeds 200,000,000 frs. (£8,000,000)." *Ibid.* Vol. 638, p. 638.

Passing from individuals to commerce, the effect of railways has been very marked, and is warmly acknowledged by the principal French writers. The following table shows the progress of French trade·—

INCREASE OF EXPORTS AND IMPORTS.

| Year. | Total Exports and Imports. | Increase per Cent. | Increase per Cent. per Annum. |
|---|---|---|---|
| 1840 | £82,520,000 | | |
| 1845 | 97,080,000 | 15., | 3. |
| 1850 | 102,204,000 | 5. | 1. |
| 1855 | 173,076,000 | 50. | 10. |
| 1860 | 232,192,000 | 34. | 6.8 |
| 1865 | 293,144,000 | 26.25 | 5.25 |

The revolution of 1848 accounts for the small increase between 1845 and 1850, but it is plain that the great increase in French commerce was between 1850 and 1860, contemporaneously with the great development of railways. *When traveling in France I have always heard railways assigned as the cause of their present commercial prosperity.*

The proportion which the exports and imports bore to the means of communication is shown in the following table:—

PROPORTION OF EXPORTS AND IMPORTS TO RAILWAYS AND NAVIGATION.

| Year. | Navigations (7,700 miles), and Railways. | Exports and Imports. | Exports and Imports per Mile Open. |
|---|---|---|---|
| 1840 | 8,264 | £82,520,000 | £9,985 |
| 1845 | 8,547 | 97,080,000 | 11,358 |
| 1850 | 9,507 | 102,204,000 | 10,750 |
| 1855 | 11,015 | 173,076,000 | 15,712 |
| 1860 | 13,286 | 232,192,000 | 17,476 |
| 1865 | 15,830 | 293,144,000 | 18,518 |

Here there is a steady rise in the amount per mile, checked only by the revolution of 1848. But the principle that there is a distinct correspondence between means of communication and the exports and imports is already shown.

The effect of railways on the condition of the working classes has also been very beneficial. The extreme lowness of fares enables them to travel cheaply, and the opportunity is largely used. The number of third class passengers in France is 75 per cent. of the total passengers, against only 58 per cent. in England (M. Flachat, p. 60). The result of these facilities of motion has been an equalization of wages throughout the country, to the great benefit of the rural populations. M. Flachat says:—

"Railways found in France great inequality in the wages of laborers; but they are constantly remedying it. Wherever they were constructed in a district of low wages, employment was eagerly sought. The working classes rapidly learnt to deserve high wages by the greater quantity of work done. Agriculture had been unable to draw out the capabilities of its workmen, and was for the moment paralyzed by want of hands; but industry developed fresh resources. The total amount of work done was considerably increased all over the country. The difficulties of agriculture were removed by obtaining in return for higher wages a larger amount of work than before, and also because machines began to be used in cultivation. Everywhere it was evident that increased energy accompanied increased remuneration. This is the point in which railways have most powerfully increased the wealth of France. The moral result of this improvement in the means of existence of the working class has been to diminish the distance which separates the man who works only for himself from the man who labors for a master. In the education of the workman's children, in his clothing, in his domestic life, and even in his amusements, there is now an improvement which raises him nearer to his master."—pp. 78 and 79.

I am sure we shall all we rejoiced at this evidence of the benefits conferred by railways upon the working classes of that great neighboring nation. I wish there was time to give you additional extracts, showing the immense services of railways to the industry of France, showing that France was kept back by the difficulty of communication, by the immense distances to be traversed and the impossibility of conveying cheaply and rapidly the raw materials of manufactures. Railways have supplied this want, and have given an impetus to production and new outlets for the produce.

Turning to the shareholders, there are some curious facts which surprised me not a little. The popular notion is that in France railway traffic bears a much higher proportion to capital expended than in England. The phrase "They manage these things better in France" is forever on the lips of the British shareholder when he talks of his own paltry 4½ per cent. dividend, or of the 3 per cent. gross receipts. The world in general believe that all 10 or 12 per cent. French lines like the Orleans of France, really have a traffic of at least that amount. But this is an entire mistake. The gross traffic receipts of France are from 9.6 per cent. on the share and debenture capital, or 4 per cent. more than in England. And the net receipts, after deduction of 45 per cent. working expenses, are now 5.28 per cent. on the total share and debenture capital, being .82 or about four-fifths per cent. higher than in England. Yet the French companies pay an average dividend of 10 per cent., while the English pay only the natural dividend of 4½. Here are the figures, for the benefit of the sceptical:—

AVERAGE RECEIPTS AND DIVIDENDS PER CENT.

| Name of Company. | 1859. | 1859. | 1861. | 1865. |
|---|---|---|---|---|
| Gross receipts | 10.5 | 10.5 | 9.61.0 | 9.6 |
| Net profits | 6.3.7 | 6.3.7 | 5.26.2 | 5.28 |
| *Dividends of Great Companies :* | | | | |
| Nord | 16.5 | 16.5 | 17 86.5 | 17.37 |
| Orleans | 18 | 20.8 | 11.20. | 11.2 |
| Midi | 10.4 | 10.4 | 8.10. | 8 |
| Ouest | 7.5 | 8.5.5 | 7.58.5 | 7.5 |
| Est | 3.13 | 8.3.13 | 6.68. | 6.6 |
| Mediterrannee | 10.6 | 10.6 | 12.15. | 12 |
| Average | 10.54 | 10.53 | 10.53 | 10.53 |

Compare these figures with those for the English lines given above. You will see the remarkable correspondence between the gross and net receipts and the very remarkable dissimilarity in the dividends. How is this accounted for?

Look at the table of capital expended. Disregarding the £27,500,000 subventions, as corresponding to the divergence tax paid by the companies, there is £233,000,000 share and debenture capital, out of which a portion of the debentures are charged to capital under the conventions for the extension lines. Being for several ways they have not yet been transferred to the revenue account. Hence the interest-bearing capital reduced, and the interest itself increased.

The large amount of debentures now comes into play, so rapidly there which there is paid from 5 to 5 per cent., leaving an overplus or surplus for the late for the shares, so raising the interest the shares to nearly 7 per cent.

But this is not enough. In 1863 the State bound itself to contribute to certain lines annual subventions, which in 1865 came to £551,000, and the State also paid during the same year, in respect of their guarantees of the debentures in the *nouveau reseau*, £1,320,000, making a total subvention in 1865 of £1,871,000, an amount sufficient to pay more than 3 per cent. on the share capital of £54,800,000. The guarantee of £1,220,000 on the *nouveau reseau*, however, is not an absolute subvention, as it will be repayable gradually by the companies when their income exceeds a fixed amount. It is therefore a loan by the State, repayable on the occurrence of a contingency, and at an uncertain date.

Thus the original interest of 5.28 per cent. on the share and debenture capital becomes 10 per cent. to the shareholder. It is a wonderfully clever arrangement and would be exceedingly palatable to Great Eastern or even Great Northern shareholders.

But consider the difference which this shows in the ideas of the two countries. In England it would never be borne for an instant that six great companies, say the London and Northwestern, Great Western, Midland, and others, should receive 10 per cent. dividend and yet obtain from the State annual subventions and guarantees amounting to £1,800,-000. No ministry dare propose such a job. The reform agitation would be nothing to the clamor with which it would be greeted; and yet in France it is the most natural thing possible. Nobody says a word against it. Nay, the feeling of the French companies and the popular opinion is that these poor 10 per cent. shareholders have been badly used, and that their legitimate 12 or 15 per cent. from the trunk lines ought not to have been lessened.

One characteristic of the French systems is the absence of competition, and this is opposed to all our ideas of freedom of communication. The Northern Company monopolizes the whole traffic between Calais and Paris. The Mediterranean Company monopolizes the whole traffic between Paris and Marseilles, a traffic of extraordinary importance and value. An attempt made two years ago by another company to abtain an extension to Marseilles and to establish an alternative route was rejected by a government commission after a very long inquiry. The consequence of this system is a great concentration of traffic in a small number of trains, to the profit of the companies and to the inconvenience of the traveler. There are in England, between places like Liverpool and London, about three times as many trains as there are in France, between Marseilles and Paris. And besides this, goods are sent less rapidly in France and delivered with less punctuality.

But there is a great deal to be said in defence of the French system. It avoids the duplicate lines necessary for competition, which France could not well afford. It keeps the companies prosperous and able to aid the government in railway extension. It is not an irresponsible monopoly, able to charge high prices to its customers, but a strictly regulated monopoly, with its tariff fixed by government at the lowest prices that will be remunerative. It is like the system of our own Metropolitan Gas and Water Companies, which enjoy a monopoly within defined districts, on terms settled by the law and revised from time to time in the interest of the public. The French government appoints commissioners of inquiry to examine into any defect or to consider improvements, and they report to the minister of public works, who has the power of making regulations which are binding on the companies. The last commission is a good instance. In February, 1864, the minister of public works issued to the companies a circular suggesting several points which required improvement, and the commission was appointed to consider their answers. The points discussed were:—

1. The adoption of a means of communication between the guard and engine-driver. This was made obligatory on the companies.

2. A means of communication between passengers and the guard. This was accepted by the companies.

3. The consumption by the locomotives of their own smoke. This was ordered to be carried out within two years.

4. The addition of second and third class carriages to express trains. The recommendation of the commission was accepted by the companies.

5. Separate carriages for unprotected females.

6. The commission demanded that on the great lines the speed of goods trains should be increased from 60 miles to 120 miles, without any increase of tariff. This very important question was referred to a sub-committee for further examination and for hearing objections.

From these details it is evident that the interests of the public are well looked after.

I should add that there is a continuous audit of the accounts of the companies by government accountants, who attend from week to week at the companies' offices for that purpose.

I will at present mention only one other point in French railway law—that the government has the power of purchasing any line of railway after fifteen years from its first concession. The price is to be fixed by taking the amount of the net profits of the seven preceding years, deducting the two lowest years and striking the average of the remaining five years. The government is then to pay to the company for the remainder of the concession an annual rent-charge or annuity equal to the average so determined, but not less than the profits of the last of the seven years. This mode of purchase appears preferable to the English law, since it does not require the creation of any new rentes or consols, and I commend it to the notice of Mr. Galt.

I have mentioned these prominent features of the French law in the hope that they may be useful in suggesting improvements in the English system.

Why should we not vest in the President of the Board of Trade a power of making and enforcing regulations for the public safety and convenience? Why should we not introduce more frequent railway commissions to consider important questions and recommend to the President of the Board of Trade or to Parliament? Why should we not have a modified system of audit, and a registration of shares and debentures?

## VIII.—Railways in Belgium and Holland.

Belgium is one of the most striking instances of the benefit of railways. In 1830 she separated from Holland, a country which possessed a much larger commerce and superior means of communication with other nations by sea and by canals. Five years later the total exports and imports of Belgium were only £10,800,000, while those of Holland were double that amount. But in 1833 the Belgian government resolved to adopt the railway system, and employed George Stephenson to plan railways between all the large towns. *The law authorizing their construction at the expense of the State passed in 1834, and no time was lost in carrying it out. Trade at once received a new impetus, and its progress since that time has been more rapid than in any other country in Europe.* The following table shows the activity with which the lines were constructed. We must remember that Belgium contains only one-tenth of the area of the United Kingdom, and that to make a fair comparison with our own progress we must multiply the table by ten.

MILES CONSTRUCTED.

| Year. | Miles Open. | Increase per annum Miles. |
|---|---|---|
| 1839 | 185 | |
| | | 25 |
| 1845 | 335 | |
| | | 48 |
| 1853 | 720 | |
| | | 45 |
| 1860 | 1,037 | |
| | | 78 |
| 1864 | 1,350 | |

Hence, the progress for a State of the size of the United Kingdom would have been—

| | Miles a Year. |
|---|---|
| 1839 to 1845 | 250 |
| 1845 to 1853 | 480 |
| 1853 to 1860 | 450 |
| 1860 to 1864 | 750 |

a rate of increase which is as great or greater than our own.

The results on commerce are shown in the following table:—

INCREASE OF EXPORTS AND IMPORTS.

| Year. | Exports and Imports | Increase per Cent. | Increase per Cent. per Annum |
|---|---|---|---|
| 1835 | £10,760,000 | | |
| | | 45.72 | 11.43 |
| 1839 | 15,680,000 | | |
| | | 71.4 | 11.9 |
| 1845 | 26,920,000 | | |
| | | 77.41 | 9.67 |
| 1853 | 47,760,000 | | |
| | | 51. | 7.3 |
| 1860 | 72,120,000 | | |
| | | 35.88 | 9. |
| 1864 | 97,280,000 | | |

I need scarcely point out the extraordinary character of this increase, which is enormous in the first ten years, and far beyond either England or France, and is not inferior to us in the later period. In the thirty years from 1835 to 1864 Belgium increased her exports and imports nearly tenfold, while England increased hers only fivefold. If we had increased our commerce in the same ratio, the English exports and imports would now be a thousand million pounds sterling.

The proportion between exports and imports and means of communication is shown in the following table, which differs from those of England and France in the rapid increase per mile —

PROPORTION OF EXPORTS AND IMPORTS TO RAILWAYS AND NAVIGATIONS.

| Year. | Canals (910 Miles) and Railways Open. | Exports and Imports. | Exports and Imports per Mile Open |
|---|---|---|---|
| 1839 | 1,055 | £15,680,000 | £14,862 |
| 1845 | 1,205 | 26,920,000 | 22,340 |
| 1853 | 1,590 | 47,760,000 | 30,037 |
| 1860 | 1,907 | 72,120,000 | 37,818 |
| 1864 | 2,220 | 97,280,000 | 42,919 |

*This enormous increase of Belgian commerce must be ascribed to her wise system of railway development, and it is not difficult to see how it arises.* Before railways, Belgium was shut out from the continent of Europe by the expensive rates of land carriage and her want of water communication. She had no colonies and but little shipping. Railways gave her direct and rapid access to Germany, Austria, and France, and made Ostend and Antwerp great continental ports. One of her chief manufactures is that of wool, of which she imports 21,000 tons, valued at £2,250,000, from Saxony, Prussia, Silesia, Poland, Bohemia, Hungary, Moravia, and the southern Provinces of Russia; and returns a large portion in a manufactured state. She is rapidly becoming the principal workshop of the continent, and every development of railways in Europe must increase her means of access and add to her trade.

Now look at Holland, which in 1835 was so much her superior. Holland was possessed of immense advantages in the perfection of her canals, which are the finest and most numerous in the world; in the large tonnage of her shipping; in her access by the Rhine to the heart of Germany; and in the command of the German trade, which was brought to her ships at Amsterdam and Rotterdam. The Dutch relied on these advantages and neglected railways. The consequence was, that by 1850 they found themselves rapidly losing the German trade, which was being diverted to Ostend and Antwerp. The Dutch Rhenish railway was constructed to remedy this loss, and was partly opened in 1853, but not fully till 1856. It succeeded in regaining part of the former connection. But now observe the result. In 1839 the Dutch exports and imports were £28,500,000, nearly double those of Belgium. In 1862 they were £59,000,000, when those of Belgium were £78,000,000. Thus while Holland had doubled her commerce Belgium had increased fivefold, and had completely passed her in the race.

Before leaving Belgium I ought to mention the cheapness of fares on her railways, which have always been much below those on English lines; a further reduction has lately been made, and I see by a French paper that the results has been to increase the passenger receipts on the State lines for the month of April from 76,956 frs. in 1865 to 198,345 frs. in 1866, of which 168,725 frs. was from third and fourth class passengers; a fact which is in favor of the plan of Mr. Galt. But it must be remembered that Belgium is the most densely populated country in the world, having 432 inhabitants to the square mile, while the United Kingdom has only 253, and England and Wales 347. A system which will pay admirably between large cities at short distances from each other, and on lines which cost little to construct, might break down completely on lines of expensive construction in more thinly inhabited districts. Mr. Galt takes his instances from railways in dense populations, and applies the rules thus obtained to railways which are under totally different conditions, and I fear that this vitiates in a great degree the soundness of his conclusions.

## IX.—RAILWAYS IN THE UNITED STATES.

In any paper on foreign railways it is impossible to omit the United States, a country where they have attained such gigantic proportions. The increase of United States lines is as follows:

| Year. | MILES CONSTRUCTED. | Total mileage. | Inc. per annum. Miles. |
|---|---|---|---|
| 1830 | | 41 | |
| | | | 215 |
| 1840 | | 2,197 | |
| | | | 465 |
| 1845 | | 4,522 | |
| | | | 590 |
| 1850 | | 7,475 | |
| | | | 1,984 |
| 1855 | | 17,398 | |
| | | | 2,274 |
| 1860 | | 28,771 | |
| | | | 1,272 |
| 1864 | | 33,860 | |

The mileage here shown is something enormous: four time that of France, two and a half that of England, and nearly as large as the total mileage of the United Kingdom and Europe, which is about 42,000 miles.

In so young a country inland traffic gives these lines the greater part of their employment, and there are no masses of expensive manufactured goods as in England or Belgium to swell the total value of foreign trade. Foreign commerce is still in its infancy, but an infancy of herculean proportions, as the following table shows:—

### INCREASE OF EXPORTS AND IMPORTS.

| Year. | Total exports and imports. | Increase per cent. | Inc. per ct. per annum. |
|---|---|---|---|
| 1830 | £31,000,000 | | |
| | | 47.60 | 3.40 |
| 1844 | 45,759,000 | | |
| | | 50.00 | 8.33 |
| 1850 | 68,758,000 | | |
| | | 62.60 | 12.52 |
| 1855 | 111,797,000 | | |
| | | 42.00 | 8.40 |
| 1860 | 158,810,000 | | |

The advance in the annual increase is very striking, being from 3½ per cent. per annum in the infancy of railways, to 8 and 12 per cent. when their extension was proceeding rapidly. Before the introduction of railways America possessed a very extensive system of canals, which amounts to nearly 6,000 miles. At the present time both canals and railways are crowded with traffic. The following table shows the relation between the growth of trade and the increase of means of communication:—

### PROPORTION OF EXPORTS AND IMPORTS TO RAILWAYS AND CANALS.

| Year. | Canals (6,000 miles) and railways open. | Total exports and imports. | Exports and imports per mile. |
|---|---|---|---|
| 1830 | 6,040 | £31,000,000 | 5,130 |
| 1844 | 10,310 | 45,759,000 | 4,437 |
| 1850 | 13,475 | 68,758,000 | 5,102 |
| 1855 | 23,398 | 111,797,000 | 4,778 |
| 1860 | 34,770 | 158,810,000 | 4,567 |

Thus, in the United States, as well as in England, France, and Belgium, the exports and imports bear a distinct relation to the miles of communication open, but lower in amount than in the European countries, as was only likely from the thinner population.

Vast as is the mileage of the American railways it is by no means near its highest point. The lines in construction, but not yet completed, are stated to be more that 15,000 miles in length, a larger number than the whole mileage of the United Kingdom, completed and uncompleted.

The manner in which these lines are made is very remarkable. The United States are very thinly populated, not containing on an average more than 32 persons per square mile in the Northern States, and 11 in the Southern. Even the most populous Northern States have only 90 persons per square mile, while England and Wales have 347 per square mile. A less expensive railway, of smaller guage, was, therefore, necessary, and the lines are almost invariably "single tracks." Their first cost have averaged from £7,000 up to £15,000 per mile, or about one-third of the expenditure in England. Of course they are very inferior in weight of rails and in sleepers, ballasting, stations, and efficiency. Even this expense was difficult to provide for where the inhabitants are so widely scattered. But in America the greatest encouragement is given to railroads, and every facility is afforded for their extension, as they are considered the most important sources of wealth and prosperity. Shares are taken largely by the inhabitants of the district traversed, land is often voted by the State, and the cities and towns find part of the capital by giving security on their municipal bonds.

I must not omit to mention the great Pacific railways, one of which is now being constructed from the State of Missouri for a distance of 2,400 miles across Kansas, Nebraska, Utah, and Nevada to San Francisco, in California. It receives from the general government subsidies of £3,300, £6,600, or £9,900 per mile, according to the difficulty of the ground, besides enormous grants of land on each side of the line. When this railway is completed the journey from Hong Kong to England will be made in thirty-three days instead of the present time of six weeks, and it is anticipated that a large portion of our Chinese traffic will pass by this route.

The general opinion undoubtedly is, that free trade is the principal cause of the immense increase since 1842 of English commerce. We see this opinion expressed every day in newspapers and reviews, in speeches and parliamentary papers. I hold in my hand a very able memorandum, lately issued by the Board of Trade, respecting the progress of British commerce before and since the adoption of free trade, in which the same view is taken, and in which the statistics of the exports and imports since 1842 are given as mainly the result of free trade. It is true that there is a reservation, acknowledging "that the increase of productive power and other causes have materially operated in effecting this vast development." But in the newspaper quotations and reviews this reservation was left out of sight, and the striking results recorded in the memorandum were entirely ascribed to free trade.

While acknowledging to the full the great benefits and the enlightened principles of free trade, I have no hesitation in saying that this popular view is a popular exaggeration, which it is the duty of staticians to correct, and I think that my reasons will be considered satisfactory by this Society. In the first place, the development of English commerce began in 1834, before free trade, but simultaneously with railways; and between 1833 and 1842 the exports and imports increased from a stationary position at £85,500,000 to £112,000,000, or 31 per cent. In the next place, from 1842 till 1860 England was the only country which adopted free trade. If England had also been the only country that made such enormous progress we might safely conclude that free trade was the chief cause of so great a fact. But this is not the case. England is only one of several countries which made an equal advance during the same period, and none of those countries, except England, had adopted free trade. The total increase of exports and imports from 1842 to 1860 in the three first countries described in this paper, and from 1844 to 1860 in the United States, was as follows :—

| Country. | 1842. | 1860. | Increase per cent. |
|---|---|---|---|
| England | £112,000,000 | £375,000,000 | 234 |
| France | 86,280,000 | 232,200,000 | 169 |
| Belgium | 19,400,000 | 72,120.000 | 272 |
| | 1844. | | |
| United States | 45,757,000 | 158,810,000 | 305 |

Thus, the English rate of increase is only third in order, and is exceeded both by Belgium and the United States. If the latter country is objected to on account of its rapid growth in population by immigration, still Belgium remains, exceeding the English rate of increase by 36 per cent. Look at the argument by induction. Here are four countries under the same condition of civilization, and having access to the same mechanical powers and inventions, which far outstrip contemporary nations. It is a probable conclusion that the same great cause was the foundation of their success. What was that common cause? It could not be free trade, for only one of the countries had adopted a free trade policy. But there was a common cause which each and all of those four countries had pre-eminently developed—the power of steam—steam machinery, steam navigation, and steam railways. I say, then, that steam was the main cause of this prodigious progress of England as well as of the other three countries.

But I will go a step farther. Steam machinery had existed for very many years before 1830, and before the great expansion of commerce. Steam navigation had also existed for many years before 1830, and before the great expansion of commerce, and steam navigation was unable to cope with the obstacle which before 1830 was so insuperable, viz: *the slowness and expense, and limited capacity of land carriage.*

I come, then, to this further conclusion, that the railways which removed this gigantic obstacle, and gave to land carriage such extraordinary rapidity and cheapness, and such unlimited capacity, must have been the main agent, the active and immediate cause of this sudden commercial development.

This conclusion appears to become a certainty when I find, from the investigation through which we have traveled, that in every one of these four great examples, the rapid development of commerce has synchronised with an equal rapid development of railways—nay, that the development of commerce has been singularly in proportion to the increased mileage of railways—so that each expansion of the railway system has been immediately followed, as if by its shadow, by a great expansion of exports and imports.

But I will not leave the case even here. Consider what are the burdens which press upon trade and manufactures. If our merchants could be presented with that wondrous carpet of the Genii of the "Arabian Nights," which transported whatever was placed upon it in one instant through the air to its farthest destination, overleaping mountains and seas and custom-houses, without expense or delay, we should have the most perfect and unburdened intercourse. But see what barriers and burdens there are in actual fact, when we trace the journey of the raw material, such as cotton or wool, to the British manufacturer, and its export as a manufactured article.

BURDENS UPON IMPORTS AND EXPORTS.

*Raw Material—*
1. Inland carriage to the sea.
2. Voyage to England.
3. Import duty.
4. Inland carriage of the manufacturer.

both in cheapness and saving of time, was far beyond any relief by free trade in taking off moderate duties.

In a vast number of cases railways did more than cheapen trade, they rendered it possible. Railways are the nearest approach that human ingenuity has yet devised to that magic carpet of the "Arabian Nights," for which I ventured to express a wish.

For all these reasons I maintain that we ought to give railways their due credit and praise, as the chief of those mighty agents which, within the last thirty years, have changed the face of civilization.

## XI.—Railways and National Debts.

In one important point the nations of Latin race have stolen a clear march upon the nations of Teutonic origin, of England, Germany, and the United States, by their appreciation and adoption for railways of the principle of a sinking fund. The idea owes its origin to the semi-Latin, semi-Teutonic intellect of Belgium. When the Belgian government, in 1834, projected a system of State railways, to be constructed with money borrowed by the State, they provided for the extinction of the loans in fifty years by an annual sinking fund. The amount borrowed was nearly £8,000,000 sterling, and the whole will be paid off in 1884, after which date the whole profits of the State lines, 352 miles in length, will become part of the revenue of the nation. But so good an investment are these lines that their present net income is £525,000 a year, and is increasing at a rate which promises in 1884 a net revenue of £960,000, *a sum which will be sufficient to pay the interest on the whole national debt, now* £26,000,000. Besides this, the conceded lines, 1,000 miles tn length, will become amortized and become State property in 90 years from the beginning of their concessions, and the profits on a capital of more than £13,000,000 will then be available toward the State revenue.

This system was copied by France, and imitated from her by the other Latin nations, Spain, Portugal, and Italy, as well as by the non-Latin States of Austria and Holland. *All these countries, at the end of various terms of 99, 90, and 85 years will practically pay off a large portion of their national debt.* Improvident Spain will pay off about £40,000,000 out of her debt of £164,000,000. Heavily burdened Austria will practically abrogate something like £65,000,000 out of her debt of £250,000,000. Italy will wipe out a large portion of her debt of £176,000,000.

But the most remarkable example is France; and I will endeavor to explain as briefly as possible the working of the French system. In France the railways are conceded for 99 years, but it is one of the conditions of the grant that all the capital, whether in shares or debentures, shall be paid off within that term by an annual *amortissement,* or sinking fund. The small amount of this annual payment is very extraordinary. The French rate of interest is 5 per cent., and the annual sinking fund necessary to pay off 100 francs in 99 years is as nearly as possible .04. Put into the English form, for the sake of clearness, this means that the annual sinking fund necessary at 5 per cent. to redeem £100 in 99 years is only 1s. per annum. As debentures are issued in France for less than 99 years when part of the concession is run out, the amount of the sinking fund varies, but it is usually said to amount on the average to one-

3

eighth per cent. As the whole expended capital of French railways represented by shares and debentures, is £233,000,000, it follows that the total annual sinking fund paid by the French companies for the redemption of that sum is less thon £300,000. The result is marvellous. that for £300,000 the French nation will acquire, in less than 99 years, an unencumbered property of £233,000,000 sterling. But this is not all. The railways represented by that £233,000,000 sterling produced in 1865 a net revenue of about £12,500,000. Before 1872 further railways will have been completed, which will be amortized at the same rate as their parent lines, and will produce before many years a net income of £4,000,000, making a total net income of the French railways £16,500,000. But the total charge of the French national debt in 1865 was only £16,000,000. *So that France has now a system in operation which, in less than 90 years from the present time, will relieve the country from the whole burden of her national debt of nearly £500,000,000.*

Is it allowable in me to ask, why are we doing nothing of the sort? When so many other nations are paying off by means of their railways a portion, or the whole of their national debts, why are we, with all our wealth and resources, to do nothing? A scheme of amortization suited to the habits of the English people, is perfectly possible, and the peculiar position of railway companies at the present moment renders it easy to carry out. I will say nothing about debentures, because a plan is now before the government dealing with them. But, I say, respecting Share Capital, that it would be perfectly practicable for the State to become the possessor of a large proportion of this stock in a comparatively short time, and at no great expense. An annual sinking fund of 5s. per cent. will pay off £100 in seventy-two years, reckoning only 4 per cent. interest. Hence, in seventy-two years, an annual sinking fund of £500,-000 a year will pay off £200,000,000. The government duty on railways amounts to £450,000 a year, and will soon reach £500,000. My proposal would be to make this a sinking fund towards purchasing £200,000,-000 of preference and other stock, and let it be invested annually by the Board of Trade, or by commissioners appointed for the purpose, like those appointed for the national debt. Instead of canceling each share as it is purchased, let it be held in trust for the nation, and the dividends applied every year in augmentation of the sinking fund. In this manner, at the end of about seventy-two years, £200,000,000 of preference and ordinary share capital would become the property of the nation, and its dividends become applicable to the interest of the national debt. As railway dividends average 4 to 4½ per cent., the devidends on the redeemed capital would pay the interest on more than £250,000,000 consols, and be equivalent to the redemption of that amount of our national debt.

I believe that this is a practical scheme. In a slightly different form it is now being carried out in France, Belgium and other continental States. I trust that before long we shall cease to be almost the only nation in Europe which does net act on the principle *"that railways are the true sinking fund for the payment of the national debt."*

The advantages of such a sinking fund invested in consols are threefold:

1. It would be invested annually in railway capital at a higher interest, and thus accumulate more rapidly.

2. It would have a different primary object, viz: the purchase of a State interest in railways, and would, therefore, be more likely to enlist popular feeling in favor of its maintenance.

3. It would be distinct and separate from the national debt, and not under the same control, and would, therefore, be less liable to be diverted to the financial necessities of the hour.

Perhaps it will be said that a railway sinking fund is unsuited to the character and habits of the English people. But surely it is our character to be prudent and to pay off encumbrances, and to adopt the best means of accomplishing that object. Surely it is not right in a great and wealthy and enlightened nation like England to incur the reproach of being spendthrift of her resources and reckless of her debts.

## XII.—FURTHER RAILWAY EXTENSION.

England is undoubtedly the country in the world best provided with railways. The statistical comparison stood thus at the end of 1865:—

RAILWAYS COMPARED WITH AREA AND POPULATION.

| Country. | Railway Miles Open. | Square Miles per Railway Mile. | Population per Railway Mile. |
|---|---|---|---|
| England and Wales..................... | 9,251 | 6½ | 2,186 |
| 1. Belgium.................................. | 1,350 | 8 | 3,625 |
| 2. United Kingdom.------------------ | 13,289 | 9 | 2,206 |
| 3. Switzerland............................. | 778 | 19 | 3,257 |
| 4. Prussia and Germany (except | | | |
| Austria).............. ...... .............. | 8,589 | 20 | 3,525 |
| 5. Northern United States (except | | | |
| Kansas, Nebraska and Oregon)..... | 24,883 | 25 | 801 |
| 6. France.................................... | 8,134 | 26 | 4,607 |
| 7. Holland.................................. | 372 | 29 | 9,066 |
| 8. Italy...................................... | 2,389 | 41 | 9,084 |
| 9. Austria................................... | 3,735 | 63 | 9,375 |
| 10. Spain...................................... | 2,721 | 67 | 5,991 |
| 11. Portugal.................................. | 419 | 87 | 8,555 |
| 12. Southern United States............... | 10,300 | 92 | 1,025 |
| 13. Canada................................... | 2,539 | 136 | 987 |
| 14. India....................... ............ | 3,186 | 287 | 42,572 |
| Total of the 14 countries......... | 82,495 | ...... | ......... |

But England has a much greater proportion of double lines, and a larger number of trains on each line; while, on the other hand, Belgium and other continental nations have lower fares and give greater accommodation to third and fourth class passengers. Both parties have something to learn—they to admit the principle of competition and increase the number of railways; we to provide cheap conveyance for the masses, without the clumsy device of excursion trains.

But now comes the question—do England and Belgium need further railways, or are they already sufficiently provided? It may partly be answered by the fact that in England there are about 3,500 miles authorized by Parliament which have not yet been made, and that in Belgium there are 450 miles (equal to 4,500 in England) conceded but not constructed. And we may also point to the circumstance that in England and Wales there were, in 1865, 6,081 miles of double line against 3,170 miles of single, showing that there is a want of cheap lines through rural districts. A glance at the railway map will confirm this inference. The lines run in the direction of the metropolis or some great town, and there are few cross-country lines. The distance between the lines supports this conclusion. Deducting the manufacturing districts, which are crowded with a railway network, the remainder of the country gives an average of about fifteen miles between each mile of railway. The average ought not to be more than eight or ten miles.

The advantage of a railway to agriculture may be estimated by the following facts. A new line would, on an average, give fresh accommodation to three and a-half miles on each side, being a total of seven square miles, or 4,560 acres for each mile of railway. It would be a very moderate estimate to suppose that cartage would be saved on one ton of produce, manure, or other articles for each acre, and that the saving per ton would be five miles at 8d. per mile. Hence the total annual saving would be £768 per mile of railway, which is 5 per cent. interest on £15,000. Thus it is almost impossible to construct a railway through a new district of fair agricultural capabilities without saving to the landowner and farmer alone the whole cost of the line. Besides this, there is the benefit to the laborers of cheap coals and better access to the market. There is also the benefit to the small towns of being put into railway communication with larger towns and wholesale producers. And there is the possibility of opening up sources of mineral wealth.

Somebody ought to make these agricultural lines, even though they may not pay a dividend to the shareholder. But who is that somebody to be? The great companies will not take the main burden lest they should lower their own dividends. The general public will not subscribe, for they know the uncertainty of the investment turning out profitable. And notwithstanding the able letters signed "H" in the *Times* some months ago, I cannot advocate the necessarily wasteful system of contractors' lines, or believe in the principle, "Never mind who is the loser so that the public is benefited." Railway extension is not promoted in the long run by wasteful financing and ruinous projects. On the contrary, such lines injure railway extension by making railways a bye-word and depreciating railway property, and they render it impossible to find supporters for sound and beneficial schemes.

The proper parties to pay for country lines are the proprietors and inhabitants of the districts through which they pass. They are benefited even if the line does not pay a dividend. They have every motive for economical construction and management, and can make a line pay where no one else can. But they will not subscribe any large portion of the capital as individuals. Very few will make a poor investment of any magnitude for the public good, though all might be ready to take their part in a general rate. Almost every country but our own has recognized the fact, and legislated on this basis, by empowering the inhabitants of a district which would be benefited to tax themselves for the construction of a railway. I have shown that in France either the department or the commune may vote a subvention out of their public funds, and that in the United States the municipalities vote subsidies of municipal bonds. In Spain the provinces and the municipalities have the power to take shares or debentures, or, if they prefer it, to vote subventions or a guarantee of interest. In Italy the municipalities do the same thing. Why should not England follow their example, and authorize the inhabitants of parishes and boroughs to rate themselves for a railway which will improve their property, or empower them to raise loans on the security of the rates, to be paid off in a certain number of years by a sinking fund, as is done for sanitary improvements? I see no other way of raising the nucleus of funds for carrying out many rural lines which would be most beneficial to the country.

I can give a remarkable instance of the benefits caused by an unremunerative railway. In 1834 the inhabitants of Whitby projected a line from Whitby along the valley of the Esk to Pickering, half way to York. The line was engineered by George Stephenson, and was originally worked by horse-power and carriages on the model of the four-horse coaches. But though considered at that time one of the wonders of the world, the line was utterly unprofitable, and the Whitby people looked upon it as a bad speculation, much as the shareholders of the London, Chatham, and Dover look on their present property. The railway was ultimately sold to the Northeastern Company; but though the shareholders got no advantage, somebody else did. Farmers and laborers came to market in Whitby, and got coals and other necessaries at reduced rates, while they sold their produce better. Very soon rents began to rise, and I find the total rise since the construction of the railway has been from an average of 15s. per acre up to 22s., or nearly 50 per cent. But far greater consequences resulted. The cliffs at Whitby were known to contain nodules of ironstone, which were picked up and sent to ironworks on the Tyne. Soon after the opening of the railways George Stephenson and a number of Whitby gentlemen formed a company, called the Whitby Stone Company, for working stone quarries and ironstone mines at Grosmont, about six miles up the railway. At first the ironstone was very badly received by the iron founders, and it was only after long and patient perseverance that the company got a sale for what they raised. It was not till 1844 and 1846 that the merits of the Cleveland ironstone were fully acknowledged and large contracts entered into for its working throughout the district. Thus the unprofitable Whitby and Pickering railway opened up the Cleveland iron district and caused the establishment of a very large number of foundries and the employment of thousands of workmen, and has added very materially to the wealth of England.

## XIII.—Conclusion.

From the facts which have been brought forward I draw the following conclusions:

1. Railways have been a most powerful agent in the progress of commerce, in improving the condition of the working classes, and in developing the agricultural and mineral resources of the country.

2. England has a more complete and efficient system of railways than any other country, but is not so far ahead that she can afford to relax her railway progress, and to let her competitors pass her in the race.

3. England ought to improve the internal organization of her railways, both as to finance and traffic, and to constitute some central authority with power to investigate and regulate.

4. A sinking fund should be instituted to purchase for the State a portion of the railway capital, and so to lighten the charge of the national debt.

5. Power should be given to parishes and boroughs to rate themselves in aid of local railways in order to facilitate the construction of country lines.

6. England, as a manufacturing and commercial country, is benefited by every extension of the railway system in foreign countries, since every new line opens up fresh markets and diminishes the cost of transporting her manufactures.

I cannot conclude without saying a word on the future of railways. The progress of the last thirty-six years has been wonderful, since that period has witnessed the construction of about 85,000 miles of railway. The next thirty-six years are likely to witness a still greater development and the construction of far more than 85,000 miles. We may look forward to England possessing, at no distant date, more that 20,000 miles, France an equal number, and the other nations of the continent increasing their mileage until it will bear the proportion of 1 railway mile to every 10 square miles of area, instead of the very much less satisfactory proportions stated in the comparative table. We may expect the period when the immense continent of North America will boast of 100,000 miles of line, clustered in the thickly-populated Eastern States and spreading plentifully through the Western to the base of the Rocky Mountains and over to California and the Pacific. We may anticipate the time when Russia will bend her energies to consolidating her vast empire by an equally vast railway network. We may predict the day when a continuous railroad will run from Dover to the Bosphorus, from the Bosphorus down the Euphrates, across Persia and Beloochistan to India, and from India to China. We may look for the age when China, with her 350,000,000 of inhabitants, will turn her intelligence and industry to railroad communication.

But who shall estimate the consequences that will follow, the prodigious increase of commerce, the activity of national intercourse, the spread of civilization, and that advance of human intelligence foretold thousands of years ago by the prophet upon the lonely plains of Palestine, "when many shall run to and fro upon the earth, and knowledge shall be increased?"

---

FOUR HUNDRED AND FORTY MILLION DOLLARS OF SUBSIDY GRANTED BY THE BRITISH GOVERNMENT TO BUILD RAILWAYS INTO THE COTTON DISTRICTS OF INDIA.

The efforts recently made by the English government to develop the resources of its vast empire in Hindostan, evince remarkable energy and sagacity. Probably no country in the world has made more material progress within the last few years than British India. Notwithstanding the discouragements arising from the mutiny of the Sepoys, and the disasters of famine and financial collapse, the present condition and future prospects of the people have been greatly improved. Railroads have been built, highways have been thrown up, canals widened and deepened, obstructions removed from rivers, bridges constructed over rivers and mountain chasms, and the jungle has been rendered passable for the first time.

These great changes in the condition of the interior of British India were initiated, or, at least, actively commenced in accordance with a policy adopted at the commencement of our civil war. England, in place of attempting to break up our monopoly of the cotton trade by an open and formal assistance of the South, resolved to effect the same object by other and surer means. Her statesmen, with far reaching sagacity, resolved to improve the opportunity afforded by the American crisis, so as to attach the tottering Indian Empire to the imperial government by a bridge of gold.

In 1860-61, the Marquis Dalhousie, Governor General, inaugurated the extensive system of internal improvement, which was to enable the people of Hindostan to compete with America for the cotton trade of the world. The most favorable cotton regions of India were inaccessible for want of proper facilities for communication. In order to get the staple to a market, it was necessary to carry it by man and horse power over vast tracts of jungle, across mountains and ravines, and ferry it over great rivers.

To obviate these difficulties, the railroad movement inaugurated was of the most comprehensive character. The population of India subject to the English government is probably not less than two hundred millions. The country comprises an area of 1,364,000 square miles, stretching 1,800 miles in length and 1,500 miles in breadth from east to west. This great country is broken up into an almost endless geographical diversity. There are vast and impassable jungles, huge forests, mighty rivers, mountain chains and extensive plains, the whole being combined with a wonderful luxuriance of vegetation, which at every step obstructs progress and almost prevents any passage by man or beast.

It was over this country, presenting so many difficulties, that Lord Dalhousie contemplated his admirable network of railroads. The system was, of course, planned with reference to the geographical features of the country, so as to connect the extremes of the vast empire with grand trunk lines, from which branch lines, or feeders, might be constructed, according to the future requirements of local commerce. Four thousand six hundred miles of railroad were to be built, at an estimated expense of $440,000,000. *The credit of the imperial government was granted to private companies, guaranteeing a certain amount of interest on all money invested in Indian railroads.* The government wisely left all details of construction and management to the energies of the companies themselves, which had every motive for economy, as all money earned above the guaranteed dividends was clear gain. The system worked so well, that last year several Indian railways exceeded the 5 per cent. guaranteed interest. During the half year ending December 31st, the East Indian and the Great Peninsular railroad companies were able to declare surplus dividends. Half the amount of surplus income was devoted to the repayment of former advances for interest by the government, and the other half was divided among the stockholders.

*The net amount of guaranteed interest paid by the government diminishes every year.* In 1865 the amount was £1,450,000; in 1866 it was £800,000, and this year only £600,000 was required. These figures indicate the profitable character of these Indian railroad enterprises.

The original system of Indian railroads contemplated the establishment of communications between Bombay, Madras, and Calcutta, the three great centres of military and commercial power. The extremes of the empire were united, and roads were cut through the great agricultural and producing districts. The East Indian Railroad Company has now under its management 1,310 miles of railway, constructed at an expense of $100,000,000, and is the longest line of road in the world under one company. The Great Indian Peninsular road will be 1,233 miles long when completed, and next year it will be open for traffic along its entire length. In 1868, from Calcutta to Bombay, a distance of 1,458 miles, there will be an unbroken railroad communication. The branch lines connecting with the main stems are of great extent, and will cost as much money as the main roads. To show the progress of Indian railroads it may be stated that it is only fourteen years since the first line was opened in that country. At the present time there are 3,200 miles in operation, and next year a thousand additional miles will be completed.

This development of railroads in British India is of the highest importance as affecting the cotton trade. Formerly we enjoyed a monopoly of the market; now, nearly one-half of the cotton manufactured in England is derived from India alone. A late Liverpool circular estimates the quantity of American cotton now on hand and to arrive before December 31st, 1867, at 680,000 bales, while the supply of India cotton for the same period is estimated at 925,000 bales. Without expressing any opinion as to the correctness of these figures, the more important fact for us to remember is that the manufacturers of England have so altered and improved their machinery as to be able to use in much larger proportion than formerly the shorter India staple, while, at the same time, the quality of cotton from that country has been decidedly and steadily improved, and is being more carefully prepared for market. Judging then of the future from the past, it may be expected to equal the American article at no distant period.

*The establishment of railroads in India removes the chief obstacles to the growth of an almost unlimited supply of cotton.* The country is admirably adapted for it, and the teeming population has long been familiar with the staple, and exhibit great aptitude in its culture. The best cotton regions have not yet been opened to the world; the only facilties for reaching a market being the slow and expensive process of cattle teams. The new railroads, however, will convey the products òt these regions to market cheaply and expeditiously. And it is a noticeable feature of Indian railroad companies that their revenues are derived from goods rather than from passengers. Of $35,000,000 income of Indian railroads during the three years ending June, 1866, two-thirds were received from merchandise traffic.